THE

POSITIVE PHILOSOPHY

OF

AUGUSTE COMTE.

FREELY TRANSLATED AND CONDENSED

BY

HARRIET MARTINEAU

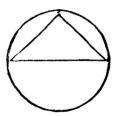

NEW YORK:
PUBLISHED BY CALVIN BLANCHARD,
82 NASSAU STREET.

1855.

Salut et Fraternité
Auguste Comte

R. CRAIGHEAD, PRINTER
53 VESEY STREET, N. Y.

PREFACE.

It may appear strange that, in these days, when the French language is almost as familiar to English readers as their own, I should have spent many months in rendering into English a work which presents no difficulties of language, and which is undoubtedly known to all philosophical students. Seldom as Comte's name is mentioned in England, there is no doubt in the minds of students of his great work that most or all of those who have added substantially to our knowledge for many years past are fully acquainted with it, and are under obligations to it which they would have thankfully acknowledged, but for the fear of offending the prejudices of the society in which they live. Whichever way we look over the whole field of science, we see the truths and ideas presented by Comte cropping out from the surface, and tacitly recognised as the foundation of all that is systematic in our knowledge. This being the case, it may appear to be a needless labor to render into our own tongue what is clearly existing in so many of the minds which are guiding and forming popular views. But it was not without reason that I undertook so serious a labor, while so much work was waiting to be done which might seem to be more urgent.

One reason, though not the chief, was that it seems to me unfair, through fear or indolence, to use the benefits conferred on us by M. Comte without acknowledgment. His fame is no doubt safe. Such a work as this is sure of receiving due honor, sooner or later. Before the end of the century, society at large will have become aware that this work is one of the chief honors of the century, and that its authors name will rank with those of the worthies who have illustrated former ages: but it does not seem to me right to assist in delaying the recognition till the author of so noble a service is beyond the reach of our gratitude and honor: and that it is demoralizing to ourselves to accept and use such a boon as he has given us in a silence which is in fact ingratitude. His honors we can not share: they are his own and incommunicable. His trials we may share, and, by sharing, lighten; and he has the strongest claim upon us for sympathy and fellowship in any popular disrepute which, in this case, as in all cases of signal social service, attends upon a first movement. Such sympathy and fellowship will, I trust, be awakened and extended in proportion to the spread among us of a popular knowledge of what M. Comte has done: and this hope was one reason, though, as I have said, not the chief, for my undertaking to reproduce his work in England in a form as popular as its nature admits.

A stronger reason was that M. Comte's work, in its original form, does no justice to its importance, even in France; and much less in England. It is in the form of lectures, the delivery of which was spread over a long course of years; and this extension of time necessitated an amount of recapitulation very injurious to its interest and philosophical aspect. M. Comte's style is singular. It is at the same time rich and diffuse. Every sentence is full fraught with meaning; yet it is overloaded with words. His scrupulous honesty leads him to guard his enunciations with epithets so constantly repeated, that though, to his own mind, they are necessary in each individual instance, they become wearisome, especially toward the end of his work, and lose their effect by constant repetition. This practice, which might be strength in a series of instructions spread over twenty years, becomes weakness when those instructions are presented as a whole; and it appeared to me worth while to condense his work, if I undertook nothing more, in order to divest it of the disadvantages arising from redundancy alone. My belief is that thus, if nothing more were done, it might be brought before the minds of many who would be deterred from the study of it by its bulk. What I have given in this volume occupies in the original six volumes averaging nearly eight hundred pages: and yet I believe it will be found that nothing essential to either statement or illustration is omitted.

My strongest inducement to this enterprise was my deep conviction of our need of this book in my own country, in a form which renders it accessible to the largest number of intelligent readers. We are living in a remarkable time, when the conflict of opinions renders a firm foundation of knowledge indispensable, not only to our intellectual, moral, and social progress, but to our holding such ground as we have gained from former ages. While our science is split up into arbitrary divisions; while abstract and concrete science are confounded together, and even mixed up with their application to the arts, and with natural history; and while the researches of the scientific world are presented as mere accretions to a heterogeneous mass of facts, there can be no hope of a scientific progress which shall satisfy and benefit those large classes of students whose business it is, not to explore, but to receive. The growth of a scientific taste among the working classes of this country is one of the most striking of the signs of the times. I believe no one can inquire into the mode of life of young men of the middle and operative classes without being struck with the desire that is shown, and the sacrifices that are made, to obtain the means of scientific study. That such a disposition should be baffled, and such study rendered almost ineffectual, by the desultory character of scientific exposition in England, while such a work as Comte's was in existence, was not to be borne, if a year or two of humble toil could help, more or less, to supply the need.

In close connection with this was another of my reasons. The supreme dread of every one who cares for the good of nation or race is that men should be adrift for want of an anchorage for their convictions. I believe that no one questions that a very large proportion of our people are now so adrift. With pain and fear, we see that a multitude, who might and should be among the wisest and best of our citizens, are alienated for ever from the kind of faith which sufficed for all in an organic period which has passed away, while no one has presented to them, and they can not obtain for themselves, any ground of conviction as firm and clear as that which sufficed for our fathers in their day. The moral dangers of such a state of fluctuation as has thus arisen are fearful in the extreme, whether the transition stage from one order of convictions to another be long or short. The work of M. Comte is unquestionably the greatest single effort that has been made to obviate this kind of danger; and my deep persuasion is that it will be found to retrieve a vast amount of wandering, of unsound speculation, of listless or reckless doubt, and of moral uncertainty and depression. Whatever else may be thought of the work, it will not be denied that it ascertains with singular sagacity and soundness the foundations of human knowledge, and its true object and scope;

and that it establishes the true filiation of the sciences within the boundaries of its own principle. Some may wish to interpolate this or that; some to amplify, and perhaps, here and there, in the most obscure recesses of the great edifice, to transpose, more or less: but any who question the general soundness of the exposition, or of the relations of its parts, are of another school, and will simply neglect the book, and occupy themselves as if it had never existed. It is not for such that I have been working, but for students who are not schoolmen; who need conviction, and must best know when their need is satisfied. When this exposition of Positive Philosophy unfolds itself in order before their eyes, they will, I am persuaded, find there at least a resting-place for their thought — a rallying-point of their scattered speculations — and possibly an immoveable basis for their intellectual and moral convictions. The time will come when the book itself will, for a while, be most discussed on account of the deficiencies which M. Comte himself presses on our notice; and when his philosophy will sustain amplifications of which he himself does not dream. It must be so, in the inevitable growth of knowledge and evolution of philosophy; and it is the fate which the philosopher himself should covet, because it is only a true book that could survive to be so treated: but, in the meantime, it gives us the basis that we demand, and the principle of action that we want, and as much instruction in the procedure, and information as to what has been already achieved, as could be given in our time; perhaps more than could have been given by any other mind of our time. Even Mathematics is here first constituted a science, venerable and unquestionable as mathematical truths have been for ages past: and we are led on, tracing as we go the clear genealogy of the sciences, till we find ourselves among the elements of Social science, as yet too crude and confused to be established, like the others, by a review of what had before been achieved; but now, by the hand of our master, discriminated, arranged, and consolidated, so as to be ready to fulfil the conditions of true science as future generations bring their contributions of knowledge and experience to build upon the foundation here laid. A thorough familiarity with the work in which all this is done would avail more to extinguish the anarchy of popular and sectional opinion in this country than any other influence that has yet been exerted, or, I believe, proposed.

It was under such convictions as these that I began, in the spring of 1851, the analysis of this work, in preparation for a translation.

A few months afterward, an unexpected aid presented itself. My purpose was related to the late Mr. Lombe, who was then residing at Florence. He was a perfect stranger to me. He told me, in a subsequent letter, that he had wished, for many years, to do what I was then attempting, and had been prevented only by ill health. My estimate of M. Comte's work, and my expectations from its introduction into England in the form of a condensed translation, were fully shared by him; and, to my utter amazement, he sent me, as the first act of our correspondence, an order on his bankers for five hundred pounds sterling. There was time, before his lamented death, for me to communicate to him my views as to the disposal of this money, and to obtain the assurance of his approbation. We planned that the larger proportion of it should be expended in getting out the work, and promoting its circulation. The last words of his last letter were an entreaty that I would let him know if more money would, in any way, improve the quality of my version, or aid the promulgation of the book. It was a matter of deep concern to me that he died before I could obtain his opinion as to the manner in which I was doing my work. All that remained was to carry out his wishes as far as possible; and to do this, no pains have been spared by myself, or by Mr. Chapman, who gave him the information that called forth his bounty.

As to the method I have pursued with my work—there will be different opinions about it, of course. Some will wish that there had been no omissions, while others would have complained of length and heaviness, if I had offered a complete translation. Some will ask why it is not a close version as far as it goes; and others, I have reason to believe, would have preferred a brief account, out of my own mind, of what Comte's philosophy is, accompanied by illustrations of my own devising. A wider expectation seems to be that I should record my own dissent, and that of some critics of much more weight, from certain of M. Comte's views. I thought long and anxiously of this; and I was not insensible to the temptation of entering my protest, here and there, against a statement, a conclusion, or a method of treatment. I should have been better satisfied still to have adduced some critical opinions of much higher value than any of mine can be. But my deliberate conclusion was that this was not the place nor the occasion for any such controversy. What I engaged to do was to present M. Comte's first great work in a useful form for English study; and it appears to me that it would be presumptuous to thrust in my own criticisms, and out of place to insert those of others. Those others can speak for themselves, and the readers of the book can criticise it for themselves. No doubt, they may be trusted not to mistake my silence for assent, nor to charge me with neglect of such criticism as the work has already evoked in this country. While I have omitted some pages of the Author's comments on French affairs, I have not attempted to alter his French view of European politics. In short, I have endeavored to bring M. Comte and his English readers face to face, with as little drawback as possible from intervention.

This by no means implies that the translation is a close one. It is a very free translation. It is more a condensation than an abridgment: but it is an abridgment too. My object was to convey the meaning of the original in the clearest way I could; and to this all other considerations were made to yield. The serious view that I have taken of my enterprise is proved by the amount of labor and of pecuniary sacrifice that I have devoted to my task. Where I have erred, it is from want of ability; for I have taken all the pains I could.

One suggestion that I made to Mr. Lombe, and that he approved, was that the three sections—Mathematics, Astronomy, and Physics—should be revised by a qualified man of science. My personal friend Professor Nichol, of Glasgow, was kind enough to undertake this service. After two careful readings, he suggested nothing material in the way of alteration, in the case of the first two sections, except the omission of Comte's speculation on the possible mathematical verification of Laplace's Cosmogony. But more had to be done with regard to the treatment of Physics. Every reader will see that that section is the weakest part of the book, in regard both to the organization and the details of the subject. In regard to the first, the author explains the fact, from the nature of the case, that Physics is rather a repository of somewhat fragmentary portions of physical science, the correlation of which is not yet clear, than a single circumscribed science. And we must say for him, in regard to the other kind of imperfection, that such advances have been made in almost every department of Physics since his second volume was published, that it would be unfair to present what he wrote under that head in 1835 as what he would have to say now. The choice lay therefore between almost rewriting this portion of M. Comte's work, or so largely abridging it that only a skeleton presentment of general principles should remain. But as the system of Positive Philosophy is much less an Expository than a Critical work, the latter alternative alone seemed open, under due consideration of justice to the Author. I have adopted therefore the plan of extensive omissions, and have retained the few short memoranda in which Professor Nichol suggested these, as notes. Although this gentleman has sanctioned my presentment of Comte's chapters on Mathematics and Physics, it must not be inferred that he agrees with his Method in Mental Philosophy, or assents to other conclusions held of main importance by the disciples of the Positive Philosophy. The contrary, indeed, is so apparent in the tenor of his own writings, that so far as his numerous readers are concerned, this remark need not have been offered. With the reservation I have made, I am bound to take the entire responsibility — the Work being absolutely and wholly my own.

It will be observed that M. Comte's later works are not referred to in any part of this book. It appears to me that they, like our English criticisms on the present Work, had better be treated of separately. Here his analytical genius has full scope; and what there is of synthesis is, in regard to social science, merely what is necessary to render his analysis possible and available. For various reasons, I think it best to stop here, feeling assured that if this Work fulfils its function, all else with which M. Comte has thought fit to follow it up will be obtained as it is demanded.

During the whole course of my long task, it has appeared to me that Comte's work is the strongest embodied rebuke ever given to that form of theological intolerance which censures Positive Philosophy for pride of reason and lowness of morals. The imputation will not be dropped, and the enmity of the religious world to the book will not slacken for its appearing among us in an English version. It can not be otherwise. The theological world can not but hate a book which treats of theological belief as a transient state of the human mind. And again, the preachers and teachers, of all sects and schools, who keep to the ancient practice, once inevitable, of contemplating and judging of the universe from the point of view of their own minds, instead of having learned to take their stand out of themselves, investigating from the universe inward, and not from within outward, must necessarily think ill of a work which exposes the futility of their method, and the worthlessness of the results to which it leads. As M. Comte treats of theology and metaphysics as destined to pass away, theologians and metaphysicians must necessarily abhor, dread, and despise his work. They merely express their own natural feelings on behalf of the objects of their reverence and the purpose of their lives, when they charge Positive Philosophy with irreverence, lack of aspiration, hardness, deficiency of grace and beauty, and so on. They are no judges of the case. Those who are — those who have passed through theology and metaphysics, and, finding what they are now worth, have risen above them — will pronounce a very different judgment on the contents of this book, though no appeal for such a judgment is made in it, and this kind of discussion is nowhere expressly provided for. To those who have learned the difficult task of postponing dreams to realities till the beauty of reality is seen in its full disclosure, while that of dreams melts into darkness, the moral charm of this work will be as impressive as its intellectual satisfactions. The aspect in which it presents Man is as favorable to his moral discipline, as it is fresh and stimulating to his intellectual taste. We find ourselves suddenly living and moving in the midst of the universe, — as a part of it, and not as its aim and object. We find ourselves living, not under capricious and arbitrary conditions, unconnected with the constitution and movements of the whole, but under great, general, invariable laws, which operate on us as a part of the whole. Certainly, I can conceive of no instruction so favorable to aspiration as that which shows us how great are our faculties, how small our knowledge, how sublime the heights which we may hope to attain, and how boundless an infinity may be assumed to spread out beyond. We find here indications in passing of the evils we suffer from our low aims, our selfish passions, and our proud ignorance; and in contrast with them, animating displays of the beauty and glory of the everlasting laws, and of the sweet serenity, lofty courage, and noble resignation, that are the natural consequence of pursuits so pure, and aims so true, as those of Positive Philosophy. Pride of intellect surely abides with those who insist on belief without evidence and on a philosophy derived from their own intellectual action, without material and corroboration from without, and not with those who are too scrupulous and too humble to transcend evidence, and to add, out of their own imaginations, to that which is, and may be, referred to other judgments. If it be desired to extinguish presumption, to draw away from low aims, to fill life with worthy occupations and elevating pleasures, and to raise human hope and human effort to the highest attainable point, it seems to me that the best resource is the pursuit of Positive Philosophy, with its train of noble truths and irresistible inducements. The prospects it opens are boundless; for among the laws it establishes that of human progress is conspicuous. The virtues it fosters are all those of which Man is capable; and the noblest are those which are more eminently fostered. The habit of truth-seeking and truth-speaking, and of true dealing with self and with all things, is evidently a primary requisite; and this habit once perfected, the natural conscience, thus disciplined, will train up all other moral attributes to some equality with it. To all who know what the study of philosophy really is—which means the study of Positive Philosophy—its effect on human aspiration and human discipline is so plain that any doubt can be explained only on the supposition that accusers do not know what it is that they are calling in question. My hope is that this book may achieve, besides the purposes entertained by its author, the one more that he did not intend, of conveying a sufficient rebuke to those who, in theological selfishness or metaphysical pride, speak evil of a philosophy which is too lofty and too simple, too humble and too generous, for the habit of their minds. The case is clear. The law of progress is conspicuously at work throughout human history. The only field of progress is now that of Positive Philosophy, under whatever name it may be known to the real students of every sect; and therefore must that philosophy be favorable to those virtues whose repression would be incompatible with progress.

CONTENTS.

INTRODUCTION.

CHAPTER I.

Account of the Aim of this Work. View of the Nature and Importance of the Positive Philosophy.

Preliminary Survey............PAGE 25
Law of Human Development........ 25
First Stage..................... 26
Second Stage.................... 26
Third Stage..................... 26
Ultimate Point of each.......... 26
Evidences of the Law............ 26
Actual.......................... 26
Theoretical..................... 27
Character of the Positive Philosophy. 28
History of the Positive Philosophy... 29
New Department of Positive Philosophy........................ 30
Social Physics.................. 30
Secondary Aim of this Work...... 30
To Review the Philosophy of the Sciences..................... 31
Glance at Speciality............ 31
Proposed new Class of Students.. 32
Advantages of the Positive Philosophy 32
1. Illustrates Intellectual Function... 32
2. Must regenerate Education..... 34
3. Advances Sciences by combining them........................ 35
4. Must reorganize Society...... 36
No hope of Reduction to a single Law 37

CHAPTER II.

View of the Hierarchy of the Positive Sciences.

Failure of Proposed Classifications... 38
True Principle of Classification...... 39
Boundaries of our Field.............. 39
Theoretical Inquiry.................. 39
Abstract Science..................... 41
Concrete Science..................... 41
Difficulty of Classification......... 42
Historical and Dogmatic Methods.... 42
True Principle of Classification..... 44
Characters........................... 44
1. Generality........................ 44
2. Independence...................... 44
Inorganic and Organic Phenomena... 44
I. INORGANIC......................... 45
1. Astronomy......................... 45
2. Physics........................... 45

3. Chemistry...............PAGE 45
II. ORGANIC.......................... 45
1. Physiology........................ 45
2. Sociology......................... 45
Five Natural Sciences: their Filiation. 46
Filiation of their Parts............. 46
Corroborations....................... 46
1. This Classification follows the Order of Disclosure of Sciences......... 46
2. Solves Heterogeneousness......... 47
3. Marks Relative Perfection of Sciences....................... 47
4. Effect on Education.............. 48
Effect on Method..................... 48
Orderly Study of Sciences............ 49
MATHEMATICS.......................... 49
A Department......................... 49
A Basis.............................. 49
An Instrument........................ 49
A Double Science..................... 50
Abstract Mathematics, an Instrument. 50
Concrete Mathematics, a Science..... 50
Mathematics Pre-eminent in the Scale 50

BOOK I.—MATHEMATICS.

CHAPTER I.

Mathematics, Abstract and Concrete.

Description of Mathematics.......... 51
Object of Mathematics............... 52
General Method...................... 52
Examples............................ 53
True Definition of Mathematics...... 54
Its Two PARTS....................... 55
Their Different Objects............. 55
Their Different Natures............. 56
Concrete Mathematics................ 56
Abstract Mathematics................ 57
Extent of its Domain................ 58
Its Universality.................... 58
Its Limitations..................... 58

CHAPTER II.

General View of Mathematical Analysis.

ANALYSIS............................ 61
True Idea of an Equation............ 61
Abstract Functions.................. 62
Concrete Functions.................. 62
Two parts of the Calculus........... 63
Algebra............................. 63
Arithmetic.......................... 63

CONTENTS.

Its Extent...........PAGE 63	Idea of Space..........PAGE 92
Its Nature...................... 64	Kinds of Extension................ 93
Algebra......................... 65	Geometrical Measurement.......... 93
Creation of New Functions......... 65	Measurement of Surfaces and Volumes 93
Finding Equations between Auxiliary Quantities........... 65	Of Curved Lines................... 93
Division of the Calculus of Functions. 66	Its Illimitable Field................. 94
SECTION 1. *Ordinary Analysis*, or Calculus of Direct Functions......... 66	Properties of Lines and Surfaces..... 95
Its Object........................ 67	Two General Methods.............. 95
Classification of Equations......... 67	Special or Ancient, and General or Modern Geometry................ 95
Algebraic Equations............... 67	Geometry of the Ancients........... 96
Algebraic Resolution of Equations... 68	Geometry of the Right Line......... 97
Our Existing Knowledge........... 68	Graphical Solutions................ 97
Numerical Resolution of Equations... 69	Descriptive Geometry.............. 97
The Theory of Equations........... 69	Algebraical Solutions.............. 98
Method of Indeterminate Coefficients. 70	Trigonometry..................... 98
SECTION 2. *Transcendental Analysis*, or Calculus of Indirect Functions.... 70	*Modern, or Analytical Geometry*.....100
Three Principal Views............. 70	Analytical Representation of Figures.100
History.......................... 71	Position..........................100
METHOD OF LEIBNITZ.............. 71	Position of a Point.................101
Generality of the Formulas......... 73	Plane Curves.....................101
Justification of the Method 73	Expression of Lines by Equations....101
NEWTON'S METHOD................ 74	Expression of Equations by Lines....102
Method of Limits.................. 74	Change in the Line Changes the Equation...........................102
Fluxions and Fluents 75	Every Definition of a Line is an Equation............................103
LAGRANGE'S METHOD............... 75	Choice of Co-ordinates..............104
Identity of the Three Methods....... 76	Determination of a Point in Space...105
Their Comparative Value........... 77	Determination of Surfaces by Equations, and of Equations by Surfaces.105
THE DIFFERENTIAL AND INTEGRAL CALCULUS........................ 79	Curves of Double Curvature.........106
Its Two Parts..................... 79	Imperfections of Analytical Geometry.106
Their Mutual Relations............. 79	Imperfections of Analysis...........107
Cases of Union of the Two......... 80	
Cases of the Differential Calculus Alone 80	
Cases of the Integral Calculus Alone.. 80	CHAPTER IV.
The Differential Calculus........... 81	*Rational Mechanics.*
Two Portions..................... 81	Its Nature........................108
Subdivisions...................... 82	Its Characters....................109
Reduction to the Elements.......... 82	Its Object........................109
Transformation of Derived Functions for New Variables................ 82	Matter not Inert in Physics.........110
Analytical Applications............. 82	Supposed Inert in Mechanics........110
The Integral Calculus............... 83	Field of Rational Mechanics.........110
Its Divisions...................... 83	Three Laws of Motion..............111
Subdivisions...................... 84	Law of Inertia....................111
One Variable or Several............ 84	Law of Equality of Action and Re-Action...........................111
Orders of Differentiation............ 84	Law of Co-Existence of Motions.....111
Quadratures...................... 85	Two Primary Divisions.............114
Algebraic Functions................ 85	STATICS and DYNAMICS.............114
Transcendental Functions........... 85	Secondary Divisions...............114
Singular Solutions................. 85	Solids and Fluids..................114
Definite Integrals.................. 86	SECTION 1. *Statics*.................115
Prospects of the Integral Calculus.... 86	Converse Methods of Treatment.....115
Calculus of Variations.............. 87	First Method.....................115
Problems Giving Rise to this Calculus. 88	Statics by Itself...................115
Other Applications................. 89	Second Method...................115
Relation to the Ordinary Calculus.... 90	Statics through Dynamics...........116
	Moments 116
CHAPTER III.	Want of Unity in the Method........116
General View of Geometry.	Virtual Velocities..................117
	Theory of Couples.................119
Its Nature........................ 92	Share of Equations in Producing Equilibrium 119
Definition........................ 92	

Connection of the Concrete with the Abstract Question..........PAGE 121
Equilibrium of Fluids..............122
Hydrostatics......................122
Liquids 123
Gases 123
SECTION 2. *Dynamics*..............124
Object............................124
Theory of Rectilinear Motion.......124
Motion of a Point.................125
Motion of a System................126
D'Alembert's Principle............126
Results...........................128
Statical Theorems.................128
Law of Repose.....................128
Stability and Instability of Equilibrium........................128
Dynamical Theorems................129
Conservation of the Motion of the Centre of Gravity................129
Principle of Areas................129
The Invariable Plain..............130
Moment of Inertia.................130
Principal Axes....................130
Conclusion 131

BOOK II.—ASTRONOMY.

CHAPTER I.

General View.

Its Nature........................132
Definition........................133
Restriction 133
Means of Exploration..............134
Its Rank..........................134
When it became a Science..........135
Reduction to a Single Law.........135
Relation to other Sciences........136
Divisions of the Science..........137
Celestial Geometry................137
Celestial Mechanics...............137

CHAPTER II.

Methods of Study of Astronomy.

SECTION 1. *Instruments*..........138
Observation 138
Shadows 139
Artificial Methods................139
The Pendulum......................140
Measurement of Angles.............140
Requisite Corrections.............141
SECTION 2. *Refraction*...........142
SECTION 3. *Parallax* 143
SECTION 4. *Catalogue of Stars*...144

CHAPTER III.

Geometrical Phenomena of the Heavenly Bodies.

SECTION 1. *Statical Phenomena*...145
Two Classes of Phenomena..........145
Planetary Distances...........PAGE 146
Form and Size.....................147
Planetary Atmospheres.............148
Earth's Form and Size.............149
Means of Discovery................149
Planetary Motions.................150
Rotation 150
Translation 151
Sidereal Revolution...............152
Motion of the Earth.............152
Evidences of the Earth's Motion...152
Ancient Conceptions...............153
How they Gave Way.................153
Earth's Rotation..................154
Influence of Centrifugal Force upon Gravity.........................154
Earth's Translation...............155
Precession of the Equinoxes.......155
Retrogradations and Stations of the Planets.........................155
Aberration of Light...............156
Influence of Scientific Fact upon Opinion.........................157
Kepler's Laws...................158
Annual Parallax...................158
Circles 158
KEPLER............................159
His Three Laws....................159
First Law.........................159
Second Law........................159
Third Law.........................160
Three Problems....................161
Predictions of Eclipses...........161
Transit of Venus..................161
Foundation of Celestial Mechanics.162
SECTION 2. *Dynamical Phenomena*..162
Gravitation.......................162
Character of Laws of Motion.......162
Their History.....................163
Newton's Demonstration............164
Old Difficulty Explained..........165
Term *Attraction* Inadmissible....165
Extent of the Demonstration.......166
Term *Gravitation* Unobjectionable.167
Gravitation is that of Molecules..167
Secondary Gravitation.............168
Domain of the Law.................168

CHAPTER IV.

Celestial Statics.

Consummation by Newton............169
Statical Considerations...........169
First Method of Inquiry into Masses..170
Second Method.....................170
Third Method......................170
SECTION 1. *Weight of the Earth*..171
SECTION 2. *Form of the Planets*..172
Difficulty of the Inquiry.........172
Geometrical Estimate..............172
Estimate from Perturbations.......173
Indirect Estimate of the Earth's Form.173
Hydrostatic Theory of Planetary Forms...........................173
SECTION 3. *The Tides*............174

Question of the Tides..........PAGE 174
Theory of the Tides..................174
Influence of the Sun................175
Of the Moon........................175
Composite Influence................175
Requisites for Exactitude..........176

CHAPTER V.
Celestial Dynamics.

Perturbations.....................177
Instantaneous.....................177
Gradual Perturbations.............178
Perturbations of Translation......179
Problem of Three Bodies...........179
Centre of the Solar System........180
Problem of the Planets............180
Of the Satellites.................180
Of the Comets.....................181
Perturbations of Rotation.........182
The Planets 182
The Satellites....................183
Device of an Invariable Plane.....183
Stability of our System...........184
Resistance of a Medium............184
Independence of the Solar System....185
Achievements of Celestial Dynamics..186

CHAPTER VI.
Sidereal Astronomy and Cosmogony.

Multiple Stars....................186
Our Cosmogony.....................188
Origin of Positive Cosmogony......188
Cosmogony of Laplace..............189
Recapitulation....................191

BOOK III.—PHYSICS.

CHAPTER I.
General View.

Imperfect Condition of the Science...192
Its Domain........................192
Compared with Chemistry...........193
Its Generality....................193
Dealing with Masses or Molecules..193
Changes of Arrangement or Composition of Molecules..............194
Description of Physics............194
Instruments.......................195
Methods of Inquiry................195
Observation 195
Experiment........................195
Application of Mathematical Analysis.195
Encyclopædic Rank of Physics......196
Relation to Astronomy.............196
To Mathematics....................197
To the other Sciences.............197
To Human Progress.................198
Human Power of Modifying Phenomena............................198
Prevision Imperfect...............199
Characteristics of each Science...199
Philosophy of Hypothesis..........199

Necessary Condition...........PAGE 200
Two Classes of Hypothesis..........200
First Class Indispensable..........200
Second Class Chimerical............201
History of the Second Class........202
In Astronomy.......................203
In Physics.........................203
Rule of Arrangement in Physics.....204
Order..............................204

CHAPTER II.
Barology.

Divisions..........................205
SECTION 1. *Statics*...............205
History............................205
Cases of Liquids...................206
First Case.........................206
Second Case........................207
Case of Gases......................208
History 208
Condition of the Problem...........209
SECTION 2. *Dynamics*..............209
History............................209
Fluids.............................211
Case of Liquids....................211
Existing State of Barology.........211

CHAPTER III.
Thermology.

Its Nature.........................212
History............................212
Relation to Mathematics............212
SECTION 1. *Mutual Thermological Influence*.........................213
Two Parts..........................213
Mutual Influence...................213
Radiation of Heat..................213
Propagation by Contact.............214
Conductibility.....................215
Permeability.......................215
Penetrability......................215
Specific Heat......................216
SECTION 2. *Constituent Changes by Heat*.216
Latent Heat........................217
Change of Volume...................217
Change in State of Aggregation.....218
Law of Engagement and Disengagement of Heat.....................218
Vapors.............................218
Temperatures of Ebullition.........219
Hygrometrical Equilibrium..........219
SECTION 3. *Thermology Connected with Analysis*......................220
SECTION 4. *Terrestrial Temperatures*..221
Interior Heat......................221
Temperature of Planetary Intervals..221
Conditions of the Problem..........222

CHAPTER IV.
Acoustics.

Its Nature.........................222
Relation to the Study of Inorganic Bodies..........................223

CONTENTS.

Relation to Physiology..........PAGE 223
To Mathematics....................223
Divisions........................225
SECTION 1. *Propagation of Sound*....226
Effect of Atmospheric Agitation.....226
SECTION 2. *Intensity of Sounds*......226
SECTION 3. *Theory of Tones*.........227
Composition of Sounds..............229
Recapitulation....................229

CHAPTER V.
Optics.

Hypotheses on the Nature of Light..230
Excessive Tendency to Systemize.....232
Divisions of Optics................233
Irrelevant Matters.................234
Theory of Vision...................234
Specific Color of Bodies...........234
SECTION 1. *Study of Direct Light*....235
Optics Proper.....................235
Imperfections.....................235
Photometry.......................235
SECTION 2. *Catoptrics*.............236
Great Law of Reflection............236
Law of Absorption not Found.......237
SECTION 3. *Dioptrics*..............237
Great Law of Refraction...........237
Newton's Discoveries on Elementary Colors..........................238
SECTION 4. *Diffraction*............239

CHAPTER VI.
Electrology.

History 240
Condition........................240
Arbitrary Hypotheses..............240
Relation to Mathematics...........241
Unsound Application...............241
Sound Application.................241
Limits...........................242
Divisions........................242
SECTION 1. *Electric Production*.....242
Causes of Electrization...........242
Chemical Action..................242
Thermological Action..............242
Friction.........................243
Pressure 243
Contact..........................243
Other Causes.....................243
Instruments......................243
SECTION 2. *Electric Statics*........244
Great Law of Distribution.........244
Electric Equilibrium..............245
SECTION 3. *Electric Dynamics*......245
Ampère's Experiments.............245
Conclusion of Physics.............247

BOOK IV.—CHEMISTRY.

CHAPTER I.

Its Nature........................249
Great Imperfection................250
Capacities.....................PAGE 250
Object of Chemistry...............250
Specific Character of its Action....251
Condition of Action..............251
Definition.......................252
Elements........................252
Combination.....................253
Rational Definition...............253
Means of Investigation............253
Observation.....................253
Experiment......................254
Comparison......................254
Chemical Analysis and Synthesis.....255
Rank of the Science...............256
Relation to Mathematics...........256
To Astronomy....................257
To Physiology...................257
To Sociology....................258
Degree of Possible Perfection.......258
Intrusion of Hypotheses...........258
Actual Imperfection..............259
Comparative Imperfection.........259
Relation to Human Progress.......260
Art of Nomenclature..............260
State of Chemical Doctrine........261
Divisions of the Science..........262
No Organic Chemistry............262
Principles of Composition and Decomposition....................263

CHAPTER II.
Inorganic Chemistry.

Mode of Beginning the Study.......264
Plurality of Elements..............264
Classification of Elements.........266
Classification of Berzelius........267
Premature Effort.................267
Requisite Preparation as to Method..267
As to Doctrine...................268
First Condition..................268
Second Condition.................269
Method of Analysis...............269
Chemical Dualism................269
Law of Double Saline Decompositions.272
Chemical Theory of Air and Water..273
Air..............................273
Water..........................274

CHAPTER III.
Doctrine of Definite Proportions.

Scope of the Doctrine.............275
As to Doctrine 276
As to Method....................276
Its History......................276
Richter's Law....................276
Berthollet's Extension.............277
Dalton's Further Extension........277
Atomic Theory...................277
Extension by Berzelius............278
Extension by Gay-Lussac..........279
Wollaston's Verification...........279
Scope of Application of Numerical Chemistry......................280

Objection of Dissolution........PAGE 281
Of Metallic Alloys.................281
Of Organic Substances............282
Application of the Principle of Dualism.282

CHAPTER IV.

The Electro-Chemical Theory.

Relation of Electricity to Chemistry..285
History of the Case................285
Nicholson's Discovery..............285
Davy's Discovery...................286
Berzelius's Extension..............286
Synthetical Process................286
Becquerel's........................286
Study of Combustion................287
Lavoisier's Theory.................287
First Division.....................288
Berthollet's Limitation............288
Second Division....................289
Difficulties of the Theory.........290
Its Position and Powers............292

CHAPTER V.

Organic Chemistry.

Confusion of two Kinds of Fact.....293
Relation of Chemistry to Anatomy...295
To Physiology......................295
Partition of Organic Chemistry.....297
Application of Dualism to Organic Compounds....................299
Summary of the Chapter.............300
Summary of the Book................300

BOOK V.—BIOLOGY.

CHAPTER I.

General View of Biology.

Opposite Starting-Points in Philosophy.301
Starting-Point of Physiology.......301
Its Present Imperfection...........302
Its Field..........................302
Relation to Medicine...............303
Its Object.........................304
Idea of Life.......................304
De Blainville's Definition.........306
Definition of Biology..............307
Means of Investigation.............308
OBSERVATION........................308
Artificial Apparatus...............308
Chemical Exploration...............308
EXPERIMENT.........................310
By Affecting the Organism..........310
By Affecting the Medium............311
Comparative Experiment.............311
Pathological Investigation.........312
COMPARISON.........................313
Five Kinds of Comparison...........315
Comparison of Parts of the same Organism.......................316
Of Phases of the same Organism.....317
Of different Organisms.............317
Relation of Biology to other Sciences.319

To Chemistry................PAGE 319
To Physics.........................320
To Astronomy.......................321
To Mathematics.....................325
Use of Scientific Fictions.........327
Condition and Prospects of the Science...........................328
Requisite Education................328
Art of Classification..............329
Influence of Biology upon the Positive Spirit........................330
Distribution of the Science........332
Where to Begin.....................334

CHAPTER II.

Anatomical Philosophy.

Development of Statical Biology....335
Process of Discovery of the Tissues...335
Combination with Comparative Anatomy..........................336
Elements and Products..............337
Vitality of the Organic Fluids.....338
Classification of the Tissues......340
ORGANIC: Cellular Tissue...........340
Dermous Tissue.....................341
Sclerous Tissue....................341
Kystous Tissue.....................341
ANIMAL: Muscular and Nervous Tissues............................341
Limitation of the Inquiry..........342

CHAPTER III.

Biotaxic Philosophy.

Comparative Anatomy of Vegetables and Animals....................343
Animal Anatomy our Subject.........344
Division of the Natural Method.....344
Natural Groups.....................344
Co-ordination of Them..............345
Three Laws of Co-ordination........345
Question of Permanence of Organic Species........................346
Two Logical Conditions of the Study.349
Subordination of Characteristics...349
Procedure in the use of the Natural Method.........................349
Division of Animal and Vegetable Kingdom........................351
Hierarchy of the Animal Kingdom...351
Attribute of Symmetry..............352
Division of Duplicate Animals......352
Division of Articulated Animals....352
Consideration of the Envelope......353
Natural Method applied to the Vegetable Kingdom..................353
Difficulty of Co-ordination........353

CHAPTER IV.

Organic or Vegetative Life.

Condition of Dynamical Biology.....355
Vital Phenomena: Their Division...356

CONTENTS.

Theory of Organic Media........PAGE 356
Exterior Conditions of Life..........357
Mechanical Conditions: Weight.....357
Pressure........................358
Motion and Rest...................358
Thermological Action..............358
Light, Electricity, etc..............359
Molecular Conditions..............359
Air and Water....................359
Study of Specifics................360
History of Physiology.............361
Philosophical Character of Physiology.362
Division of the Study.............363
Two Functions of the Organic Life...364
Results of Organic Action.........365
State of Composition and Decomposition...................365
Vital Heat........................365
Electrical State..................366
Production and Development of Living Bodies.........................366
Decline of the Organism..........367

CHAPTER V.

The Animal Life.

Transition from the Inorganic to the Organic.......................369
Primitive Nervous Properties.......370
Relation of the Animal to the Organic Life............................370
To Inorganic Philosophy...........371
Properties of Tissue..............371
Sensibility......................372
Irritability......................372
Present State of Analysis of Animality.373
Movement.........................373
Exterior Sensation................375
Interior Sensation................376
Mode of Action of Animality......376
Intermittence....................377
Sleep............................377
Habit............................378
Need of Activity.................379
Association of the Animal Functions..379

CHAPTER VI.

Intellectual and Moral or Cerebral Functions.

Short-coming of Descartes..........380
History till Gall's Time...........381
Positive Theory of Cerebral Functions.382
Its Proper Place..................382
Vices of Psychological Systems.....382
Method: Interior Observation......383
Doctrine: Relation between the Affective and Intellectual Faculties.....384
Brutes and Man...................384
Theory of the I...................385
Reason and Instinct...............385
Basis of Gall's Doctrine..........387
Divisions of the Brain............388
Subdivision.......................389
Objections........................390

Necessity of Human Actions....PAGE 390
Answered........................390
Hypothetical Distribution of Faculties.391
Needed Improvements..............392
Anatomical Basis.................392
Physiological Analysis of Faculties...393
Examination of Historical Cases....393
Pathological and Comparative Analysis.394
Laws of Action...................395
Intermittence and Continuity.......396
Association......................396
Unity of the Brain and Nervous System............................396
Imperfect State of Phrenology.....397
Present State of Biology..........397
Retrospect of Natural Philosophy....397

BOOK VI.—SOCIAL PHYSICS.

CHAPTER I.

Necessity of an Opportuneness of this New Science.

Proposal of the Subject............399
Conditions of Order and Progress....401
The Theological Polity............402
Criterion of Social Doctrine.......403
Failure of the Theological Polity.....403
The Metaphysical Polity...........406
Becomes Obstructive..............408
Dogma of Liberty of Conscience.....408
Dogma of Equality................411
Dogma of the Sovereignty of the People............................412
Dogma of National Independence....412
Inconsistency of the Metaphysical Doctrine........................413
Notion of a State of Nature.........413
Adhesion to the Worn-out..........414
Recurrence to War................415
Principle of Political Centralization..415
The Stationary Doctrine...........418
Dangers of the Critical Period......420
Intellectual Anarchy..............420
Destruction of Public Morality......422
Private Morality..................423
Political Corruption..............425
Low Aims of Political Questions....427
Fatal to Progress.................427
Fatal to Order...................428
Incompetence of Political Leaders....429
Advent of the Positive Philosophy...430
Logical Coherence of the Doctrine...431
Its Effect on Order...............432
Its Effect on Progress.............434
Anarchical Tendencies of the Scientific Class...........................437
Conclusion.......................438

CHAPTER II.

Principal Philosophical Attempts to Constitute a Social Science.

History of Social Science..........439
Aristotle's "Politics".............442

Montesquieu..................PAGE 442
Condorcet.......................444
Political Economy................446
Growth of Historical Study........449

CHAPTER III.

Characteristics of the Positive Method in its Application to Social Phenomena.

Infantile State of Social Science.....452
The Relative Superseding the Absolute.453
Presumptuous Character of the Existing Political Spirit...............454
Prevision of Social Phenomena......456
Spirit of Social Science............457
Statical Study......................457
Social Organization................458
Political and Social Concurrence....459
Interconnection of the Social Organism........................461
Order of Statical Study.............462
Dynamical Study...................463
Social Continuity..................464
Produced by Natural Laws.........464
Notion of Human Perfectibility.....467
Limits of Political Action...........469
Social Phenomena Modifiable.......469
Order of Modifying Influences......471
Means of Investigation in Social Science.473
Direct Means......................474
Observation 474
Experiment........................477
Comparison.......................478
Comparison with Inferior Animals...478
Comparison of Co-Existing States of Society.........................479
Comparison of Consecutive States....481
Promise of a Fourth Method of Investigation 484

CHAPTER IV.

Relation of Sociology to the other Departments of Positive Philosophy.

Relation to Biology.................486
Relation to Inorganic Philosophy....489
Man's action on the External World..491
Necessary Education................492
Mathematical Preparation..........492
Pretended Theory of Chances.......492
Reaction of Sociology..............494
As to Doctrine....................494
As to Method.....................495
Speculative Rank of Sociology.....497

CHAPTER V.

Social Statics, or Theory of the Spontaneous Order of Human Society.

Three Aspects.....................498
1. The Individual 498
2. The Family 502
The Sexual Relation...............504
The Parental Relation.............506
3. Society 508

Distribution of Employments....PAGE 510
Inconveniences....................511
Basis of the True Theory of Government 512
Elementary Subordination..........512
Tendency of Society to Government..514

CHAPTER VI.

Social Dynamics; or Theory of the Natural Progress of Human Society.

Scientific View of Human Progression.515
Course of Man's Social Development.516
Rate of Progress...................517
Ennui 517
Duration of Human Life...........518
Increase of Population.............519
The Order of Evolution............521
Law of the Three Periods..........522
The Theological Period.............523
Intellectual Influence of the Theological Philosophy..................525
Social Influences of the Theological Philosophy 528
Institution of a Speculative Class....529
The Positive Stage.................530
Attempted Union of the Two Philosophies 531
The Metaphysical Period...........533
Co-existence of the Three Periods in the Same Mind..................534
Corresponding Material Development.535
Primitive Military Life.............535
Primitive Slavery 536
The Military *Régime* Provisional.....536
Affinity between the Theological and Military *Régime*..................537
Affinity between the Positive and Industrial Spirit....................539
Intermediate *Régime*................540

CHAPTER VII.

Preparation of the Historical Question.— First Theological Phase: Fetichism.— Beginning of the Theological and Military System.

Limitations of the Analysis.........541
Abstract Treatment of History......542
Abstract Inquiry into Laws.........543
Co-existence of Successive States.....544
Fetichism.........................545
Starting-point of the Human Race....545
Relation of Fetichism to Morals......548
To Language......................548
To Intellect.......................549
To Society........................549
Astrolatry 550
Relation of Fetichism to Human Knowledge 551
To the Fine Arts...................552
To Industry.......................552
Political Influence..................553
Institution of Agriculture...........555
Protection to Products......... 556

Transition to Polytheism.......PAGE 557
The Metaphysical Spirit Traceable...560

CHAPTER VIII.

Second Phase: Polytheism.—Development of the Theological and Military System.

True Sense of Polytheism..........562
Its Operation on the Human Mind...562
Polytheistic Science................564
Polytheistic Art...................566
Polytheistic Industry..............571
Social Attributes of Polytheism......572
Polity of Polytheism...............573
Worship...........................574
Civilization by War................574
Sacerdotal Sanctions...............575
Two Characteristics of the Polity....576
Slavery...........................576
Concentration of Spiritual and Temporal Power....................578
Morality of Polytheism.............580
Moral Effects of Slavery...........580
Subordination of Morality to Polity..581
Personal Morality..................582
Social Morality....................583
Domestic Morality..................583
Three Phases of Polytheism.........584
The Egyptian, or Theocratic........584
Caste.............................585
The Greek, or Intellectual.........588
Science...........................589
Philosophy........................591
The Roman, or Military.............592
Conquest..........................593
Morality..........................594
Intellectual Development...........594
Preparation for Monotheism.........595
The Jews..........................598

CHAPTER IX.

Age of Monotheism.—Modification of the Theological and Military System.

Catholicism, the Form..............599
Principle of Political Rule..........600
The Great Problem..................603
Separation of Spiritual and Temporal Power........................603
Transposition of Morals and Politics..604
Function of Each...................604
The Speculative Class..............605
The Catholic System 606
Ecclesiastical Organization..........607
Elective Principle.................607
Monastic Institutions..............608
Special Education of the Clergy.....608
Restriction of Inspiration...........609
Ecclesiastical Celibacy.............610
Temporal Sovereignty of the Popes..610
Educational Function...............612
Dogmatic Conditions................614
Dogma of Exclusive Salvation.......614
Of the Fall of Man.................615
Of Purgatory......................615

Of Christ's Divinity............PAGE 615
Of the Real Presence...............615
Worship...........................616
Significance of Controversies........616
Temporal Organization of the *Régime*.617
The Germanic Invasions.............617
Rise of Defensive System...........618
Of Territorial Independence........618
Slavery Converted into Serfage......619
Intervention of the Church Throughout..............................619
Institution of Chivalry.............621
Operation of the Feudal System.....621
Moral Aspect of the *Régime*........622
Rise of Morality over Polity........622
Source of Moral Influence of Catholicism..........................623
Moral Types.......................626
Personal Morality under Catholicism.626
Domestic..........................627
Social............................628
Intellectual Aspect of the *Régime*...629
Philosophy........................630
Science...........................631
Art...............................632
Industry..........................633
Provisional Nature of the *Régime*...633
Division between Natural and Moral Philosophy......................634
The Metaphysical Spirit............634
Temporal Decline...................635
Conclusion........................636

CHAPTER X.

Metaphysical State, and Critical Period of Modern Society.

Conduct of the Inquiry.............637
Necessity of a Transitional State.....638
Its Commencement..................638
Division of the Critical Period......640
Causes of Spontaneous Decline......640
Decline under Negative Doctrine....642
Character of the Provisional Philosophy..............................642
Christian Period of the Doctrine.....643
Deistical Period...................643
Organs of the Doctrine.............645
Scholasticism.....................645
The Legists.......................646
Period of Spontaneous Spiritual Decline..............................647
Spontaneous Temporal Decline......649
True Character of the Reformation...650
The Jesuits.......................652
Final Decay of Catholicism.........652
Vices of Protestantism.............653
Temporal Dictatorship..............654
Royal and Aristocratic.............655
Rise of Ministerial Function........657
Military Decline...................657
Rise of Diplomatic Function........659
Intellectual Influence of Protestantism..............................660
Catholic Share in Protestant Results..662

Jansenism	PAGE 662
Quietism	662
Moral Influence of Protestantism	663
Three Stages of Dissolution	664
Lutheranism	664
Calvinism	664
Socinianism	665
Quakerism	665
Political Revolutions of Protestantism	665
Holland	666
England	666
America	666
Attendant Errors	667
Subjection of Spiritual Power	667
Moral Changes under Protestantism	668
Stage of Full Development of the Critical Doctrine	669
Protestantism Opposed to Progress	670
The Negative Philosophy	671
Three Periods of the Negative Philosophy	673
Systemized	673
Hobbes	673
Its Intellectual Character	673
Its Moral Character	674
Its Political Character	675
Its Propagation	676
School of Voltaire	678
Its Political Action	679
School of Rousseau	680
The Economists	681
Attendant Evils	681

CHAPTER XI.

Rise of the Elements of the Positive State.—Preparation for Social Reorganization.

Date of Modern History	683
Rise of New Elements	684
Philosophical Order of Employments	685
Classifications	685
Order of Succession	686
THE INDUSTRIAL MOVEMENT	688
Birth of Political Liberty	691
Characteristics of the Industrial Movement	692
Personal Effect	692
Domestic Effect	693
Social Effect	693
Industrial Policy	695
Relation to Catholicism	695
Relation to the Temporal Authority	696
Administration	696
Three Periods	697
Paid Armies	697
Rise of Public Credit	698
Political Alliances	698
Mechanical Inventions	698
The Compass	698
Fire-Arms	699
Printing	699
Maritime Discovery	700
Second Period	701
Colonial System	702

Slavery	PAGE 703
Third Period	704
Final Subordination of the Military Spirit	704
Spread of Industry	704
THE INTELLECTUAL MOVEMENT	706
The Æsthetic Development	706
Intellectual Originality	708
Relation of Art to Industry	709
Critical Character of Art	711
Retrograde Character	712
Relation of Art to Politics	713
Spread of Art	715
The Scientific Development	716
New Birth of Science	717
Relation to Monotheism	718
Astrology	720
Alchemy	720
First Modern Phase of Progress	721
Second Phase	723
Filiation of Discoveries	724
Relation of Science to Old Philosophy	724
Galileo	725
Social Relations of Science	725
Third Phase	726
Relations of Discoveries	727
Stage of Speciality	728
The Philosophical Development	729
Reason and Faith	729
Bacon and Descartes	731
Political Philosophy	733
The Scotch School	734
Political Philosophy	734
Idea of Progression	735
Gaps to be Supplied	736
In Industry	736
In Art	738
In Philosophy	738
In Science	738
Existing Needs	738

CHAPTER XII.

Review of the Revolutionary Crisis.—Ascertainment of the Final Tendency of Modern Society.

France first Revolutionized	739
Precursory Events	740
First Stage.—The Constituent Assembly	740
Second Stage.—The National Convention	742
Alliance of Foes	742
Constitutional Attempt	744
Military Ascendency	745
Napoleon Bonaparte	745
Restoration of the Bourbons	747
Fall of the Bourbons	748
The Next Reign	748
Extension of the Movement	749
Completion of the Theological Decay	750
Decay of the Military System	752
Recent Industrial Progress	754
Recent Æsthetic Progress	755
Recent Scientific Progress	756

Abuses..........................PAGE 757
Recent Philosophical Progress.......761
The Law of Evolution..............763
Speculative Preparation...........764
The Spiritual Authority...........765
Its Educational Function..........769
Regeneration of Morality..........771
International Duty................772
Basis of Assent...................773
The Temporal Authority............774
Public and Private Function.......774
Principle of Co-Ordination........775
Speculative Classes Highest.......776
The Practical Classes.............776
Privileges and Compensations......777
Practical Privacy.................778
Practical Freedom.................779
Popular Claims....................779
Reciprocal Effects................781
Preparatory Stage.................781
Promotion of Order and Progress...782
National Participation............784
France............................784
Italy.............................784
Germany...........................785
England...........................785
Spain.............................786
Co-operation of Thinkers..........787
Summary of Results under the Sociological Theory..................787

CHAPTER XIII.

Final Estimate of the Positive Method.

Principle of Unity................788
Which Element shall Prevail.......788
First General Conclusion..........789
The Mathematical Element..........790
The Sociological Element..........793
Solves Antagonisms................796
Spirit of the Method..............798
Nature of the Method..............799
Inquiry into Laws.................799
Accordance with Common Sense......799
Conception of Natural Laws........801

Logical Method..................PAGE 803
Scientific Method.................803
Stability of Opinions.............805
Destination of the Method.........806
The Individual....................806
The Race..........................807
Speculative Life..................808
Practical Life....................808
Liberty of Method.................809
Extension of the Positive Method..810
Abstract of Concrete Science......811
Relations of Phases...............813
Mathematics.......................813
Astronomy.........................813
Physics and Chemistry.............814
Biology...........................815
Sociology.........................816

CHAPTER XIV.

Estimate of the Results of Positive Doctrine in its Preparatory Stage.

The Mathematical Element..........818
Application to Sociology..........819
The Astronomical Element..........821
The Physical......................822
The Chemical......................823
The Biological....................824
The Sociological..................825

CHAPTER XV.

Estimate of the Final Action of the Positive Philosophy.

The Scientific Action.............828
Abstract Speculation..............829
Concrete Research.................830
The Moral Action..................831
Personal Morality.................832
Domestic Morality.................832
Social Morality...................833
Political Action..................834
Double Government.................834
The Æsthetic Action...............836
The Five Nations..................838

POSITIVE PHILOSOPHY.

INTRODUCTION.

CHAPTER I.

ACCOUNT OF THE AIM OF THIS WORK.—VIEW OF THE NATURE AND IMPORTANCE OF THE POSITIVE PHILOSOPHY.

A GENERAL statement of any system of philosophy may be either a sketch of a doctrine to be established, or a summary of a doctrine already established. If greater value belongs to the last, the first is still important, as characterizing from its origin the subject to be treated. In a case like the present, where the proposed study is vast and hitherto indeterminate, it is especially important that the field of research should be marked out with all possible accuracy. For this purpose, I will glance at the considerations which have originated this work, and which will be fully elaborated in the course of it.

In order to understand the true value and character of the Positive Philosophy, we must take a brief general view of the progressive course of the human mind, regarded as a whole; for no conception can be understood otherwise than through its history.

From the study of the development of human intelligence, in all directions, and through all times, the discovery arises of a great fundamental law, to which it is necessarily subject, and which has a solid foundation of proof, both in the facts of our organization and in our historical experience. The law is this:—that each of our leading conceptions—each branch of our knowledge—passes successively through three different theoretical condititions: the Theological, or fictitious; the Metaphysical, or abstract; and the Scientific, or positive. In other words, the human mind, by its nature, employs in its progress three methods of philosophizing, the character of which is essentially different, and even radically opposed: viz., the theological method, the metaphysical, and the positive. Hence arise three philosophies, or general systems of conceptions on the aggregate of phenomena, *Law of human progress.*

each of which excludes the others. The first is the necessary point of departure of the human understanding; and the third is its fixed and definite state. The second is merely a state of transition.

<small>First Stage.</small> In the theological state, the human mind, seeking the essential nature of beings, the first and final causes (the origin and purpose) of all effects—in short, Absolute knowledge—supposes all phenomena to be produced by the immediate action of supernatural beings.

<small>Second Stage.</small> In the metaphysical state, which is only a modification of the first, the mind supposes, instead of supernatural beings, abstract forces, veritable entities (that is, personified abstractions) inherent in all beings, and capable of producing all phenomena. What is called the explanation of phenomena is, in this stage, a mere reference of each to its proper entity.

<small>Third Stage.</small> In the final, the positive state, the mind has given over the vain search after Absolute notions, the origin and destination of the universe, and the causes of phenomena, and applies itself to the study of their laws—that is, their invariable relations of succession and resemblance. Reasoning and observation, duly combined, are the means of this knowledge. What is now understood when we speak of an explanation of facts is simply the establishment of a connection between single phenomena and some general facts, the number of which continually diminishes with the progress of science.

<small>Ultimate point of each.</small> The Theological system arrived at the highest perfection of which it is capable when it substituted the providential action of a single Being for the varied operations of the numerous divinities which had been before imagined. In the same way, in the last stage of the Metaphysical system, men substitute one great entity (Nature) as the cause of all phenomena, instead of the multitude of entities at first supposed. In the same way, again, the ultimate perfection of the Positive system would be (if such perfection could be hoped for) to represent all phenomena as particular aspects of a single general fact—such as Gravitation, for instance.

The importance of the working of this general law will be established hereafter. At present, it must suffice to point out some of the grounds of it.

<small>Evidences of the law.</small> There is no science which, having attained the positive stage, does not bear marks of having passed through the others. Some time since it was (whatever it might be) composed, as we can now perceive, of metaphysical abstractions; and, further back in the course of time, it took its form from theological conceptions. <small>Actual.</small> We shall have only too much occasion to see, as we proceed, that our most advanced sciences still bear very evident marks of the two earlier periods through which they have passed.

The progress of the individual mind is not only an illustration, but an indirect evidence of that of the general mind. The point of departure of the individual and of the race being the same, the phases of the mind of a man correspond to the epochs of the mind of the race. Now, each of us is aware, if he looks back upon his own history, that he was a theologian in his childhood, a metaphysician in his youth, and a natural philosopher in his manhood. All men who are up to their age can verify this for themselves.

Besides the observation of facts, we have theoretical reasons in support of this law.

The most important of these reasons arises from the necessity that always exists for some theory to which to refer our facts, combined with the clear impossibility that, at the outset of human knowledge, men could have formed theories out of the observation of facts. All good intellects have repeated, since Bacon's time, that there can be no real knowledge but that which is based on observed facts. This is incontestable, in our present advanced stage; but, if we look back to the primitive stage of human knowledge, we shall see that it must have been otherwise then. If it is true that every theory must be based upon observed facts, it is equally true that facts can not be observed without the guidance of some theory. Without such guidance, our facts would be desultory and fruitless; we could not retain them: for the most part we could not even perceive them.

Theoretical.

Thus, between the necessity of observing facts in order to form a theory, and having a theory in order to observe facts, the human mind would have been entangled in a vicious circle, but for the natural opening afforded by Theological conceptions. This is the fundamental reason for the theological character of the primitive philosophy. This necessity is confirmed by the perfect suitability of the theological philosophy to the earliest researches of the human mind. It is remarkable that the most inaccessible questions—those of the nature of beings, and the origin and purpose of phenomena—should be the first to occur in a primitive state, while those which are really within our reach are regarded as almost unworthy of serious study. The reason is evident enough:—that experience alone can teach us the measure of our powers; and if men had not begun by an exaggerated estimate of what they can do, they would never have done all that they are capable of. Our organization requires this. At such a period there could have been no reception of a positive philosophy, whose function is to discover the laws of phenomena, and whose leading characteristic it is to regard as interdicted to human reason those sublime mysteries which theology explains, even to their minutest details, with the most attractive facility. It is just so under a practical view of the nature of the researches with which men first occupied themselves. Such inquiries offered the powerful charm of unlimited empire over the external world—a world destined wholly for our use, and involved in every way with our existence. The theological philosophy, presenting this view, administered exactly the stimulus necessary to incite the human mind to the irksome labor without which it could make no progress. We can now scarcely conceive of such a state of things, our reason having become sufficiently mature to enter upon laborious scientific researches, without needing any such stimulus as wrought upon the imaginations of astrologers and alchemists. We have motive enough in the hope of discovering the laws of phenomena, with a view to the confirmation or rejection of a theory. But it could not be so in the earliest days; and it is to the chimeras of astrology and alchemy that we owe the long series of observations and experiments on which our positive science is based. Kepler felt this on behalf of astronomy, and Berthollet on behalf of chemistry. Thus was a spontaneous philosophy, the theological, the only possible beginning, method, and provisional system, out of which the Positive philosophy could grow. It is easy, after this, to perceive how Metaphysical methods and doctrines must have afforded the means of transition from the one to the other.

The human understanding, slow in its advance, could not step at once from the theological into the positive philosophy. The two are so radically opposed, that an intermediate system of conceptions has been necessary to render the transition possible. It is only in doing this, that metaphysical conceptions have any utility whatever. In contemplating phenomena, men substitute for supernatural direction a corresponding entity. This entity may have been supposed to be derived from the supernatural action: but it is more easily lost sight of, leaving attention free from the facts themselves, till, at length, metaphysical agents have ceased to be anything more than the abstract names of phenomena. It is not easy to say by what other process than this our minds could have passed from supernatural considerations to natural; from the theological system to the positive.

The law of human development being thus established, let us consider what is the proper nature of the Positive Philosophy.

Character of the Positive Philosophy. As we have seen, the first characteristic of the Positive Philosophy is that it regards all phenomena as subjected to invariable natural *Laws*. Our business is, —seeing how vain is any research into what are called *Causes*, whether first or final,—to pursue an accurate discovery of these Laws, with a view to reducing them to the smallest possible number. By speculating upon causes, we could solve no difficulty about origin and purpose. Our real business is to analyse accurately the circumstances of phenomena, and to connect them by the natural relations of succession and resemblance. The best illustration of this is in the case of the doctrine of Gravitation. We say that the general phenomena of the universe are *explained* by it, because it connects under one head the whole immense variety of astronomical facts; exhibiting the constant tendency of atoms toward each other in direct proportion to their masses, and in inverse proportion to the squares of their distance; while the general fact itself is a mere extension of one which is perfectly familiar to us, and which we therefore say that we know;—the weight of bodies on the surface of the earth. As to what weight and attraction are, we have nothing to do with that, for it is not a matter of knowledge at all. Theologians and metaphysicians may imagine and refine about such questions; but positive philosophy rejects them. When any attempt has been made to explain them, it has ended only in saying that attraction is universal weight, and that weight is terrestrial attraction: that is, that the two orders of phenomena are identical; which is the point from which the question set out. Again, M. Fourier, in his fine series of researches on Heat, has given us all the most important and precise laws of the phenomena of heat, and many large and new truths, without once inquiring into its nature, as his predecessors had done when they disputed about calorific matter and the action of a universal ether. In treating his subject in the Positive method, he finds inexhaustible material for all his activity of research, without betaking himself to insoluble questions.

Before ascertaining the stage which the Positive Philosophy has reached, we must bear in mind that the different kinds of our knowledge have passed through the three stages of progress at different rates, and have not therefore arrived at the same time. *History of the Positive Philosophy.* The rate of advance depends on the nature of the knowledge in question, so distinctly that, as we shall see hereafter, this consideration constitutes an accessary to the fundamental law of progress. Any kind of knowledge reaches the positive stage early in proportion to its generality, simplicity, and independence of other departments. Astronomical science, which is above all made up of facts that are general, simple, and independent of other sciences, arrived first; then terrestrial Physics; then Chemistry; and, at length, Physiology.

It is difficult to assign any precise date to this revolution in science. It may be said, like everything else, to have been always going on; and especially since the labors of Aristotle and the school of Alexandria; and then from the introduction of natural science into the West of Europe by the Arabs. But, if we must fix upon some marked period, to serve as a rallying point, it must be that,—about two centuries ago,—when the human mind was astir under the precepts of Bacon, the conceptions of Descartes, and the discoveries of Galileo. Then it was that the spirit of the Positive philosophy rose up in opposition to that of the superstitious and scholastic systems which had hitherto obscured the true character of all science. Since that date, the progress of the Positive philosophy, and the decline of the other two, have been so marked that no rational mind now doubts that the revolution is destined to go on to its completion,—every branch of knowledge being, sooner or later, brought within the operation of Positive philosophy. This is not yet the case. Some are still lying outside: and not till they are brought in will the Positive philosophy possess that character of universality which is necessary to its definitive constitution.

In mentioning just now the four principal categories of phenomena,—astronomical, physical, chemical, and physiological,—there was an omission which will have been noticed. Nothing was said of Social phenomena. Though involved with the physiological, Social phenomena demand a distinct classification, both on account of their importance and of their difficulty. They are the most individual, the most complicated, the most dependent on all others; and therefore they must be the latest,—even if they had no special obstacle to encounter. This branch of science has not hitherto entered into the domain of Positive philosophy. Theological and metaphysical methods, exploded in other departments, are as yet exclusively applied, both in the way of inquiry and discussion, in all treatment of Social subjects, though the best minds are heartily weary of eternal disputes about divine right and the sovereignty of the people. This is the great, while it is evidently the only gap which has to be filled, to constitute, solid and entire, the Positive Philosophy. Now that the human mind has grasped celestial and terrestrial physics,—mechanical and chemical; organic physics, both vegetable and animal,—there remains one science, to fill up the series of sciences of observation,—Social physics. This is what men have now most need of: and this it is the principal aim of the present work to establish.

<small>New department of Positive philosophy.</small>

<small>Social Physics.</small>
It would be absurd to pretend to offer this new science at once in a complete state. Others, less new, are in very unequal conditions of forwardness. But the same character of positivity which is impressed on all the others will be shown to belong to this. This once done, the philosophical system of the moderns will be in fact complete, as there will then be no phenomenon which does not naturally enter into some one of the five great categories. All our fundamental conceptions having become homogeneous, the Positive state will be fully established. It can never again change its character, though it will be for ever in course of development by additions of new knowledge. Having acquired the character of universality which has hitherto been the only advantage resting with the two preceding systems, it will supersede them by its natural superiority, and leave to them only an historical existence.

<small>Secondary aim of this work.</small>
We have stated the special aim of this work. Its secondary and general aim is this:—to review what has been effected in the Sciences, in order to show that they are not radically separate, but all branches from the same trunk. If we had confined ourselves to the first and special object of the work, we should have produced merely a study of Social physics: whereas, in introducing the second and general, we offer a study of Positive philosophy, passing in review all the positive sciences already formed.

The purpose of this work is not to give an account of the Natural Sciences. Besides that it would be endless, and that it would require a scientific preparation such as no one man possesses, it would be apart from our object, which is to go through a course of not Positive Science, but Positive Philosophy. We have only to consider each fundamental science in its relation to the whole positive system, and to the spirit which characterizes it; that is, with regard to its methods and its chief results. *To review the philosophy of the Sciences.*

The two aims, though distinct, are inseparable; for, on the one hand, there can be no positive philosophy without a basis of social science, without which it could not be all-comprehensive; and, on the other hand, we could not pursue Social science without having been prepared by the study of phenomena less complicated than those of society, and furnished with a knowledge of laws and anterior facts which have a bearing upon social science. Though the fundamental sciences are not all equally interesting to ordinary minds, there is no one of them that can be neglected in an inquiry like the present; and, in the eye of philosophy, all are of equal value to human welfare. Even those which appear the least interesting have their own value, either on account of the perfection of their methods, or as being the necessary basis of all the others.

Lest it should be supposed that our course will lead us into a wilderness of such special studies as are at present the bane of a true positive philosophy, we will briefly advert to the existing prevalence of such special pursuit. In the primitive state of human knowledge there is no regular division of intellectual labor. Every student cultivates all the sciences. As knowledge accrues, the sciences part off; and students devote themselves each to some one branch. It is owing to this division of employment, and concentration of whole minds upon a single department, that science has made so prodigious an advance in modern times; and the perfection of this division is one of the most important characteristics of the Positive philosophy. But, while admitting all the merits of this change, we can not be blind to the eminent disadvantages which arise from the limitation of minds to a particular study. It is inevitable that each should be possessed with exclusive notions, and be therefore incapable of the general superiority of ancient students, who actually owed that general superiority to the inferiority of their knowledge. We must consider whether the evil can be avoided without losing the good of the modern arrangement; for the evil is becoming urgent. We all acknowledge that the divisions established for the convenience of scientific pursuit are radically artificial; and yet there are very few who can embrace in idea the whole of any one science: each science moreover being itself only a part of a great whole. Almost every one is busy about his own particular section, without much thought about its relation to the general system of positive knowledge. We must not be blind to the evil, nor slow *Speciality.*

in seeking a remedy. We must not forget that this is the weak side of the positive philosophy, by which it may yet be attacked, with some hope of success, by the adherents of the theological and metaphysical systems. As to the remedy, it certainly does not lie in a return to the ancient confusion of pursuits, which would be mere retrogression, if it were possible, which it is not. It lies in perfecting the division of employments itself,—in carrying it one degree higher,—in constituting one more speciality from the study of scientific generalities. Let us have a new class of students, suitably prepared, whose business it shall be to take the respective sciences as they are, determine the spirit of each, ascertain their relations and mutual connection, and reduce their respective principles to the smallest number of general principles, in conformity with the fundamental rules of the Positive Method. At the same time, let other students be prepared for their special pursuit by an education which recognises the whole scope of positive science, so as to profit by the labors of the students of generalities, and so as to correct reciprocally, under that guidance, the results obtained by each. We see some approach already to this arrangement. Once established, there would be nothing to apprehend from any extent of division of employments. When we once have a class of learned men, at the disposal of all others, whose business it shall be to connect each new discovery with the general system, we may dismiss all fear of the great whole being lost sight of in the pursuit of the details of knowledge. The organization of scientific research will then be complete; and it will henceforth have occasion only to extend its development, and not to change its character. After all, the formation of such a new class as is proposed would be merely an extension of the principle which has created all the classes we have. While science was narrow, there was only one class: as it expanded, more were instituted. With a further advance a fresh need arises, and this new class will be the result.

The general spirit of a course of Positive Philosophy having been thus set forth, we must now glance at the chief advantages which may be derived, on behalf of human progression, from the study of it. Of these advantages, four may be especially pointed out.

I. The study of the Positive Philosophy affords the only rational means of exhibiting the logical laws of the human mind, which have hitherto been sought by unfit methods. To explain what is meant by this, we may refer to a saying of M. de Blainville, in his work on Comparative Anatomy, that every active, and especially every living being, may be regarded under two relations—the Statical and the Dynamical; that is, under conditions or in action. It is clear that all considerations range themselves under the one or the other of these heads. Let us apply this classification to the intellectual functions.

If we regard these functions under their Statical aspect—that is, if we consider the conditions under which they exist—we must determine the organic circumstances of the case, which inquiry involves it with anatomy and physiology. If we look at the Dynamic aspect, we have to study simply the exercise and results of the intellectual powers of the human race, which is neither more nor less than the general object of the Positive Philosophy. In short, looking at all scientific theories as so many great logical facts, it is only by the thorough observation of these facts that we can arrive at the knowledge of logical laws. These being the only means of knowledge of intellectual phenomena, the illusory psychology, which is the last phase of theology, is excluded. It pretends to accomplish the discovery of the laws of the human mind by contemplating it in itself; that is, by separating it from causes and effects. Such an attempt, made in defiance of the physiological study of our intellectual organs, and of the observation of rational methods of procedure, can not succeed at this time of day.

The Positive Philosophy, which has been rising since the time of Bacon, has now secured such a preponderance, that the metaphysicians themselves profess to ground their pretended science on an observation of facts. They talk of external and internal facts, and say that their business is with the latter. This is much like saying that vision is explained by luminous objects painting their images upon the retina. To this the physiologists reply that another eye would be needed to see the image. In the same manner, the mind may observe all phenomena but its own. It may be said that a man's intellect may observe his passions, the seat of the reason being somewhat apart from that of the emotions in the brain; but there can be nothing like scientific observation of the passions, except from without, as the stir of the emotions disturbs the observing faculties more or less. It is yet more out of the question to make an intellectual observation of intellectual processes. The observing and observed organ are here the same, and its action can not be pure and natural. In order to observe, your intellect must pause from activity; yet it is this very activity that you want to observe. If you can not effect the pause, you can not observe: if you do effect it, there is nothing to observe. The results of such a method are in proportion to its absurdity. After two thousand years of psychological pursuit, no one proposition is established to the satisfaction of its followers. They are divided, to this day, into a multitude of schools, still disputing about the very elements of their doctrine. This interior observation gives birth to almost as many theories as there are observers. We ask in vain for any one discovery, great or small, which has been made under this method. The psychologists have done some good in keeping up the activity of our understandings, when there was no better work for our faculties to do; and they may have added something to our stock of knowledge. If they have done so, it is by practising the Positive method—by observing the progress of the human mind in the light of science; that is, by ceasing, for the moment, to be psychologists.

The view just given in relation to logical Science becomes yet more striking when we consider the logical Art.

The Positive Method can be judged of only in action. It can not be looked at by itself, apart from the work on which it is employed. At all events, such a contemplation would be only a dead study, which could produce nothing in the mind which loses time upon it. We may talk for ever about the method, and state it in terms very wisely, without knowing half so much about it as the man who has once put it in practice upon a single particular of actual research, even without any philosophical intention. Thus it is that psychologists, by dint of reading the precepts of Bacon and the discourses of Descartes, have mistaken their own dreams for science.

Without saying whether it will ever be possible to establish, *à priori*, a true method of investigation, independent of a philosophical study of the sciences, it is clear that the thing has never been done yet, and that we are not capable of doing it now. We can not, as yet, explain the great logical procedures, apart from their applications. If we ever do, it will remain as necessary then as now to form good intellectual habits by studying the regular application of the scientific methods which we shall have attained.

This, then, is the first great result of the Positive Philosophy—the manifestation by experiment of the laws which rule the Intellect in the investigation of truth; and, as a consequence, the knowledge of the general rules suitable for that object.

Must regenerate Education.   II. The second effect of the Positive Philosophy, an effect not less important and far more urgently wanted, will be to regenerate Education.

The best minds are agreed that our European education, still essentially theological, metaphysical, and literary, must be superseded by a Positive training, conformable to our time and needs. Even the governments of our day have shared, where they have not originated, the attempts to establish positive instruction; and this is a striking indication of the prevalent sense of what is wanted. While encouraging such endeavors to the utmost, we must not however, conceal from ourselves that everything yet done is inadequate to the object. The present exclusive speciality of our pursuits, and the consequent isolation of the sciences, spoil our teaching. If any student desires to form an idea of natural philosophy as a whole, he is compelled to go through each department as it is now taught, as if he were to be only an astronomer, or only a chemist; so that, be his intellect what it may, his training must remain very imperfect. And yet his object requires that he should obtain general positive conceptions of all the classes of natural phenomena. It is such an aggregate of conceptions, whether on a great or on a small scale, which must henceforth be the permanent basis of all human combinations. It will constitute the mind of future generations. In order to this regeneration of our intellectual system, it is necessary that the sciences, considered as branches from one trunk, should yield us, as a whole, their chief methods and their most important results. The specialities of science can be pursued by those whose vocation lies in that direction. They are indispensable; and they are not likely to be neglected; but they can never of themselves renovate our system of Education; and, to be of their full use, they must rest upon the basis of that general instruction which is a direct result of the Positive Philosophy.

III. The same special study of scientific generalities must also aid the progress of the respective positive sciences: and this constitutes our third head of advantages. Advances sciences by combining them.

The divisions which we establish between the sciences are, though not arbitrary, essentially artificial. The subject of our researches is one: we divide it for our convenience, in order to deal the more easily with its difficulties. But it sometimes happens—and especially with the most important doctrines of each science—that we need what we can not obtain under the present isolation of the sciences—a combination of several special points of view; and for want of this, very important problems wait for their solution much longer than they otherwise need do. To go back into the past for example: Descartes' grant conception with regard to analytical geometry is a discovery which has changed the whole aspect of mathematical science, and yielded the germ of all future progress; and it issued from the union of two sciences which had always before been separately regarded and pursued. The case of pending questions is yet more impressive; as, for instance, in Chemistry, the doctrine of Definite Proportions. Without entering upon the discussion of the fundamental principle of this theory, we may say with assurance that, in order to determine it—in order to determine whether it is a law of nature that atoms should necessarily combine in fixed numbers—it will be indispensable that the chemical point of view should be united with the physiological. The failure of the theory with regard to organic bodies indicates that the cause of this immense exception must be investigated; and such an inquiry belongs as much to physiology as to chemistry. Again, it is as yet undecided whether azote is a simple or a compound body. It was concluded by almost all chemists that azote is a simple body; the illustrious Berzilius hesitated, on purely chemical considerations; but he was also influenced by the physiological observation that animals which receive no azote in their food have as much of it in their tissues as carnivorous animals. From this we see how physiology must unite with chemistry to inform us whether azote is simple or compound, and to institute a new series of researches upon the relation between the composition of living bodies and their mode of alimentation.

Such is the advantage which, in the third place, we shall owe to

Positive philosophy—the elucidation of the respective sciences by their combination. In the fourth place

<small>Must reorganize society.</small> IV. The Positive Philosophy offers the only solid basis for that Social Reorganization which must succeed the critical condition in which the most civilized nations are now living.

It can not be necessary to prove to anybody who reads this work that Ideas govern the world, or throw it into chaos; in other words, that all social mechanism rests upon Opinions. The great political and moral crisis that societies are now undergoing is shown by a rigid analysis to arise out of intellectual anarchy. While stability in fundamental maxims is the first condition of genuine social order, we are suffering under an utter disagreement which may be called universal. Till a certain number of general ideas can be acknowledged as a rallying-point of social doctrine, the nations will remain in a revolutionary state, whatever palliatives may be devised; and their institutions can be only provisional. But whenever the necessary agreement on first principles can be obtained, appropriate institutions will issue from them, without shock or resistance; for the causes of disorder will have been arrested by the mere fact of the agreement. It is in this direction that those must look who desire a natural and regular, a normal state of society.

Now, the existing disorder is abundantly accounted for by the existence, all at once, of three incompatible philosophies—the theological, the metaphysical, and the positive. Any one of these might alone secure some sort of social order; but while the three co-exist, it is impossible for us to understand one another upon any essential point whatever. If this is true, we have only to ascertain which of the philosophies must, in the nature of things, prevail; and, this ascertained, every man, whatever may have been his former views, can not but concur in its triumph. The problem once recognised, can not remain long unsolved; for all considerations whatever point to the Positive Philosophy as the one destined to prevail. It alone has been advancing during a course of centuries, throughout which the others have been declining. The fact is incontestable. Some may deplore it, but none can destroy it, nor therefore neglect it but under penalty of being betrayed by illusory speculations. This general revolution of the human mind is nearly accomplished. We have only to complete the Positive Philosophy by bringing Social phenomena within its comprehension, and afterward consolidating the whole into one body of homogeneous doctrine. The marked preference which almost all minds, from the highest to the commonest, accord to positive knowledge over vague and mystical conceptions, is a pledge of what the reception of this philosophy will be when it has acquired the only quality that it now wants—a character of due generality. When it has become complete, its supremacy will take place spontaneously, and will re-establish order throughout society. There is, at present, no conflict but between the theological and the metaphysical philosophies. They are contending for the task of reorganizing society; but it is a work too mighty for either of them. The positive philosophy has hitherto intervened only to examine both, and both are abundantly discredited by the process. It is time now to be doing something more effective, without wasting our forces in needless controversy. It is time to complete the vast intellectual operation begun by Bacon, Descartes, and Galileo, by constructing the system of general ideas which must henceforth prevail among the human race. This is the way to put an end to the revolutionary crisis which is tormenting the civilized nations of the world.

Leaving these four points of advantage, we must attend to one precautionary reflection.

Because it is proposed to consolidate the whole of our acquired knowledge into one body of homogeneous doctrine, it must not be supposed that we are going to study this vast variety as proceeding from a single principle, and as subjected to a single law. *No hope of reduction to a single law.* There is something so chimerical in attempts at universal explanation by a single law, that it may be as well to secure this Work at once from any imputation of the kind, though its development will show how undeserved such an imputation would be. Our intellectual resources are too narrow, and the universe is too complex, to leave any hope that it will ever be within our power to carry scientific perfection to its last degree of simplicity. Moreover, it appears as if the value of such an attainment, supposing it possible, were greatly overrated. The only way, for instance, in which we could achieve the business, would be by connecting all natural phenomena with the most general law we know —which is that of gravitation, by which astronomical phenomena are already connected with a portion of terrestrial physics. Laplace has indicated that chemical phenomena may be regarded as simple atomic effects of the Newtonian attraction, modified by the form and mutual position of the atoms. But supposing this view proveable (which it can not be while we are without data about the constitution of bodies), the difficulty of its application would doubtless be found so great that we must still maintain the existing division between astronomy and chemistry, with the difference that we now regard as natural that division which we should then call artificial. Laplace himself presented his idea only as a philosophic device, incapable of exercising any useful influence over the progress of chemical science. Moreover, supposing this insuperable difficulty overcome, we should be no nearer to scientific unity, since we then should still have to connect the whole of physiological phenomena with the same law, which certainly would not be the least difficult part of the enterprise. Yet, all things considered, the hypothesis we have glanced at would be the most favorable to the desired unity.

The consideration of all phenomena as referable to a single origin is by no means necessary to the systematic formation of science, any more than to the realization of the great and happy consequences that we anticipate from the positive philosophy. The only necessary unity is that of Method, which is already in great part established. As for the doctrine, it need not be *one;* it is enough that it be *homogeneous*. It is, then, under the double aspect of unity of method and homogeneousness of doctrine that we shall consider the different classes of positive theories in this work. While pursuing the philosophical aim of all science, the lessening of the number of general laws requisite for the explanation of natural phenomena, we shall regard as presumptuous every attempt, in all future time, to reduce them rigorously to one.

Having thus endeavored to determine the spirit and influence of the Positive Philosophy, and to mark the goal of our labors, we have now to proceed to the exposition of the system; that is, to the determination of the universal, or encyclopædic order, which must regulate the different classes of natural phenomena, and consequently the corresponding positive sciences.

CHAPTER II.

VIEW OF THE HIERARCHY OF THE POSITIVE SCIENCES.

IN proceeding to offer a Classification of the Sciences, we must leave on one side all others that have as yet been attempted. Such scales as those of Bacon and D'Alembert are constructed upon an arbitrary division of the faculties of the mind; whereas, our principal faculties are often engaged at the same time in any scientific pursuit. As for other classifications, they have failed, through one fault or another, to command assent: so that there are almost as many schemes as there are individuals to propose them. The failure has been so conspicuous, that the best minds feel a prejudice against this kind of enterprise, in any shape.

<small>Failure of proposed classifications.</small>

Now, what is the reason of this?—For one reason, the distribution of the sciences, having become a somewhat discredited task, has of late been undertaken chiefly by persons who have no sound knowledge of any science at all. A more important and less personal reason, however, is the want of homogeneousness in the different parts of the intellectual system,—some having successively become positive, while others remain theological or metaphysical. Among such incoherent materials, classification is of course impossible. Every attempt at a distribution has failed from this cause, without the distributor being able to see why:—without his discovering that a radical contrariety existed between the materials

PRINCIPLE OF CLASSIFICATION OF THE SCIENCES. 39

he was endeavoring to combine. The fact was clear enough, if it had but been understood, that the enterprise was premature; and that it was useless to undertake it till our principal scientific conceptions should all have become positive. The preceding chapter seems to show that this indispensable condition may now be considered fulfilled: and thus the time has arrived for laying down a sound and durable system of scientific order.

We may derive encouragement from the example set by recent botanists and zoologists, whose philosophical labors have exhibited the true principle of classification; viz., that the classification must proceed from the study of the things to be classified, and must by no means be determined by *à priori* considerations. The real affinities and natural connections presented by objects being allowed to determine their order, the classification itself becomes the expression of the most general fact. And thus does the positive method apply to the question of classification itself, as well as to the objects included under it. It follows that the mutual dependence of the sciences,—a dependence resulting from that of the corresponding phenomena,—must determine the arrangement of the system of human knowledge. Before proceeding to investigate this mutual dependence, we have only to ascertain the real bounds of the classification proposed: in other words, to settle what we mean by human knowledge, as the subject of this work. *True principle of classification.*

The field of human labor is either speculation or action: and thus, we are accustomed to divide our knowledge into the theoretical and the practical. It is obvious that, in this inquiry, we have to do only with the theoretical. We are not going to treat of all human notions whatever, but of those fundamental conceptions of the different orders of phenomena which furnish a solid basis to all combinations, and are not founded on any antecedent intellectual system. In such a study, speculation is our material, and not the application of it,—except where the application may happen to throw back light on its speculative origin. This is probably what Bacon meant by that First Philosophy which he declared to be an extract from the whole of Science, and which has been so differently and so strangely interpreted by his metaphysical commentators. *Boundaries of our field.*

There can be no doubt that Man's study of nature must furnish the only basis of his action upon nature; for it is only by knowing the laws of phenomena, and thus being able to foresee them, that we can, in active life, set them to modify one another for our advantage. Our direct natural power over everything about us is extremely weak, and altogether disproportioned to our needs. Whenever we effect anything great, it is through a knowledge of natural laws, by which we can set one agent to work upon another, —even very weak modifying elements producing a change in the results of a large aggregate of causes. The relation of science to art may be summed up in a brief expression:

From Science comes Prevision: from Prevision comes Action.

We must not, however, fall into the error of our time, of regarding Science chiefly as a basis of Art. However great may be the services rendered to Industry by science, however true may be the saying that Knowledge is Power, we must never forget that the sciences have a higher destination still; and not only higher, but more direct—that of satisfying the craving of our understanding to know the laws of phenomena. To feel how deep and urgent this need is, we have only to consider for a moment the physiological effects of *consternation*, and to remember that the most terrible sensation we are capable of, is that which we experience when any phenomenon seems to arise in violation of the familiar laws of nature. This need of disposing facts in a comprehensible order (which is the proper object of all scientific theories) is so inherent in our organization, that if we could not satisfy it by positive conceptions, we must inevitably return to those theological and metaphysical explanations which had their origin in this very fact of human nature. It is this original tendency which acts as a preservative, in the minds of men of science, against the narrowness and incompleteness which the practical habits of our age are apt to produce. It is through this that we are able to maintain just and noble ideas of the importance and destination of the sciences; and if it were not thus, the human understanding would soon, as Condorcet has observed, come to a stand, even as to the practical applications for the sake of which higher things had been sacrificed; for, if the arts flow from science, the neglect of science must destroy the consequent arts. Some of the most important arts are derived from speculations pursued during long ages with a purely scientific intention. For instance, the ancient Greek geometers delighted themselves with beautiful speculations on Conic Sections; those speculations wrought, after a long series of generations, the renovation of astronomy; and out of this has the art of navigation attained a perfection which it never could have reached otherwise than through the speculative labors of Archimedes and Apollonius: so that, to use Condorcet's illustration, "the sailor who is preserved from shipwreck by the exact observation of the longitude, owes his life to a theory conceived two thousand years before by men of genius who had in view simply geometrical speculations."

Our business, it is clear, is with theoretical researches, letting alone their practical application altogether. Though we may conceive of a course of study which should unite the generalities of speculation and application, the time is not come for it. To say nothing of its vast extent, it would require preliminary achievements which have not yet been attempted. We must first be in possession of appropriate Special conceptions, formed according to scientific theories; and for these we have yet to wait. Meantime, an intermediate class is rising up, whose particular destination is to organize the relation of theory and practice; such as the engineers, who do not labor in the advancement of science, but who study it in its existing state, to apply it to practical purposes. Such classes are furnishing us with the elements of a future body of doctrine on the theories of the different arts. Already, Monge, in his view of descriptive geometry, has given us a general theory of the arts of construction. But we have as yet only a few scattered instances of this nature. The time will come when out of such results, a department of Positive philosophy may arise; but it will be in a distant future. If we remember that several sciences are implicated in every important art,—that, for instance, a true theory of Agriculture requires a combination of physiological, chemical, mechanical, and even astronomical and mathematical science,—it will be evident that true theories of the arts must wait for a large and equable development of these constituent sciences.

One more preliminary remark occurs, before we finish the prescription of our limits,—the ascertainment of our field of inquiry. We must distinguish between the two classes of Natural science;—the abstract or general, which have for their object the discovery of the laws which regulate phenomena in all conceivable cases: and the concrete, particular, or descriptive, which are sometimes called Natural sciences in a restricted sense, whose function it is to apply these laws to the actual history of existing beings. *Abstract science.* *Concrete science.* The first are fundamental; and our business is with them alone, as the second are derived, and however important, not rising into the rank of our subjects of contemplation. We shall treat of physiology, but not of botany and zoology, which are derived from it. We shall treat of chemistry, but not of mineralogy, which is secondary to it.—We may say of Concrete Physics, as these secondary sciences are called, the same thing that we said of theories of the arts,—that they require a preliminary knowledge of several sciences, and an advance of those sciences not yet achieved; so that, if there were no other reason, we must leave these secondary classes alone. At a future time Concrete Physics will have made progress, according to the development of Abstract Physics, and will afford a mass of less incoherent materials than those which it now presents. At present, too few of the students of these secondary sciences appear to be even aware that a due acquaintance with the primary sciences is requisite to all successful prosecution of their own.

We have now considered,

First, that science being composed of speculative knowledge and of practical knowledge, we have to deal only with the first; and

Second, that theoretical knowledge, or science properly so called, being divided into general and particular, or abstract and concrete science, we have again to deal only with the first.

Being thus in possession of our proper subject, duly prescribed, we may proceed to the ascertainment of the true order of the fundamental sciences.

Difficulty of classification. This classification of the sciences is not so easy a matter as it may appear. However natural it may be, it will always involve something, if not arbitrary, at least artificial; and in so far, it will always involve imperfection. It is impossible to fulfil, quite rigorously, the object of presenting the sciences in their natural connection, and according to their mutual dependence, so as to avoid the smallest danger of being involved in a vicious circle. It is easy to show why.

Historical and dogmatic methods. Every science may be exhibited under two methods or procedures, the Historical and the Dogmatic. These are wholly distinct from each other, and any other method can be nothing but some combination of these two. By the first method knowledge is presented in the same order in which it was actually obtained by the human mind, together with the way in which it was obtained. By the second, the system of ideas is presented as it might be conceived of at this day, by a mind which, duly prepared and placed at the right point of view, should begin to reconstitute the science as a whole. A new science must be pursued historically, the only thing to be done being to study in chronological order the different works which have contributed to the progress of the science. But when such materials have become recast to form a general system, to meet the demand for a more natural logical order, it is because the science is too far advanced for the historical order to be practicable or suitable. The more discoveries are made, the greater becomes the labor of the historical method of study, and the more effectual the dogmatic, because the new conceptions bring forward the earlier ones in a fresh light. Thus, the education of an ancient geometer consisted simply in the study, in their due order, of the very small number of original treatises then existing on the different parts of geometry. The writings of Archimedes and Apollonius were, in fact, about all. On the contrary, a modern geometer commonly finishes his education without having read a single original work dating further back than the most recent discoveries, which can not be known by any other means. Thus the Dogmatic Method is for ever superseding the Historical, as we advance to a higher position in science. If every mind had to pass through all the stages that every predecessor in the study had gone through, it is clear that, however easy it is to learn rather than invent, it would be impossible to effect the purpose of education,—to place the student on the vantage-ground gained by the labors of all the men who have gone before. By the dogmatic method this is done, even though the living student may have only an ordinary intellect, and the dead may have been men of lofty genius. By the dogmatic method therefore must every advanced science be attained, with so much of the historical combined with it as is rendered necessary by discoveries too recent to be studied elsewhere than in their own records. The only objection to the preference of the Dogmatic method is that it does not show how the science was attained; but a moment's reflection will show that this is the case also with the Historical method. To pursue a science historically is quite a different thing from learning the history of its progress. This last pertains to the study of human history, as we shall see when we reach the final division of this work. It is true that a science can not be completely understood without a knowledge of how it arose; and again, a dogmatic knowledge of any science is necessary to an understanding of its history; and therefore we shall notice, in treating of the fundamental sciences, the incidents of their origin, when distinct and illustrative; and we shall use their history, in a scientific sense, in our treatment of Social Physics; but the historical study, important, even essential, as it is, remains entirely distinct from the proper dogmatic study of science. These considerations, in this place, tend to define more precisely the spirit of our course of inquiry, while they more exactly determine the conditions under which we may hope to succeed in the construction of a true scale of the aggregate fundamental sciences. Great confusion would arise from any attempt to adhere strictly to historical order in our exposition of the sciences, for they have not all advanced at the same rate; and we must be for ever borrowing from each some fact to illustrate another, without regard to priority of origin. Thus, it is clear that, in the system of the sciences, astronomy must come before physics, properly so called: and yet, several branches of physics, above all, optics, are indispensable to the complete exposition of astronomy. Minor defects, if inevitable, can not invalidate a classification which, on the whole, fulfils the principal conditions of the case. They belong to what is essentially artificial in our division of intellectual labor. In the main, however, our classification agrees with the history of science; the more general and simple sciences actually occurring first and advancing best in human history, and being followed by the more complex and restricted, though all were, since the earliest times, enlarging simultaneously.

A simple mathematical illustration will precisely represent the difficulty of the question we have to resolve, while it will sum up the preliminary considerations we have just concluded.

We propose to classify the fundamental sciences. They are six, as we shall soon see. We can not make them less; and most scientific men would reckon them as more. Six objects admit of 720 different dispositions, or, in popular language, changes. Thus we have to choose the one right order (and there can be but one right) out of 720 possible ones. Very few of these have ever been proposed; yet we might venture to say that there is probably not one in favor of which some plausible reason might not be assigned; for we see the wildest divergences among the schemes which have been proposed,—the sciences which are placed by some at the head of the scale being sent by others to the further extremity. Our problem is, then, to find the one rational order, among a host of possible systems.

True principle of classification. Now we must remember that we have to look for the principle of classification in the comparison of the different orders of phenomena, through which Science discovers the laws which are her object. What we have to determine is the real dependence of scientific studies. Now, this dependence can result only from that of the corresponding phenomena. All observable phenomena may be included within a very few natural categories, so arranged as that the study of each category may be grounded on the principal laws of the preceding, and serve as the basis of the next ensuing. This order is determined by the degree of simplicity, or, what comes to the same thing, of generality of their phenomena. Hence results their successive dependence, and the greater or lesser facility for being studied.

Generality.

Dependence.

It is clear, *à priori*, that the most simple phenomena must be the most general; for whatever is observed in the greatest number of cases is of course the most disengaged from the incidents of particular cases. We must begin then with the study of the most general or simple phenomena, going on successively to the more particular or complex. This must be the most methodical way, for this order of generality or simplicity fixes the degree of facility in the study of phenomena, while it determines the necessary connection of the sciences by the successive dependence of their phenomena. It is worthy of remark in this place that the most general and simple phenomena are the furthest removed from Man's ordinary sphere, and must thereby be studied in a calmer and more rational frame of mind than those in which he is more nearly implicated; and this constitutes a new ground for the corresponding sciences being developed more rapidly.

We have now obtained our rule. Next we proceed to our classification.

Inorganic and Organic phenomena. We are first struck by the clear division of all natural phenomena into two classes—of inorganic and of organic bodies. The organized are evidently, in fact, more complex and less general than the inorganic, and depend upon them, instead of being depended on by them. Therefore it is that physiological study should begin with inorganic phenomena; since the organic include all the qualities belonging to them, with a special order added, viz., the vital phenomena, which belong to organization. We have not to investigate the nature of either; for the positive philosophy does not inquire into natures. Whether their natures be supposed different or the same, it is evidently necessary to separate the two studies of inorganic matter and of living bodies. Our classification will stand through any future decision as to the way in which living bodies are to be regarded; for, on any supposition, the general laws of inorganic physics must be established before we can proceed with success to the examination of a dependent class of phenomena.

INORGANIC AND ORGANIC PHYSICS.

Each of these great halves of natural philosophy has subdivisions. Inorganic physics must, in accordance with our rule of generality and the order of dependence of phenomena, be divided into two sections—of celestial and terrestrial phenomena. Thus we have Astronomy, geometrical and mechanical, and Terrestrial Physics. The necessity of this division is exactly the same as in the former case.

I. INORGANIC.
1. Astronomy.

Astronomical phenomena are the most general, simple, and abstract of all; and therefore the study of natural philosophy must clearly begin with them. They are themselves independent, while the laws to which they are subject influence all others whatsoever. The general effects of gravitation preponderate, in all terrestrial phenomena, over all effects which may be peculiar to them, and modify the original ones. It follows that the analysis of the simplest terrestrial phenomenon, not only chemical, but even purely mechanical, presents a greater complication than the most compound astronomical phenomenon. The most difficult astronomical question involves less intricacy than the simple movement of even a solid body, when the determining circumstances are to be computed. Thus we see that we must separate these two studies, and proceed to the second only through the first, from which it is derived.

In the same manner, we find a natural division of Terrestrial Physics into two, according as we regard bodies in their mechanical or their chemical character. Hence we have Physics, properly so called, and Chemistry. Again, the second class must be studied through the first. Chemical phenomena are more complicated than mechanical, and depend upon them, without influencing them in return. Every one knows that all chemical action is first submitted to the influence of weight, heat, electricity, etc., and presents moreover something which modifies all these. Thus, while it follows Physics, it presents itself as a distinct science.

2. Physics.
3. Chemistry.

Such are the divisions of the sciences relating to inorganic matter. An analogous division arises in the other half of Natural Philosophy—the science of organized bodies.

II. ORGANIC.

Here we find ourselves presented with two orders of phenomena; those which relate to the individual, and those which relate to the species, especially when it is gregarious. With regard to Man, especially, this distinction is fundamental. The last order of phenomena is evidently dependent on the first, and is more complex. Hence we have two great sections in organic physics—Physiology, properly so called, and Social Physics, which is dependent on it. In all Social phenomena we perceive the working of the physiological laws of the individual; and moreover something which modifies their effects, and which belongs to the influence of individuals over each other—singularly complicated in the case of the human race by the

1. Physiology.
2. Sociology.

influence of generations on their successors. Thus it is clear that our social science must issue from that which relates to the life of the individual. On the other hand, there is no occasion to suppose, as some eminent physiologists have done, that Social Physics is only an appendage to physiology. The phenomena of the two are not identical, though they are homogeneous; and it is of high importance to hold the two sciences separate. As social conditions modify the operation of physiological laws, Social Physics must have a set of observations of its own.

It would be easy to make the divisions of the Organic half of Science correspond with those of the Inorganic, by dividing physiology into vegetable and animal, according to popular custom. But this distinction, however important in Concrete Physics (in that secondary and special class of studies before declared to be inappropriate to this work), hardly extends into those Abstract Physics with which we have to do. Vegetables and animals come alike under our notice, when our object is to learn the general laws of life—that is, to study physiology. To say nothing of the fact that the distinction grows ever fainter and more dubious with new discoveries, it bears no relation to our plan of research; and we shall therefore consider that there is only one division in the science of organized bodies.

Five Natural Sciences. Thus we have before us Five fundamental Sciences in successive dependence—Astronomy, Physics, Chemistry, Physiology, and finally Social Physics. The first considers the most general, simple, abstract, and remote phenomena known to us, and those which affect all others without being affected by them. The last considers the most particular, compound, concrete phenomena, and those which are the most interesting to Man. Between these two, the degrees of speciality, of complexity, and individuality, are in regular proportion to the place of the respective sciences in the scale exhibited. This—casting out everything *Their filiation.* arbitrary—we must regard as the true filiation of the sciences; and in it we find the plan of this work.

Filiation of their parts. As we proceed, we shall find that the same principle which gives this order to the whole body of science arranges the parts of each science; and its soundness will therefore be freshly attested as often as it presents itself afresh. There is no refusing a principle which distributes the interior of each science after the same method with the aggregate sciences. But this is not the place in which to do more than indicate what we shall contemplate more closely hereafter. We must now rapidly review some of the leading properties of the hierarchy of science that has been disclosed.

Corroborations. This gradation is in essential conformity with the order which has spontaneously taken place among the *1. This classification follows the order of disclosure of sciences.* branches of natural philosophy, when pursued separately, and without any purpose of establishing such order. Such an accordance is a strong presumption that the arrangement is natural. Again, it coincides with the actual development of natural philosophy. If no leading science can be effectually pursued otherwise than through those which precede it in the scale, it is evident that no vast development of any science could take place prior to the great astronomical discoveries to which we owe the impulse given to the whole. The progression may since have been simultaneous; but it has taken place in the order we have recognised.

This consideration is so important that it is difficult to understand without it the history of the human mind. *2. Solves heterogeneousness.* The general law which governs this history, as we have already seen, can not be verified, unless we combine it with the scientific gradation just laid down: for it is according to this gradation that the different human theories have attained in succession the theological state, the metaphysical, and finally the positive. If we do not bear in mind the law which governs progression, we shall encounter insurmountable difficulties; for it is clear that the theological or metaphysical state of some fundamental theories must have temporarily coincided with the positive state of others which precede them in our established gradation, and actually have at times coincided with them; and this must involve the law itself in an obscurity which can be cleared up only by the classification we have proposed.

Again, this classification marks, with precision, the relative perfection of the different sciences, which consists in the degree of precision of knowledge, and in *3. Marks relative perfection in sciences.* the relation of its different branches. It is easy to see that the more general, simple, and abstract any phenomena are, the less they depend on others, and the more precise they are in themselves, and the more clear in their relations with each other. Thus, organic phenomena are less exact and systematic than inorganic; and of these again terrestrial are less exact and systematic than those of astronomy. This fact is completely accounted for by the gradation we have laid down; and we shall see as we proceed, that the possibility of applying mathematical analysis to the study of phenomena is exactly in proportion to the rank which they hold in the scale of the whole.

There is one liability to be guarded against, which we may mention here. We must beware of confounding the degree of precision which we are able to attain in regard *Defects are in us, not in science.* to any science, with the certainty of the science itself. The certainty of science, and our precision in the knowledge of it, are two very different things, which have been too often confounded; and are so still, though less than formerly. A very absurd proposition may be very precise; as if we should say, for instance, that the sum of the angles of a triangle is equal to three right angles; and a very certain proposition may be wanting in precision in our statement of it; as, for instance, when we assert that every man will die. If the different sciences offer to us a varying degree of precision, it is from no want of certainty in themselves, but of our mastery of their phenomena.

4. Effect on Education. The most interesting property of our formula of gradation is its effect on education, both general and scientific. This is its direct and unquestionable result. It will be more and more evident as we proceed, that no science can be effectually pursued without the preparation of a competent knowledge of the anterior sciences on which it depends. Physical philosophers can not understand Physics without at least a general knowledge of Astronomy; nor Chemists, without Physics and Astronomy; nor Physiologists, without Chemistry, Physics, and Astronomy; nor, above all, the students of Social philosophy, without a general knowledge of all the anterior sciences. As such conditions are, as yet, rarely fulfilled, and as no organization exists for their fulfilment, there is among us, in fact, no rational scientific education. To this may be attributed, in great part, the imperfection of even the most important sciences at this day. If the fact is so in regard to scientific education, it is no less striking in regard to general education. Our intellectual system can not be renovated till the natural sciences are studied in their proper order. Even the highest understandings are apt to associate their ideas according to the order in which they were received: and it is only an intellect here and there, in any age, which in its utmost vigor can, like Bacon, Descartes, and Leibnitz, make a clearance in their field of knowledge, so as to reconstruct from the foundation their system of ideas.

Effect on Method. Such is the operation of our great law upon scientific education through its effect on Doctrine. We can not appreciate it duly without seeing how it affects Method.

As the phenomena which are homogeneous have been classed under one science, while those which belong to other sciences are heterogeneous, it follows that the Positive Method must be constantly modified in a uniform manner in the range of the same fundamental science, and will undergo modifications, different and more and more compound, in passing from one science to another. Thus, under the scale laid down, we shall meet with it in all its varieties; which could not happen if we were to adopt a scale which should not fulfil the conditions we have admitted. This is an all-important consideration; for if, as we have already seen, we can not understand the positive method in the abstract, but only by its application, it is clear that we can have no adequate conception of it but by studying it in its varieties of application. No one science, however well chosen, could exhibit it. Though the Method is always the same, its procedure is varied. For instance, it should be Observation with regard to one kind of phenomena, and Experiment with regard to another; and different kinds of experiment, according to the case. In the same way, a general precept, derived from one fundamental science, however applicable to another, must have its spirit preserved by a reference to its origin; as in the case of the theory of Classifications. The best idea of the Positive Method would, of course, be obtained by the study of the most primitive and exalted of the sciences, if we were confined to one; but this isolated view would give no idea of its capacity of application to others in a modified form. Each science has its own proper advantages; and without some knowledge of them all, no conception can be formed of the power of the Method.

One more consideration must be briefly adverted to. It is necessary, not only to have a general knowledge of all the sciences, but to study them in their order. What can come of a study of complicated phenomena, if the student have not learned, by the contemplation of the simpler, what a Law is, what it is to Observe; what a Positive conception is; and even what a chain of reasoning is? Yet this is the way our young physiologists proceed every day—plunging into the study of living bodies, without any other preparation than a knowledge of a dead language or two, or at most a superficial acquaintance with Physics and Chemistry, acquired without any philosophical method, or reference to any true point of departure in Natural philosophy. In the same way, with regard to Social phenomena, which are yet more complicated, what can be effected but by the rectification of the intellectual instrument, through an adequate study of the range of anterior phenomena? There are many who admit this: but they do not see how to set about the work, nor understand the Method itself, for want of the preparatory study; and thus, the admission remains barren, and social theories abide in the theological or metaphysical state, in spite of the efforts of those who believe themselves positive reformers. *Orderly study of sciences.*

These, then, are the four points of view under which we have recognised the importance of a Rational and Positive Classification.

It can not but have been observed, that in our enumeration of the sciences there is a prodigious omission. We have said nothing of Mathematical science. The omission was intentional; and the reason is no other than the vast importance of mathematics. This science will be the first of which we shall treat. Meantime, in order not to omit from our sketch a department so prominent, we may indicate here the general results of the study we are about to enter upon. *Mathematics.*

In the present state of our knowledge, we must regard Mathematics less as a constituent part of natural philosophy than as having been, since the time of Descartes and Newton, the true basis of the whole of natural philosophy; though it is, exactly speaking, both the one and the other. To us it is of less value for the knowledge of which it consists, substantial and valuable as that knowledge is, than as being the most powerful instrument that the human mind can employ in the investigation of the laws of natural phenomena. *A department. A basis. An instrument.*

A double science.   In due precision, Mathematics must be divided into two great sciences, quite distinct from each other—Abstract Mathematics, or the Calculus (taking the word in its most extended sense), and Concrete Mathematics, which is composed of General Geometry and of Rational Mechanics. The Concrete part is necessarily founded on the Abstract, and it becomes in its turn the basis of all natural philosophy; all the phenomena of the universe being regarded, as far as possible, as geometrical or mechanical.

Abstract mathematics an instrument.   The Abstract portion is the only one which is purely instrumental, it being simply an immense extension of natural logic to a certain order of deductions. Geometry and mechanics must, on the contrary, be regarded as true natural sciences, founded, like all others, on observation, though, by the extreme simplicity of their phenomena, they can be systematized to much greater perfection. It is this capacity which has caused the experimental character of their first principles to be too much lost sight of. But these two physical sciences have this peculiarity, that they are now, and will be more and more, employed rather as method than as doctrine.

Concrete mathematics a science.

It needs scarcely to be pointed out that, in placing Mathematics at the head of Positive Philosophy, we are only extending the application of the principle which has governed our whole Classification. We are simply carrying back our principle to its first manifestation. Geometrical and Mechanical phenomena are the most general, the most simple, the most abstract of all,—the most irreducible to others, the most independent of them; serving, in fact, as a basis to all others. It follows that the study of them is an indispensable preliminary to that of all others. Therefore must Mathematics hold the first place in the hierarchy of the sciences, and be the point of departure of all Education, whether general or special. In an empirical way, this has hitherto been the custom,—a custom which arose from the great antiquity of mathematical science. We now see why it must be renewed on a rational foundation.

Mathematics preeminent in the scale.

We have now considered, in the form of a philosophical problem, the rational plan of the study of the Positive Philosophy. The order that results is this; an order which of all possible arrangements is the only one that accords with the natural manifestation of all phenomena. MATHEMATICS, ASTRONOMY, PHYSICS, CHEMISTRY, PHYSIOLOGY, SOCIAL PHYSICS.

BOOK I.

MATHEMATICS.

CHAPTER I.

MATHEMATICS, ABSTRACT AND CONCRETE.

WE are now to enter upon the study of the first of the Six great Sciences: and we begin by establishing the importance of the Positive Philosophy in perfecting the character of each science in itself.

Though Mathematics is the most ancient and the most perfect science of all, the general idea of it is far from being clearly determined. The definition of the science, and its chief divisions, have remained up to this time vague and uncertain. The plural form of the name (grammatically used as singular) indicates the want of unity in its philosophical character, as commonly conceived. In fact, it is only since the beginning of the last century that it could be conceived of as a whole; and since that time geometers have been too much engaged on its different branches, and in applying it to the most important laws of the universe, to have much attention left for the general system of the science. Now, however, the pursuit of its specialities is no longer so engrossing as to exclude us from the study of Mathematics in its unity. It has now reached a degree of consistency which admits of the effort to reduce its parts into a system, in preparation for further advance. The latest achievements of mathematicians have prepared the way for this by evidencing a character of unity in its principal parts which was not before known to exist. Such is eminently the spirit of the great author of the Theory of Functions and of Analytical Mechanics.

The common description of Mathematics, as *the science of Magnitudes,* or somewhat more positively, *the science which relates to the Measurement of Magnitudes,* is too vague and unmeaning to have been used but for want of a better. Yet the idea contained in it is just at bottom, and is even suffi-

Description of Mathematics.

ciently extensive, if properly understood; but it needs precision and depth. It is important in such matters not to depart unnecessarily from notions generally admitted; and we will therefore see how, from this point of view, we can rise to such a definition of Mathematics as will be adequate to the importance, extent, and difficulty of the science.

Object of Mathematics. Our first idea of *measuring* a magnitude is simply that of comparing the magnitude in question with another supposed to be known, which is taken for the *unit* of comparison among all others of the same kind. Thus, when we define mathematics as being the measurement of magnitudes, we give a very imperfect idea of it, and one which seems to bear no relation, in this respect, to any science whatever. We seem to speak only of a series of mechanical procedures, like a superposition of lines, for obtaining the comparison of magnitudes, instead of a vast chain of reasonings, inexhaustible by the intellect. Nevertheless, this definition has no other fault than not being deep enough. It does not mistake the real aim of mathematics, but it presents as direct an object which is usually indirect; and thus it misleads us as to the nature of the science. To rectify this, we must attend to a general fact, which is easily established—that the direct measurement of a magnitude is often an impossible operation; so that if we had no other means of doing what we want, we must often forego the knowledge we desire. We can rarely even measure a right line by another right line; and this is the simplest measurement of all. The very first condition of this is, that we should be able to traverse the line from one end to the other; and this can not be done with the greater number of the distances which interest us the most. We can not do it with the heavenly bodies, nor with the earth and any heavenly body, nor even with many distances on the earth; and again, the length must be neither too great nor too small, and it must be conveniently situated; and a line which could be easily measured if it were horizontal, becomes impracticable if vertical. There are so few lines capable of being directly measured with precision, that we are compelled to resort to artificial lines, created to admit of a direct determination, and to be the point of reference for all others. If there is difficulty about the measurement of lines, the embarrassment is much greater when we have to deal with surfaces, volumes, velocities, times, forces, etc., and in general with all other magnitudes susceptible of estimate, and, by their nature, difficult of direct measurement. It is the general fact of this difficulty, inherent in almost every case, which necessitates the formation of mathematical science; for, finding direct measurement so often impossible, we are compelled to devise means of doing it indirectly. Hence arose Mathematics.

General method. The general method employed, and the only conceivable one, is to connect the magnitudes in question with some that can be directly determined, and thus to ascertain the former, through their relations with the latter. Such is the precise object of Mathematics, regarded as a whole. To form anything like a worthy idea of it, we must remember that the indirect determination of magnitudes may have many degrees of indirectness. It often happens that the magnitudes to which undetermined magnitudes are to be referred can not themselves be measured directly, and must themselves be made the subject of a prior process, and so on through a whole series; and thus, the mind is often obliged to establish a long course of intermediaries between the one and the other point of the inquiry—points which may appear at the outset to have no connection whatever.

If this appears too abstract, it may become plain by a few examples. In observing a falling body, we are aware that two quantities are involved: the height from which the body falls, and the time occupied in its descent. These two quantities are connected, as they vary together, and together remain fixed. In the language of mathematicians, they are *functions* of each other. The measurement of one being impracticable, it is supplied by that of the other. By observing the time occupied by a stone in falling down a precipice, we can ascertan the height of the precipice as accurately as if we could measure it with a horizontal line. In another case, we may be able to know the height whence a body has fallen, and unable to observe the time with precision, and then we must have recourse to the inverse question,— to determine the time by the distance; as, for instance, if we were to inquire how long it would take a body to fall from the moon. In these cases, the question is very simple, supposing we do not complicate it with considerations of intensity of gravity, resistance of a fluid medium, etc. But, to enlarge the question, we must contemplate the phenomenon in its greatest generality by supposing the fall to be oblique, and taking into account all the principal circumstances. Then, instead of two variable quantities, simply connected, the phenomenon will present a considerable number,—the space traversed, whether in a vertical or horizontal direction; the time employed in traversing it; the velocity of the body at each point of its course; and even the intensity and direction of the impulse which sent it forth; and finally, in some cases, the resistance of the medium, and the intensity of gravity. All these quantities are so connected that each, in its turn, may be determined indirectly by means of the others, and thus we shall have as many mathematical inquiries as there are magnitudes coexisting in the phenomenon considered. Such a very simple change as this in the physical conditions of a problem may place a mathematical question, originally quite elementary, in the rank of those difficult questions whose complete and rigorous solution transcends the power of the human understanding.

Examples.

Again—we may take a geometrical example. We want to determine a distance not directly measurable. We shall conceive of it as making a part of some *figure*, or system of lines of some sort, of which the other parts are directly measurable; let us say a triangle (for this is the simplest, and to it all others are reducible). The distance in question is supposed to form a portion of a triangle, in which we are able to determine directly, either another side and two angles, or two sides and one angle. The knowledge required is obtained by the mathematical labor of deducing the unknown distance from the observed elements, by means of the relation between them. The process may, and commonly does, become highly complicated by the elements supposed to be known being themselves determinable only in an indirect manner, by the aid of fresh auxiliary systems, the number of which may be very considerable. The distance, once ascertained, will often enable us to obtain new quantities which will offer occasion for new mathematical questions. Thus, when we once know the distance of any object, the observation, simple and always possible, of its apparent diameter, may disclose to us, with certainty, however indirectly, its real dimensions; and at length, by a series of analogous inquiries, its surface, its volume, even its weight, and a multitude of other qualities which might have seemed out of the reach of our knowledge for ever. It is by such labors that Man has learned to know, not only the distances of the planets from the earth and from each other, but their actual magnitude—their true form, even to the inequalities on their surface, and (what seems much more out of his reach) their respective masses, their mean densities, and the leading circumstances of the fall of heavy bodies on their respective surfaces, etc. Through the power of mathematical theories, all this and very much more has been obtained by means of a very small number of straight lines, properly chosen, and a larger number of angles. We might even say, to describe the general bearing of the science in a sentence, that but for the fear of multiplying mathematical operations unnecessarily, and for the consequent necessity of reserving them for the determination of quantities which could not be measured directly, the knowledge of all magnitudes susceptible of precise estimate which can be offered by the various orders of phenomena, would be finally reducible to the immediate measurement of a single straight line, and of a suitable number of angles.

True definition of mathematics. We can now define Mathematical science with precision. It has for its object the *indirect* measurement of magnitudes, and it proposes *to determine magnitudes by each other, according to the precise relations which exist between them.* Preceding definitions have given to Mathematics the character of an Art; this raises it at once to the rank of a true Science. According to this definition, the spirit of Mathematics consists in regarding as mutually connected all the quantities which can be presented by any phenomenon whatsoever, in order to deduce all from each other. Now, there is evidently no phenomenon which may not be regarded as affording such considerations. Hence results the naturally indefinite extent, and the rigorous logical universality of Mathematical science. As for its actual practical extent, we shall see what that is hereafter.

These explanations justify the name of Mathematics, applied to the science we are considering. By itself it signifies SCIENCE. The Greeks had no other, and we may call it *the* science; for its definition is neither more nor less (if we omit the specific notion of magnitudes) than the definition of all science whatsoever. All science consists in the co-ordination of facts; and no science could exist among isolated observations. It might even be said that Mathematics might enable us to dispense with all direct observation, by empowering us to deduce from the smallest possible number of immediate data the largest possible amount of results. Is not this the real use, both in speculation and in action, of the *laws* which we discover among natural phenomena? If so, Mathematics merely urges to the ultimate degree, in its own way, researches which every real science pursues, in various inferior degrees of its own sphere. Thus it is only through Mathematics that we can thoroughly understand what true science is. Here alone can we find in the highest degree simplicity and severity of scientific law, and such abstraction as the human mind can attain. Any scientific education setting forth from any other point, is faulty in its basis.

Thus far, we have viewed the science as a whole. We must now consider its primary division. The secondary divisions will be laid down afterward.

Every mathematical solution spontaneously separates into two parts. The inquiry being, as we have seen, the determination of unknown magnitudes, through their relation to the known, the student must, in the first place, ascertain what these relations are, in the case under his notice. This first is the *Concrete* part of the inquiry. When it is accomplished, what remains is a pure question of numbers, consisting simply in a determination of unknown numbers when we know by what relation they are connected with known numbers. This second operation is the *Abstract* part of the inquiry. The primary division of Mathematics is therefore into two great sciences:—ABSTRACT MATHEMATICS, and CONCRETE MATHEMATICS. This division exists in all complete mathematical questions whatever, whether more or less simple. *Its two parts. Their different objects.*

Recurring to the simplest case of a falling body, we must begin by learning the relation between the height from which it falls, and the time occupied in falling. As Geometers say, we must find the *equation* which exists between them. Till this is done, there is no basis for a computation. This ascertainment may be extremely difficult, and it is incomparably the superior part of the problem. The true scientific science is so modern, that, as far as we know, no one before Galileo had remarked the acceleration of velocity in a falling body, the natural supposition having been that the height was in uniform proportion to the time. This first inquiry issued in the discovery of the law of Galileo. The Concrete part being accomplished, the Abstract remains. We have ascertained that the spaces traversed in each second increase as the series of odd numbers, and we now have only the task of the computation of the height from the time, or of the time from the height; and this consists in finding that, by the established law, the first of these two quantities is a known multiple of the second power of the other; whence we may finally determine the value of the one when that of the other is given. In this instance the concrete question is the more difficult of the two. If the same phenomenon were taken in its greatest generality, the reverse would be the case. Take the two together, and they may be regarded as exactly equivalent in difficulty. The mathematical law may be easy to ascertain, and difficult to work; or it may be difficult to ascertain, and easy to work. In importance, in extent, and in difficulty, these two great sections of Mathematical Science will be seen hereafter to be equivalent.

Their different natures. We have seen the difference in their objects. They are no less different in their nature.

The Concrete must depend on the character of the objects examined, and must vary when new phenomena present themselves: whereas, the Abstract is wholly independent of the nature of the objects, and is concerned only with their numerical relations. Thus, a great variety of phenomena may be brought under one geometrical solution. Cases which appear as unlike each other as possible may stand for one another under the Abstract process, which thus serves for all, while the Concrete process must be new in each case. Thus the Concrete process is Special, and the Abstract is General. The character of the Concrete is experimental, physical, phenomenal: while the Abstract is purely logical, rational. The Concrete part of every mathematical question is necessarily founded on consideration of the external world; while the Abstract part consists of a series of logical deductions. The equations being once found, in any case, it is for the understanding, without external aid, to educe the results which these equations contain.

We see how natural and complete this main division is. We will briefly prescribe the limits of each section.

Concrete Mathematics. As it is the business of Concrete Mathematics to discover the equations of phenomena, we might suppose that it must comprehend as many distinct sciences as there are distinct categories of phenomena; but we are very far indeed from having discovered mathematical laws in all orders of phenomena. In fact, there are as yet only two great categories of phenomena whose equations are constantly known—Geometrical and Mechanical phenomena. Thus, the Concrete part of Mathematics consists of GEOMETRY and RATIONAL MECHANICS.

There is a point of view from which all phenomena might be included under these two divisions. All natural effects, considered statically or dynamically, might be referred to laws of extension or laws of motion. But this point of view is too high for us at present; and it is only in the regions of Astronomy, and, partially, of terrestrial Physics, that this vast transformation has taken place.

We will then proceed on the supposition that Geometry and Mechanics are the constituents of Concrete Mathematics.

The nature of Abstract Mathematics is precisely determined. It is composed of what is called the *Calculus*, taking this word in its widest extension, which reaches from the simplest numerical operations to the highest combinations of transcendental analysis. Its proper object is to resolve all questions of numbers. Its starting-point is that which is the limit of Concrete Mathematics—the knowledge of the precise relations—that is, the equations—between different magnitudes which are considered simultaneously. The object of the Calculus, however indirect or complicated the relations may be, is to discover unknown quantities by the known. This science, though more advanced than any other, is, in reality, only at its beginning yet; but it is necessary, in order to define the nature of any science, to suppose it perfect. And the true character of the Calculus is what we have said.

Abstract Mathematics.

From an historical point of view, Mathematical Analysis appears to have arisen out of the contemplation of geometrical and mechanical facts; but it is not the less independent of these sciences, logically speaking. Analytical ideas are, above all others, universal, abstract, and simple; and geometrical and mechanical conceptions are necessarily founded on them. Mathematical Analysis is therefore the true rational basis of the whole system of our positive knowledge. We can now also explain why it not only gives precision to our actual knowledge, but establishes a far more perfect co-ordination in the study of phenomena which allow of such an application. If a single analytical question, brought to an abstract solution, involves the *implicit* solution of a multitude of physical questions, the mind is enabled to perceive relations between phenomena apparently isolated, and to extract from them the quality which they have in common. To the wonder of the student, unsuspected relations arise between problems which, instead of being, as they appeared before, wholly unconnected, turn out to be identical. There appears to be no connection between the determination of the direction of a curve at each of its points and that of the velocity of a body at each moment of its variable motion; yet, in the eyes of the geometer, these questions are but one.

When we have seized the true general character of Mathematical Analysis, we easily see how perfect it is, in comparison with all other branches of our positive science. The perfection consists in the simplicity of the ideas contemplated; and not, as Condillac and others have supposed, to the conciseness and generality of the signs used as instruments of reasoning. The signs are of admirable use to work out the ideas, when once obtained; but, in fact, all the great analytical conceptions were formed without any essential aid from the signs. Subjects which are by their nature inferior in simplicity and generality can not be raised to logical perfection by any artifice of scientific language.

Extent of its domain. We have now seen what is the object and what is the character of Mathematical Science. It remains for us to consider the extent of its domain.

Its universality. We must first admit that, in a logical view, this science is necessarily and rigorously universal. There is no inquiry which is not finally reducible to a question of Numbers; for there is none which may not be conceived of as consisting in the determination of quantities by each other, according to certain relations. The fact is, we are always endeavoring to arrive at numbers, at fixed quantities, whatever may be our subject, however uncertain our methods, and however rough our results. Nothing can appear less like a mathematical inquiry than the study of living bodies in a state of disease; yet, in studying the cure of disease, we are endeavoring to ascertain the quantities of the different agents which are to modify the organism, in order to bring it to its natural state, admitting, as geometers do, for some of these quantities, in certain cases, values which are equal to zero, negative, or even contradictory. It is not meant that such a method can be actually followed in the case of complicated phenomena; but the logical extension of the science, which is what we are now considering, comprehends such instances as this.

Kant has divided human ideas into the two categories of quantity and quality, which, if true, would destroy the universality of Mathematics; but Descartes's fundamental conception of the relation of the concrete to the abstract in Mathematics abolishes this division, and proves that all ideas of quality are reducible to ideas of quantity. He had in view geometrical phenomena only; but his successors have included in this generalization, first, mechanical phenomena, and, more recently, those of heat. There are now no geometers who do not consider it of universal application, and admit that every phenomenon may be as logically capable of being represented by an equation as a curve or a motion, if only we were always capable (which we are very far from being) of first discovering, and then resolving it.

Its limitations. The limitations of Mathematical science are not, then, in its nature. The limitations are in our intelligence: and by these we find the domain of the science remarkably restricted, in proportion as phenomena, in becoming special, become complex.

Though, as we have seen, every question may be conceived of as reducible to numbers, the reduction can not be made by us except in the case of the simplest and most general phenomena. The difficulty of finding the equation in the case of special, and therefore complex phenomena, soon becomes insurmountable, so that, at the utmost, it is only the phenomena of the first three classes,—that is, only those of Inorganic Physics,—that we can even hope to subject to the process. The properties of inorganic bodies are nearly invariable; and therefore, with regard to them, the first condition of mathematical inquiry can be fulfilled: the different quantities which they present may be resolved into fixed numbers; but the variableness of the properties of organic bodies is beyond our management. An inorganic body, possessing solidity, form, consistency, specific gravity, elasticity, etc., presents qualities which are within our estimate, and can be treated mathematically; but the case is altered when Chemical action is added to these. Complications and variations then enter into the question which at present baffle mathematical analysis. Hereafter, it may be discovered what fixed numbers exist in chemical combinations: but we are as yet very far from having any practical knowledge of them. Still further are we from being able to form such computations amidst the continual agitation of atoms which constitutes what we call *life*, and therefore from being able to carry mathematical analysis into the study of Physiology. By the rapidity of their changes, and their incessant numerical variations, vital phenomena are, practically, placed in opposition to mathematical processes. If we should desire to compute, in a single case, the most simple facts of a living body,—such as its mean density, its temperature, the velocity of its circulation, the proportion of elements which at any moment compose its solids or its fluids, the quantity of oxygen which it consumes in a given time, the amount of its absorptions or its exhalation,—and, yet more, the energy of its muscular force, the intensity of its impressions, etc., we must make as many observations as there are species or races, and varieties in each; we must measure the changes which take place in passing from one individual to another, and in the same individual, according to age, health, interior condition, surrounding circumstances perpetually varying, such as the constitution of the atmosphere, etc. It is clear that no mathematical precision can be attained amidst a complexity like this. Social phenomena, being more complicated still, are even more out of the question, as subjects for mathematical analysis. It is not that a mathematical basis does not exist in these cases, as truly as in phenomena which exhibit, in all clearness, the law of gravitation: but that our faculties are too limited for the working of problems so intricate. We are baffled by various phenomena of inorganic bodies, when they are very complex. For instance, no one doubts that meteorological phenomena are subject to mathematical laws, however little we yet know about them; but their multiplicity renders their observed results as variable and irregular as if each cause were free of all such conditions.

We find a second limitation in the number of conditions to be studied, even if we were sure of the mathematical law which governs each agent. Our feeble faculties could not grasp and wield such an aggregate of conditions, however certain might be our knowledge of each. In the simplest cases in which we desire to approximate the abstract to the concrete conditions, with any completeness,—as in the phenomenon of the flow of a fluid from a given orifice, by virtue of its gravity alone,—the difficulty is such that we are, as yet, without any mathematical solution of this very problem. The same is the case with the yet more simple instance of the movement of a solid projectile through a resisting medium.

To the popular mind it may appear strange, considering these facts, that we know so much as we do about the planets. But in reality, that class of phenomena is the most simple of all within our cognizance. The most complex problem which they present is the influence of a third body acting in the same way on two which are tending toward each other in virtue of gravitation; and this is a more simple question than any terrestrial problem whatever. We have, however, attained only approximate solutions in this case. And the high perfections to which solar astronomy has been brought by the use of mathematical science is owing to our having profited by those facilities that we may call accidental, which the favorable constitution of our planetary system presents. The planets which compose it are few; their masses are very unequal, and much less than that of the sun; they are far distant from each other; their forms are nearly spherical; their orbits are nearly circular, and only slightly inclined in relation to each other; and so on. Their perturbations are, in consequence, inconsiderable, for the most part; and all we have to do is usually to take into the account, together with the influence of the sun on each planet, the influence of one other planet, capable, by its size and its nearness, of occasioning perceptible derangements. If any of the conditions mentioned above had been different, though the law of gravitation had existed as it is, we might not at this day have discovered it. And if we were now to try to investigate Chemical phenomena by the same law, we should find a solution as impossible as it would be in astronomy, if the conditions of the heavenly bodies were such as we could not reduce to an analysis.

In showing that Mathematical analysis can be applied only to Inorganic Physics, we are not restricting its domain. Its rigorous universality, in a logical view, has been established. To pretend that it is practically applicable to the same extent would be merely to lead away the human mind from the true direction of scientific study, in pursuit of an impossible perfection. The most difficult sciences must remain, for an indefinite time, in that preliminary state which prepares for the others the time when they too may become capable of mathematical treatment. Our business is to study phenomena, in the characters and relations in which they present themselves to us, abstaining from introducing considerations of quantities, and mathematical laws, which it is beyond our power to apply.

We owe to Mathematics both the origin of Positive Philosophy and its Method. When this Method was introduced into the other sciences, it was natural that it should be urged too far. But each science modified the method by the operation of its own peculiar phenomena. Thus only could that true definitive character be brought out, which must prevent its being ever confounded with that of any other fundamental science.

The aim, character, and general relations of Mathematical Science have now been exhibited as fully as they could be in such a sketch as this. We must next pass in review the three great sciences of which it is composed,—the Calculus, Geometry, and Rational Mechanics.

CHAPTER II.

GENERAL VIEW OF MATHEMATICAL ANALYSIS.

THE historical development of the Abstract portion of Mathematical science has, since the time of Descartes, *Analysis.* been for the most part determined by that of the Concrete. Yet the Calculus in all its principal branches must be understood before passing on to Geometry and Mechanics. The Concrete portions of the science depend on the Abstract, which are wholly independent of them. We will now, therefore, proceed to a rapid review of the leading conceptions of the Analysis.

First, however, we must take some notice of the general idea of an *equation*, and see how far it is from *True idea of an equation.* being the true one on which geometers proceed in practice; for without settling this point we can not determine, with any precision, the real aim and extent of abstract mathematics.

The business of concrete mathematics is to discover the equations which express the mathematical laws of the phenomenon under consideration; and these equations are the starting-point of the calculus, which must obtain from them certain quantities by means of others. It is only by forming a true idea of an equation that we can lay down the real line of separation between the concrete and the abstract part of mathematics.

It is giving much too extended a sense to the notion of an equation to suppose that it means every kind of relation of equality between *any* two functions of the magnitudes under consideration; for, if every equation is a relation of equality, it is far from being the case that, reciprocally, every relation of equality must be an equation of the kind to which analysis is, by the nature of the case, applicable. It is evident that this confusion must render it almost impossible to explain the difficulty we find in establishing the relation of the concrete to the abstract which meets us in every great mathematical question, taken by itself. If the word equation meant what we are apt to suppose, it is not easy to see what difficulty there could be, in general, in establishing the equations of any problem whatever. This ordinary notion of an equation is widely unlike what geometers understand in the actual working of the science.

According to my view, functions must themselves be divided into

Abstract and Concrete; the first of which alone can enter into true equations. Every equation is a relation of equality between two abstract functions of the magnitudes in question, including with the primary magnitudes all the auxiliary magnitudes which may be connected with the problem, and the introduction of which may facilitate the discovery of the equations sought.

This distinction may be established by both the *à priori* and *à posteriori* methods; by characterizing each kind of function, and by enumerating all the abstract functions yet known,—at least with regard to their elements.

Abstract functions. *A priori*; Abstract functions express a mode of dependence between magnitudes which may be conceived between numbers alone, without the need of pointing out any phenomena in which it may be found realized; while Concrete functions are those whose expression requires a specified actual case of physics, geometry, mechanics, etc.

Concrete functions.

Most functions were concrete in their origin,—even those which are at present the most purely abstract; and the ancients discovered only through geometrical definitions elementary algebraic properties of functions, to which a numerical value was not attached till long afterward, rendering abstract to us what was concrete to the old geometers. There is another example which well exhibits the distinction just made—that of circular functions, both direct and inverse, which are still sometimes concrete, sometimes abstract, according to the point of view from which they are regarded.

A posteriori; the distinguishing character, abstract or concrete, of a function having been established, the question of any determinate function being abstract, and therefore able to enter into true analytical equations, becomes a simple question of fact, as we are acquainted with the elements which compose all the abstract functions at present known. We say we know them all, though analytical functions are infinite in number, because we are here speaking, it must be remembered, of the elements—of the simple, not of the compound. We have ten elementary formulas; and, few as they are, they may give rise to an infinite number of analytical combinations. There is no reason for supposing that there can never be more. We have more than Descartes had, and even Newton and Leibnitz; and our successors will doubtless introduce additions, though there is so much difficulty attending their augmentation, that we can not hope that it will proceed very far.

It is the insufficiency of this very small number of analytical elements which constitutes our difficulty in passing from the concrete to the abstract. In order to establish the equations of phenomena, we must conceive of their mathematical laws by the aid of functions composed of these few elements. Up to this point the question has been essentially concrete, not coming within the domain of the calculus. The difficulty of the passage from the concrete to the abstract in general consists in our having only these few analytical elements with which to represent all the precise relations which the whole range of natural phenomena afford to us. Amid their infinite variety, our conceptions must be far below the real difficulty; and especially because these elements of our analysis have been supplied to us by the mathematical consideration of the simplest phenomena of a geometrical origin, which can afford us, *à priori*, no rational guaranty of their fitness to represent the mathematical laws of all other classes of phenomena. We shall hereafter see how this difficulty of the relation of the concrete to the abstract has been diminished, without its being necessary to multiply the number of analytical elements.

Thus far we have considered the Calculus as a whole. We must now consider its divisions. These divisions we must call the *Algebraic Calculus*, or *Algebra*, and the *Arithmetical Calculus*, or *Arithmetic*, taking care to give them the most extended logical sense, and not the restricted one in which the terms are usually received. Two parts of the Calculus.

It is clear that every question of Mathematical Analysis presents two successive parts, perfectly distinct in their nature. The first stage is the transformation of the proposed equations, so as to exhibit the mode of formation of unknown quantities by the known. This constitutes the *algebraic* question. Then ensues the task of finding the *values* of the *formulas* thus obtained. The values of the numbers sought are already represented by certain explicit functions of given numbers: these values must be determined; and this is the *arithmetical* question. Thus the algebraic and the arithmetical calculus differ in their object. They differ also in their view of quantities,—Algebra considering quantities in regard to their *relations*, and Arithmetic in regard to their *values*. In practice, it is not always possible, owing to the imperfection of the science of the calculus, to separate the processes entirely in obtaining a solution; but the radical difference of the two operations should never be lost sight of. *Algebra*, then, is the *Calculus of Function*, and *Arithmetic* the *Calculus of Values*. Algebra.
Arithmetic.

We have seen that the division of the Calculus is into two branches. It remains for us to compare the two, in order to learn their respective extent, importance, and difficulty.

The Calculus of Values, Arithmetic, appears at first to have as wide a field as Algebra, since as many questions might seem to arise from it as we can conceive different algebraic formulas to be valued. But a very simple reflection will show that it is not so. Functions being divided into simple and compound, it is evident that when we become able to determine the value of simple functions, there will be no difficulty with the compound. In the algebraic relation, a compound function plays a very different part from that of the elementary functions which constitute it; and this is the source of our chief analytical difficulties. But it is quite otherwise with the Arithmetical Calculus. Thus, the number of distinct arithmetical Arithmetic.

Its Extent.

operations is indicated by that of the abstract elementary functions, which we have seen to be very few. The determination of the values of these ten functions necessarily affords that of all the infinite number comprehended in the whole of mathematical analysis: and there can be no new arithmetical operations otherwise than by the creation of new analytical elements, which must, in any case, for ever be extremely small. The domain of arithmetic then is, by its nature, narrowly restricted, while that of algebra is rigorously indefinite. Still, the domain of arithmetic is more extensive than is commonly represented; for there are many questions treated as incidental in the midst of a body of analytical researches, which, consisting of determinations of values, are truly arithmetical. Of this kind are the construction of a table of logarithms, and the calculation of trigonometrical tables, and some distinct and higher procedures; in short, every operation which has for its object the determination of the values of functions. And we must also include that part of the science of the Calculus which we call the Theory of Numbers, the object of which is to discover the properties inherent in different numbers, in virtue of their values, independent of any particular system of numeration. It constitutes a sort of transcendental arithmetic. Though the domain of arithmetic is thus larger than is commonly supposed, this Calculus of values will yet never be more than a point, as it were, in comparison with the calculus of functions, of which mathematical science essentially consists. This is evident, when we look into the real nature of arithmetical questions.

Determinations of values are, in fact, nothing else than real *transformations* of the functions to be valued. These

<small>Its nature.</small>

transformations have a special end; but they are essentially of the same nature as all taught by analysis. In this view, the Calculus of values may be regarded as a supplement, and a particular application of the Calculus of functions, so that arithmetic disappears, at it were, as a distinct section in the body of abstract mathematics. To make this evident, we must observe that when we desire to determine the value of an unknown number whose mode of formation is given, we define and express that value in merely announcing the arithmetical question, already defined and expressed under a certain form; and that, in determining its value, we merely express it under another determinate form to which we are in the habit of referring the idea of each particular number by making it re-enter into the regular system of numeration. This is made clear by what happens when the mode of numeration is such that the question is its own answer; as, for instance, when we want to add together seven and thirty, and call the result seven-and-thirty. In adding other numbers, the terms are not so ready, and we transform the question; as when we add together twenty-three and fourteen: but not the less is the operation merely one of transformation of a question already defined and expressed. In this view, the *calculus of values* might be regarded as a particular ap-

ALGEBRA. 65

plication of the *calculus of functions*, arithmetic thereby disappearing, as a distinct section, from the domain of abstract mathematics.—And here we have done with the Calculus of values, and pass to the Calculus of functions, of which abstract mathematics is essentially composed.

We have seen that the difficulty of establishing the relation of the concrete to the abstract is owing to the insufficiency of the very small number of analytical elements that we are in possession of. The obstacle has been surmounted in a great number of important cases: and we will now see how the establishment of the equations of phenomena has been achieved. *Algebra.*

The first means of remedying the difficulty of the small number of analytical elements seems to be to create new ones. But a little consideration will show that this resource is illusory. A new analytical element would not serve unless we could immediately determine its value: but how can we determine the value of a function which is simple; that is, which is not formed by a combination of those already known? This appears almost impossible: but the introduction of another elementary abstract function into analysis supposes the simultaneous creation of a new arithmetical operation; which is certainly extremely difficult. If we try to proceed according to the method which procured us the elements we possess, we are left in entire uncertainty; for the artifices thus employed are evidently exhausted. We have thus no idea how to proceed to create new elementary abstract functions. Yet, we must not therefore conclude that we have reached the limit appointed by the powers of our understanding. Special improvements in mathematical analysis have yielded us some partial substitutes, which have increased our resources: but it is clear that the augmentation of these elements can not proceed but with extreme slowness. It is not in this direction, then, that the human mind has found its means of facilitating the establishment of equations. *Creation of new functions.*

This first method being discarded, there remains only one other. As it is impossible to find the equations directly, we must seek for corresponding ones between other auxiliary quantities, connected with the first according to a certain determinate law, and from the relation between which we may ascend to that of the primitive magnitudes. This is the fertile conception which we term the *transcendental analysis*, and use as our finest instrument for the mathematical exploration of natural phenomena. *Finding equations between auxiliary quantities.*

This conception has a much larger scope than even profound geometers have hitherto supposed; for the auxiliary quantities resorted to might be derived, according to any law whatever, from the immediate elements of the question. It is well to notice this; because our future improved analytical resources may perhaps be found in a new mode of *derivation*. But, at present, the only aux-

5

iliary quantities habitually substituted for the primitive quantities in *transcendental analysis* are what are called—

1st, *infinitely small* elements, the *differentials* of different orders of those quantities, if we conceive of this analysis in the manner of Leibnitz: or

2d, the *fluxions*, the *limits* of the ratios of the simultaneous increments of the primitive quantities, compared with one another; or, more briefly, the *prime* and *ultimate* ratios of these increments, if we adopt the conception of Newton: or

3d, the *derivatives*, properly so called, of these quantities; that is, the coefficients of the different terms of their respective increments, according to the conception of Lagrange.

These conceptions, and all others that have been proposed, are by their nature identical. The various grounds of preference of each of them will be exhibited hereafter.

Division of the Calculus of functions. We now see that the Calculus of functions, or Algebra, must consist of two distinct branches. The one has for its object the *resolution* of equations when they are directly established between the magnitudes in question: the other, setting out from equations (generally much more easy to form) between quantities indirectly connected with those of the problem, has to deduce, by invariable analytical procedures, the corresponding equations between the direct magnitudes in question;—bringing the problem within the domain of the preceding calculus.—It might seem that the transcendental analysis ought to be studied before the ordinary, as it provides the equations which the other has to resolve. But, though the transcendental is logically independent of the ordinary, it is best to follow the usual method of study, taking the ordinary first; for, the proposed questions always requiring to be completed by ordinary analysis, they must be left in suspense if the instrument of resolution had not been studied beforehand.

To ordinary analysis I propose to give the name of CALCULUS OF DIRECT FUNCTIONS. To transcendental analysis (which is known by the name of Infinitesimal Calculus, Calculus of fluxions and of fluents, Calculus of Vanishing quantities, the Differential and Integral Calculus, etc., according to the view in which it has been conceived) I shall give the title of CALCULUS OF INDIRECT FUNCTIONS. I obtain these terms by generalizing and giving precision to the ideas of Lagrange, and employ them to indicate the exact character of the two forms of analysis.

SECTION I.

ORDINARY ANALYSIS, OR CALCULUS OF DIRECT FUNCTIONS.

ALGEBRA is adequate to the solution of mathematical questions which are so simple that we can form directly the equations between the magnitudes considered, without its being necessary to bring into the problem, either in substitution or alliance, any system of auxiliary quantities derived from the primary. It is true, in the majority of important cases, its use requires to be preceded and prepared for by that of the calculus of indirect functions, by which the establishment of equations is facilitated: but though algebra then takes the second place, it is not the less a necessary agent in the solution of the question; so that the Calculus of direct functions must continue to be, by its nature, the basis of mathematical analysis. We must now, then, notice the rational composition of this calculus, and the degree of development it has attained.

Its object being the resolution of equations (that is, the discovery of the mode of formation of unknown quantities by the known, according to the equations which exist between them), it presents as many parts as we can imagine distinct classes of equations; and its extent is therefore rigorously indefinite, because the number of analytical functions susceptible of entering into equations is illimitable, though, as we have seen, composed of a very small number of primitive elements. *Its object.*

The rational classification of equations must evidently be determined by the nature of the analytical elements of which their numbers are composed. Accordingly, analysts first divide equations with one or more variables into two principal classes, according as they contain functions of only the first three of the ten couples, or as they include also either exponental or circular functions. Though the names of algebraic and transcendental functions given to these principal groups are inapt, the division between the corresponding equations is really enough, in so far as that the resolution of equations containing the transcendental functions is more difficult than that of algebraic equations. Hence the study of the first is extremely imperfect, and our analytical methods relate almost exclusively to the elaboration of the second. *Classification of Equations.*

Our business now is with these Algebraic equations only. In the first place, we must observe that, though they may often contain *irrational* functions of the unknown quantities, as well as *rational* functions, the first case can always be brought under the second, by transformation more or less easy; so that it is only with the latter that analysts have had to occupy themselves, to resolve all the algebraic equations. As to their classification, the early method of classing them according to the number of their terms has been retained only for equations with two terms, which are, in fact, susceptible of a resolution proper to themselves. The classification by their degrees, long universally established, is eminently natural; for this distinction rigorously determines the greater or less difficulty of their resolution. The gradation can be independently, as well as practically exhibited: for the most general equation of each degree necessarily comprehends all those of the different inferior degrees, as must also the formula which determines the unknown quantity: and therefore, however slight we may, *à priori*, suppose the difficulty to be of the degree *Algebraic equations.*

under notice, it must offer more and more obstacles, in proportion to the rank of the degree, because it is complicated in the execution with those of all the preceding degrees.

<small>Algebraic resolution of equations.</small> This increase of difficulty is so great, that the resolution of algebraic equations is as yet known to us only in the first four degrees. In this respect, algebra has advanced but little since the labors of Descartes and the Italian analysts of the sixteenth century; though there has probably not been a single geometer for two centuries past who has not striven to advance the resolution of equations. The general equation of the fifth degree has itself, thus far, resisted all attempts. The formula of the fourth degree is so difficult as to be almost inapplicable; and analysts, while by no means despairing of the resolution of equations of the fifth, and even higher degrees, being obtained, have tacitly agreed to give up such researches.

The only question of this kind which would be of eminent importance, at least in its logical relations, would be the general resolution of algebraic equations of any degree whatever. But the more we ponder this subject, the more we are led to suppose, with Lagrange, that it exceeds the scope of our understandings. Even if the requisite formula could be obtained, it could not be usefully applied, unless we could simplify it, without impairing its generality, by the introduction of a new class of analytical elements, of which we have as yet no idea. And, besides, if we had obtained the resolution of algebraic equations of any degree whatever, we should still have treated only a very small part of algebra, properly so called; that is, of the calculus of direct functions, comprehending the resolution of all the equations that can be formed by the analytical functions known to us at this day. Again, we must remember that by a law of our nature, we shall always remain below the difficulty of science, our means of conceiving of new questions being always more powerful than our resources for resolving them; in other words, the human mind being more apt at imagining than at reasoning. Thus, if we had resolved all the analytical equations now known, and if to do this, we had found new analytical elements, these again would introduce classes of equations of which we now know nothing; and so, however great might be the increase of our knowledge, the imperfection of our algebraic science would be perpetually reproduced.

<small>Our existing knowledge.</small> The methods that we have are, the complete resolution of the equations of the first four degrees; of any binomial equations; of certain special equations of the superior degrees; and of a very small number of exponential, logarithmic, and circular equations. These elements are very limited; but geometers have succeeded in treating with them a great number of important questions in an admirable manner. The improvements introduced within a century into mathematical analysis have contributed more to render the little knowledge that we have immeasurably useful, than to increase it.

To fill up the vast gap in the resolution of algebraic equations of the higher degrees, analysts have had recourse to a new order of questions,—to what they call the *numerical resolution* of equations. Not being able to obtain the real algebraic formula, they have sought to determine at least the *value* of each unknown quantity for such or such a designated system of particular values attributed to the given quantities. This operation is a mixture of algebraic with arithmetical questions; and it has been so cultivated as to be rendered possible in all cases, for equations of any degree and even of any form. The methods for this are now sufficiently general; and what remains is to simplify them so as to fit them for regular application. While such is the state of algebra, we have to endeavor so to dispose the questions to be worked as require finally only this *numerical* resolution of the equations. We must not forget however that this is very imperfect algebra; and it is only isolated, or truly final questions (which are very few), that can be brought finally to depend upon only the *numerical* resolution of equations. Most questions are only preparatory,—a first stage of the solution of other questions; and in these cases it is evidently not the *value* of the unknown quantity that we want to discover, but the *formula* which exhibits its derivation. Even in the most simple questions, when this numerical resolution is strictly sufficient, it is not the less a very imperfect method. Because we cannot abstract and treat separately the algebraic part of the question, which is common to all the cases which result from the mere variation of the given numbers, we are obliged to go over again the whole series of operations for the slightest change that may take place in any one of the quantities concerned.

Thus is the calculus of direct functions at present divided into two parts, as it is employed for the algebraic or the numerical resolution of equations. The first, the only satisfactory one, is unfortunately very restricted, and there is little hope that it will ever be otherwise: the second, usually insufficient, has at least the advantage of a much greater generality. They must be carefully distinguished in our minds, on account of their different objects, and therefore of the different ways in which quantities are considered by them. Moreover, there is, in regard to their methods, an entirely different procedure in their rational distribution. In the first part, we have nothing to do with the *values* of the unknown quantities, and the division must take place according to the nature of the equations which we are able to resolve; whereas in the second, we have nothing to do with the *degrees* of the equations, as the methods are applicable to equations of any degree whatever; but the concern is with the numerical character of the *values* of the unknown quantities.

These two parts, which constitute the immediate object of the Calculus of direct functions, are subordinated to a third, purely speculative, from which both derive their most effectual resources, and which has been very exactly designated by the general name of *Theory of Equations*, though it relates, as yet, only to *algebraic* equations. The numerical resolution of equations has, on account of its generality, special need of this rational foundation.

Two orders of question divide this important department of algebra between them; first, those which relate to the composition of equations, and then those that relate to their transformation; the business of these last being to modify the roots of an equation without knowing them, according to any given law, provided this law is uniform in relation to all these roots.

One more theory remains to be noticed, to complete our rapid exhibition of the different essential parts of the calculus of direct functions. This theory, which relates to the transformation of functions into series by the aid of what is called the Method of indeterminate Coefficients, is one of the most fertile and important in algebra. This eminently analytical method is one of the most remarkable discoveries of Descartes. The invention and development of the infinitesimal calculus, for which it might be very happily substituted in some respects, has undoubtedly deprived it of some of its importance; but the growing extension of the transcendental analysis has, while lessening its necessity, multiplied its applications and enlarged its resources; so that, by the useful combination of the two theories, the employment of the method of indeterminate coefficients has become much more extensive than it was even before the formation of the calculus of indirect functions.

Method of indeterminate Coefficients.

I have now completed my sketch of the Calculus of Direct Functions. We must next pass on to the more important and extensive branch of our science, the Calculus of Indirect Functions.

SECTION II.

TRANSCENDENTAL ANALYSIS, OR CALCULUS OF INDIRECT FUNCTIONS.

Three principal views.

We referred (p. 65) in a former section to the views of the transcendental analysis presented by Leibnitz, Newton, and Lagrange. We shall see that each conception has advantages of its own, that all are finally equivalent, and that no method has yet been found which unites their respective characteristics. Whenever the combination takes place, it will probably be by some method founded on the conception of Lagrange. The other two will then offer only an historical interest; and meanwhile, the science must be regarded as in a merely provisional state, which requires the use of all the three conceptions at the same time; for it is only by the use of them all that an adequate idea of the analysis and its applications can be formed. The vast extent and difficulty of this part of mathematics, and its recent formation, should prevent our being at all surprised at the existing want of system. The conception which will doubtless give a fixed and uniform character to the science has come into the hands of only one new generation of geometers since its creation; and the intellectual habits requisite to perfect it have not been sufficiently formed.

The first germ of the infinitesimal method (which can be conceived of independently of the Calculus) may be recognised in the old Greek *Method of Exhaustions*, employed to pass from the properties of straight lines to those of curves. The method consisted in substituting for the curve the auxiliary consideration of a polygon, inscribed or circumscribed, by means of which the curve itself was reached, the limits of the primitive ratios being suitably taken. There is no doubt of the filiation of ideas in this case; but there was in it no equivalent for our modern methods; for the ancients had no logical and general means for the determination of these limits, which was the chief difficulty of the question. The task remaining for modern geometers was to generalize the conception of the ancients, and, considering it in an abstract manner, to reduce it to a system of calculation, which was impossible to them.

History.

Lagrange justly ascribes to the great geometer Fermat the first idea in this new direction. Fermat may be regarded as having initiated the direct formation of transcendental analysis by his method for the determination of *maxima* and *minima*, and for the finding of *tangents*, in which process he introduced auxiliaries which he afterward suppressed as null when the equations obtained had undergone certain suitable transformations. After some modifications of the ideas of Fermat in the intermediate time, Leibnitz stripped the process of some complications, and formed the analysis into a general and distinct calculus, having its own notation: and Leibnitz is thus the creator of transcendental analysis, as we employ it now. This pre-eminent discovery was so ripe, as all great conceptions are at the hour of their advent, that Newton had at the same time, or rather earlier, discovered a method exactly equivalent, regarding the analysis from a different point of view, much more logical in itself, but less adapted than that of Leibnitz to give all practicable extent and facility to the fundamental method. Lagrange afterward, discarding the heterogeneous considerations which had guided Leibnitz and Newton, reduced the analysis to a purely algebraic system, which only wants more aptitude for application.

We will notice the three methods in their order.

The method of Leibnitz consists in introducing into the calculus, in order to facilitate the establishment of equations, the infinitely small elements or *differentials* which are supposed to constitute the quantities whose relations we are seeking. There are relations between these differentials which are simpler and more discoverable than those of the primitive quantities; and by these we may afterward (through a special calculus employed to eliminate these auxiliary infinitesimals) recur to the equations sought, which it would usually have been impossible to obtain directly.

METHOD OF LEIBNITZ.

This indirect analysis may have various degrees of indirectness; for, when there is too much difficulty in forming the equation between the differentials of the magnitudes under notice, a second application of the method is required, the differentials being now treated as new primitive quantities, and a relation being sought between *their* infinitely small elements, or *second differentials*, and so on; the same transformation being repeated any number of times, provided the whole number of auxiliaries be finally eliminated.

It may be asked by novices in these studies, how these auxiliary quantities can be of use while they are of the same species with the magnitudes to be treated, seeing that the greater or less value of any quantity can not affect any inquiry which has nothing to do with value at all. The explanation is this. We must begin by distinguishing the different orders of infinitely small quantities, obtaining a precise idea of this by considering them as being either the successive powers of the same primitive infinitely small quantity, or as being quantities which may be regarded as having finite ratios with these powers; so that, for instance, the second or third or other differentials of the same variable are classed as infinitely small quantities of the second, third or other order, because it is easy to exhibit in them finite multiples of the second, third, or other powers of a certain first differential. These preliminary ideas being laid down, the spirit of the infinitesimal analysis consists in constantly neglecting the infinitely small quantities in comparison with finite quantities; and generally the infinitely small quantities of any order whatever in comparison with all those of an inferior order. We see at once how such a power must facilitate the formation of equations between the differentials of quantities, since we can substitute for these differentials such other elements as we may choose, and as will be more simple to treat, only observing the condition that the new elements shall differ from the preceding only by quantities infinitely small in relation to them. It is thus that it becomes possible in geometry to treat curved lines as composed of an infinity of rectilinear elements, and curved surfaces as formed of plane elements; and, in mechanics, varied motions as an infinite series of uniform motions, succeeding each other at infinitely small intervals of time. Such a mere hint as this of the varied application of this method may give some idea of the vast scope of the conception of transcendental analysis, as formed by Leibnitz. It is, beyond all question, the loftiest idea ever yet attained by the human mind.

It is clear that this conception was necessary to complete the basis of mathematical science, by enabling us to establish, in a broad and practical manner, the relation of the concrete to the abstract. In this respect, we must regard it as the necessary complement of the great fundamental idea of Descartes on the general analytical representation of natural phenomena; an idea which could not be duly estimated or put to use till after the formation of the infinitesimal analysis.

This analysis has another property besides that of facilitating the study of the mathematical laws of all phenomena, and perhaps not less important than that. The differential formulas exhibit an extreme generality, expressing in a single equation each determinate phenomenon, however varied may be the subjects to which it belongs. Thus, one such equation gives the tangents of all curves, another their rectifications, a third their quadratures; and, in the same way, one invariable formula expresses the mathematical law of all variable motion; and one single equation represents the distribution of heat in any body, and for any case. This remarkable generality is the basis of the loftiest views of the geometers. Thus, this analysis has not only furnished a general method for forming equations indirectly which could not have been directly discovered, but it has introduced a new order of more natural laws for our use in the mathematical study of natural phenomena, enabling us to rise at times to a perception of positive approximations between classes of wholly different phenomena, through the analogies presented by the differential expressions of their mathematical laws. In virtue of this second property of the analysis, the entire system of an immense science, like geometry or mechanics, has submitted to a condensation into a small number of analytical formulas, from which the solution of all particular problems can be deduced, by invariable rules.

Generality of the formulas.

This beautiful method is, however, imperfect in its logical basis. At first, geometers were naturally more intent upon extending the discovery and multiplying its applications than upon establishing the logical foundation of its processes. It was enough, for some time, to be able to produce, in answer to objections, unhoped-for solutions of the most difficult problems. It became necessary, however, to recur to the basis of the new analysis, to establish the rigorous exactness of the processes employed, notwithstanding their apparent breaches of the ordinary laws of reasoning. Leibnitz himself failed to justify his conception, giving, when urged, an answer which represented it as a mere approximative calculus, the successive operations of which might, it is evident, admit an augmenting amount of error. Some of his successors were satisfied with showing that its results accorded with those obtained by ordinary algebra, or the geometry of the ancients, reproducing by these last some solutions which could be at first obtained only by the new method. Some, again, demonstrated the conformity of the new conception with others; that of Newton especially, which was unquestionably exact. This afforded a practical justification: but, in a case of such unequalled importance, a logical justification is also required,—a direct proof of the necessary rationality of the infinitesimal method. It was Carnot who furnished this at last, by showing that the method was founded on the principle of the necessary compensation of errors. We can not say that all the logical scaffolding of the infinitesimal method may not have a merely provisional existence, vicious as it is in its nature:

Justification of the Method.

but, in the present state of our knowledge, Carnot's principle of the necessary compensation of errors is of more importance, in legitimating the analysis of Leibnitz, than is even yet commonly supposed. His reasoning is founded on the conception of infinitesimal quantities indefinitely decreasing, while those from which they are derived are fixed. The infinitely small errors introduced with the auxiliaries can not have occasioned other than infinitely small errors in all the equations; and when the relations of finite quantities are reached, these relations must be rigorously exact, since the only errors then possible must be finite ones, which can not have entered: and thus the finite equations become perfect. Carnot's theory is doubtless more subtile than solid; but it has no other radical logical vice than that of the infinitesimal method itself, of which it is, as it seems to me, the natural development and general explanation; so that it must be adopted as long as that method is directly employed.

The philosophical character of the transcendental analysis has now been sufficiently exhibited to allow of my giving only the principal idea of the other two methods.

NEWTON'S METHOD. Newton offered his conception under several different forms in succession. That which is now most commonly adopted, at least on the continent, was called by himself, sometimes the *Method of prime and ultimate Ratios*, sometimes the *Method of Limits*, by which last term it is now usually known.

Method of limits. Under this method, the auxiliaries introduced are the limits of the ratios of the simultaneous increments of the primitive quantities; or, in other words, the final ratios of these increments; limits or final ratios which we can easily show to have a determinate and finite value. A special calculus, which is the equivalent of the infinitesimal calculus, is afterward employed, to rise from the equations between these limits to the corresponding equations between the primitive quantities themselves.

The power of easy expression of the mathematical laws of phenomena given by this analysis arises from the calculus applying, not to the increments themselves of the proposed quantities, but to the limits of the ratios of those increments; and from our being therefore able always to substitute for each increment any other magnitude more easy to treat, provided their final ratio is the ratio of equality; or, in other words, that the limit of their ratio is unity. It is clear, in fact, that the calculus of limits can be in no way affected by this substitution. Starting from this principle, we find nearly the equivalent of the facilities offered by the analysis of Leibnitz, which are merely considered from another point of view. Thus, curves will be regarded as the limits of a series of rectilinear polygons, and variable motions as the limits of an aggregate of uniform motions of continually nearer approximation, etc., etc. Such is, in substance, Newton's conception; or rather, that which Maclaurin and D'Alembert have offered as the most rational basis of the transcendental analysis, in the endeavor to fix and arrange Newton's ideas on the subject.

Newton had another view, however, which ought to be presented here, because it is still the special form of the calculus of indirect functions commonly adopted by English geometers; and also, on account of its ingenious clearness in some cases, and of its having furnished the notation best adapted to this manner of regarding the transcendental analysis. I mean the Calculus of *fluxions* and of *fluents*, founded on the general notion of *velocities*.

Fluxions and fluents.

To facilitate the conception of the fundamental idea, let us conceive of every curve as generated by a point affected by a motion varying according to any law whatever. The different quantities presented by the curve, the abscissa, the ordinate, the arc, the area, etc., will be regarded as simultaneously produced by successive degrees during this motion. The *velocity* with which each one will have been described will be called the *fluxion* of that quantity, which inversely would have been called its *fluent*. Henceforth, the transcendental analysis will according to this conception, consist in forming directly the equations between the fluxions of the proposed quantities, to deduce from them afterward, by a special Calculus, the equations between the fluents themselves. What has just been stated respecting curves may evidently be transferred to any magnitudes whatever, regarded, by the help of a suitable image, as some being produced by the motion of others. This method is evidently the same with that of limits complicated with the foreign idea of motion. It is, in fact, only a way of representing, by a comparison derived from mechanics, the method of prime and ultimate ratios, which alone is reducible to a calculus. It therefore necessarily admits of the same general advantages in the various principal applications of the transcendental analysis, without its being requisite for us to offer special proofs of this.

Lagrange's conception consists in its admirable simplicity, in considering the transcendental analysis to be a great algebraic artifice, by which, to facilitate the establishment of equations, we must introduce, in the place of or with the primitive functions, their *derived* functions; that is, according to the definition of Lagrange, the coefficient of the first term of the increment of each function, arranged according to the ascending powers of the increment of its variable. The Calculus of indirect functions, properly so called, is destined here, as well as in the conceptions of Leibnitz and Newton, to eliminate these derivatives, employed as auxiliaries, to deduce from their relations the corresponding equations between the primitive magnitudes. The transcendental analysis is then only a simple, but very considerable extension of ordinary analysis. It has long been a common practice with geometers to introduce, in analytical investigations, in the place of the magnitudes in question, their different powers, or their logarithms, or their sines, etc., in order to simplify the equations, and even to ob-

LAGRANGE'S METHOD.

tain them more easily. Successive *derivation* is a general artifice of the same nature, only of greater extent, and commanding in consequence, much more important resources for this common object.

But, though we may easily conceive, à *priori*, that the auxiliary use of these derivatives *may* facilitate the study of equations, it is not easy to explain why it *must* be so under this method of derivation, rather than any other transformation. This is the weak side of Lagrange's great idea. We have not yet become able to lay hold of its precise advantages, in an abstract manner, and without recurrence to the other conceptions of the transcendental analysis. These advantages can be established only in the separate consideration of each principal question; and this verification becomes laborious, in the treatment of a complex problem.

Other theories have been proposed, such as Euler's *Calculus of vanishing quantities*: but they are merely modifications of the three just exhibited. We must next compare and estimate these methods; and in the first place observe their perfect and necessary conformity.

Identity of the three methods. Considering the three methods in regard to their destination, independently of preliminary ideas, it is clear that they all consist in the same general logical artifice; that is, the introduction of a certain system of auxiliary magnitudes uniformly correlative with those under investigation; the auxiliaries being substituted for the express object of facilitating the analytical expression of the mathematical laws of phenomena, though they must be finally eliminated by the help of a special calculus. It was this which determined me to define the transcendental analysis as the Calculus of indirect functions, in order to mark its true philosophical character, while excluding all discussion about the best manner of conceiving and applying it. Whatever may be the method employed, the general effect of this analysis is to bring every mathematical question more speedily into the domain of the calculus, and thus to lessen considerably the grand difficulty of the passage from the concrete to the abstract. We can not hope that the Calculus will ever lay hold of all questions of natural philosophy—geometrical, mechanical, thermological, etc.—from their birth. That would be a contradiction. In every problem there must be a certain preliminary operation before the calculus can be of any use, and one which could not by its nature be subjected to abstract and invariable rules:—it is that which has for its object the establishment of equations, which are the indispensable point of departure for all analytical investigations. But this preliminary elaboration has been remarkably simplified by the creation of the transcendental analysis, which has thus hastened the moment at which general and abstract processes may be uniformly and exactly applied to the solution, by reducing the operation to finding the equations between auxiliary magnitudes, whence the Calculus leads to equations directly relating to the proposed magnitudes, which had formerly to be established directly. Whether these indirect equations are *differential* equations, according to Leibnitz, or equations of *limits*, according to Newton, or *derived* equations, according to Lagrange, the general procedure is evidently always the same. The coincidence is not only in the result but in the process; for the auxiliaries introduced are really identical, being only regarded from different points of view. The conceptions of Leibnitz and Newton consist in making known in any case two general necessary properties of the derived function of Lagrange. The transcendental analysis, then, examined abstractly and in its principle, is always the same, whatever conception is adopted; and the processes of the Calculus of indirect functions are necessarily identical in these different methods, which must therefore, under any application whatever, lead to rigorously uniform results.

If we endeavor to estimate their comparative value, we shall find in each of the three conceptions advantages and inconveniencies which are peculiar to it, and which prevent geometers from adhering to any one of them, as exclusive and final. *Their comparative value.*

The method of Leibnitz has eminently the advantage in the rapidity and ease with which it effects the formation of equations between auxiliary magnitudes. We owe to its use the high perfection attained by all the general theories of geometry and mechanics. Whatever may be the speculative opinions of geometers as to the infinitesimal method, they all employ it in the treatment of any new question. Lagrange himself, after having reconstructed the analysis on a new basis, rendered a candid and decisive homage to the conception of Leibnitz, by employing it exclusively in the whole system of his 'Analytical Mechanics.' Such a fact needs no comment. Yet we are obliged to admit, with Lagrange, that the conception of Leibnitz is radically vicious in its logical relations. He himself declared the notion of infinitely small quantities to be a *false idea:* and it is in fact impossible to conceive of them clearly, though we may sometimes fancy that we do. This false idea bears, to my mind, the characteristic impress of the metaphysical age of its birth and tendencies of its originator. By the ingenious principle of the compensation of errors, we may, as we have already seen, explain the necessary exactness of the processes which compose the method; but it is a radical inconvenience to be obliged to indicate, in Mathematics, two classes of reasonings so unlike, as that the one order are perfectly rigorous, while by the others we designedly commit errors which have to be afterward compensated. There is nothing very logical in this; nor is anything obtained by pleading, as some do, that this method can be made to enter into that of limits, which is logically irreproachable. This is eluding the difficulty, and not resolving it; and besides, the advantages of this method, its ease and rapidity, are almost entirely lost under such a transformation. Finally, the infinitesimal method exhibits the very serious defect of breaking the unity of abstract mathematics by creating a transcendental analysis founded upon principles widely different from those which serve as a basis to ordinary analysis. This division of analysis into two systems, almost wholly independent, tends to prevent the formation of general analytical conceptions. To estimate the consequences duly, we must recur, in thought, to the state of the science before Lagrange had established a general and complete harmony between these two great sections.

Newton's conception is free from the logical objections imputable to that of Leibnitz. The notion of *limits* is in fact remarkable for its distinctness and precision. The equations are, in this case, regarded as exact from their origin; and the general rules of reasoning are as constantly observed as in ordinary analysis. But it is weak in resources, and embarrassing in operation, compared with the infinitesimal method. In its applications, the relative inferiority of this theory is very strongly marked. It also separates the ordinary and transcendental analysis, though not so conspicuously as the theory of Leibnitz. As Lagrange remarked, the idea of *limits*, though clear and exact, is not the less a foreign idea, on which analytical theories ought not to be dependent.

This perfect unity of analysis, and a purely abstract character in the fundamental ideas, are found in the conception of Lagrange, and there alone. It is, therefore, the most philosophical of all. Discarding every heterogeneous consideration, Lagrange reduced the transcendental analysis to its proper character,—that of presenting a very extensive class of analytical transformations, which facilitate, in a remarkable degree, the expression of the conditions of the various problems. This exhibits the conception as a simple extension of ordinary analysis. It is a superior algebra. All the different parts of abstract mathematics, till then so incoherent, might be from that moment conceived of as forming a single system. This philosophical superiority marks it for adoption as the final theory of transcendental analysis; but it presents too many difficulties, in its application, in comparison with the others, to admit of its exclusive preference at present. Lagrange himself had great difficulty in rediscovering, by his own method, the principal results already obtained by the infinitesimal method, on general questions in geometry and mechanics; and we may judge by that what obstacles would occur in treating in the same way questions really new and important. Though Lagrange, stimulated by difficulty, obtained results in some cases which other men would have despaired of, it is not the less true that his conception has thus far remained, as a whole, essentially unsuited to applications.

The result of such a comparison of these three methods is the conviction that, in order to understand the transcendental analysis thoroughly, we should not only study it in its principles according to all these conceptions, but should accustom ourselves to employ them all (and especially the first and last) almost indifferently, in the solution of all important questions, whether of the calculus of indirect functions in itself, or of its applications. In all the other departments of mathematical science, the consideration of different methods for a single class of questions may be useful, apart from the historical interest which it presents; but it is not indispensable. Here, on the contrary, it is strictly indispensable. Without it there can be no philosophical judgment of this admirable creation of the human mind; nor any success and facility in the use of this powerful instrument.

THE DIFFERENTIAL AND INTEGRAL CALCULUS.

The Calculus of Indirect Functions is necessarily divided into two parts; or rather, it is composed of two distinct calculi, having the relation of converse action. *Its Two Parts.* By the one we seek the relations between the auxiliary magnitudes, by means of the relations between the corresponding primitive magnitudes; by the other we seek, conversely, these direct equations by means of the indirect equations first established. This is the double object of the transcendental analysis.

Different names have been given to the two systems, according to the point of view from which the entire analysis has been regarded. The infinitesimal method, properly so called, being most in use, almost all geometers employ the terms *Differential Calculus* and *Integral Calculus* established by Leibnitz. Newton, in accordance with his method, called the first the *Calculus of Fluxions*, and the second the *Calculus af Fluents*, terms which were till lately commonly adopted in England. According to the theory of Lagrange, the one would be called the *Calculus of Derived Functions*, and the other the *Calculus of Primitive Functions*. I shall make use of the terms of Leibnitz, as the fittest for the formation of secondary expressions, though we must, as has been shown, employ all the conceptions concurrently, approaching as nearly as may be to that of Lagrange.

The differential calculus is obviously the rational basis of the integral. We have seen that ten simple *Their mutual relations.* functions constitute the elements of our analysis. We cannot know how to integrate directly any other differential expressions than those produced by the differentiation of those ten functions. The art of integration consists therefore in bringing all the other cases, as far as possible, to depend wholly on this small number of simple functions.

It may not be apparent to all minds what can be the proper utility of the differential calculus, independently of this necessary connection with the integral calculus, which seems as if it must be in itself the only directly indispensable one; in fact, the elimination of the *infinitesimals* or the *derivatives*, introduced as auxiliaries, being the final object of the calculus of indirect functions, it is natural to think that the calculus which teaches us to deduce the equations between the primitive magnitudes from those between the auxiliary magnitudes must meet all the general needs of the transcendental analysis, without our seeing at first what special and constant part the solution of the inverse question can have in such an analysis. A common answer is assigning to the differential calculus the office of forming the differential equations; but this is clearly an error; for the primitive formation of differential equations is not the business of any calculus, for it is, on the contrary, the point of departure of any calculus whatever. The very use of the differential calculus is enabling us to *differentiate* the various equations; and it cannot therefore be the process for establishing them. This common error arises from confounding the infinitesimal *calculus* with the infinitesimal *method*, which last facilitates the formation of equations, in every application of the transcendental analysis. The calculus is the indispensable complement of the method; but it is perfectly distinct from it. But again, we should much misconceive the peculiar importance of this first branch of the calculus of indirect functions if we saw in it only a preliminary process, designed merely to prepare an indispensable basis for the integral calculus. A few words will show that a primary direct and necessary office is always assigned to the differential calculus. In

Cases of union of the two. forming differential equations, we rarely restrict ourselves to introducing differentially only those magnitudes whose relations are sought. It would often be impossible to establish equations without introducing other magnitudes whose relations are, or are supposed to be, known. Now in such cases it is necessary that the differentials of these intermediaries should be eliminated before the equations are fit for integration. This elimination belongs to the differential calculus; for it must be done by determining, by means of the equations between the intermediary functions, the relations of their differentials; and this is merely a question of differentiation. This is the way in which the differential calculus not only prepares a basis for the integral, but makes it available in a multitude of cases which could not otherwise be treated. There are some questions, few, but highly

Cases of the Differential calculus alone. important, which admit of the employment of the differential calculus alone. They are those in which the magnitudes sought enter directly, and not by their differentials, into the primitive differential equations, which then contain differentially only the various known functions employed, as we saw just now, as intermediaries. This calculus is here entirely sufficient for the elimination of the infinitesimals, without the question giving rise to any integration. There are also questions, few, but highly important, which are the converse of the last, requiring the employment of the integral calculus alone. In these, the

Cases of the Integral calculus alone. differential equations are found to be immediately ready for integration, because they contain, at their first formation, only the infinitesimals which relate to the functions sought, or to the really independent variables, without the introduction, differentially, of any intermediaries being required. If intermediary functions are introduced, they will, by the hypothesis, enter directly, and not by their differentials; and then, ordinary algebra will serve for their elimination, and to bring the question to depend on the integral calculus only. The differential calculus is, in such cases, not essential to the solution of the problem, which will depend entirely on the integral calculus. Thus, all questions to which the analysis is applicable are contained in three classes. The first class comprehends the problems which may be resolved by the differential calculus alone. The second, those which may be resolved by the integral calculus alone. These are only exceptional; the third constituting the normal case; that in which the differential and integral calculus have each a distinct and necessary part in the solution of problems.

The Differential Calculus.

The entire system of the differential calculus is simple and perfect, while the integral calculus remains extremely imperfect.

We have nothing to do here with the applications of either calculus, which are quite a different study from that of the abstract principles of differentiation and integration. The consequence of the common practice of confounding these principles with their application, especially in geometry, is that it becomes difficult to conceive of either analysis or geometry. It is in the department of Concrete Mathematics that the application should be studied. *The Differential Calculus.*

The first division of the differential calculus is grounded on the condition whether the functions to be differentiated are *explicit* or *implicit*; the one giving rise to the differentiation of *formulas*, and the other to the differentiation of *equations*. *Two portions.* This classification is rendered necessary by the imperfection of ordinary analysis; for if we knew how to resolve all equations algebraically, it would be possible to render every implicit function explicit; and, by differentiating it only in that state, the second part of the differential calculus would be immediately included in the first, without giving rise to any new difficulty. But the algebraic resolution of equations is, as we know, still scarcely past its infancy, and unknown for the greater number of cases; and we have to differentiate a function without knowing it, though it is determinate. Thus we have two classes of questions, the differentiation of implicit functions being a distinct case from that of explicit functions, and much more complicated. We have to begin by the differentiation of formulas, and we may then refer to this first case the differentiation of equations, by certain analytical considerations which we are not concerned with here. There is another view in which the two general cases of differentiation are distinct. The relation obtained between the differentials is always more indirect, in comparison with that of the finite quantities, in the differentiation of implicit, than in that of explicit functions. We shall meet with this consideration in the case of the integral calculus, where it acquires a preponderant importance.

Subdivisions. Each of these parts of the differential calculus is again divided: and this subdivision exhibits two very distinct theories, according as we have to differentiate functions of a single variable, or functions of several independent variables,—the second branch being of far greater complexity than the first, in the case of explicit functions, and much more in that of implicit. One more distinction remains, to complete this brief sketch of the parts of the differential calculus. The case in which it is required to differentiate at once different implicit functions *combined* in certain primitive equations must be distinguished from that in which all these functions are *separate*. The same imperfection of ordinary analysis which prevents our converting every implicit function into an equivalent explicit one, renders us unable to separate the functions which enter simultaneously into any system of equations; and the functions are evidently still more implicit in the case of combined than of separate functions: and in differentiating, we are not only unable to resolve the primitive equations, but even to effect the proper elimination among them.

Reduction to the elements. We have now seen the different parts of this calculus in their natural connection and rational distribution. The whole calculus is finally found to rest upon the differentiation of explicit functions with a single variable,—the only one which is ever executed directly. Now, it is easy to understand that this first theory, this necessary basis of the whole system, simply consists of the differentiation of the elementary functions, ten in number, which compose all our analytical combinations; for the differentiation of compound functions is evidently deduced, immediately and necessarily, from that of their constituent simple functions. We find, then, the whole system of differentiation reduced to the knowledge of the ten fundamental differentials, and to that of the two general principles, by one of which the differentiation of implicit functions is deduced from that of explicit, and by the other, the differentiation of functions of several variables is reduced to that of functions of a single variable. Such is the simplicity and perfection of the system of the differential calculus.

Transformation of derived functions for new variables. The *transformations of derived Functions for new variables* is a theory which must be just mentioned, to avoid the omission of an indispensable complement of the system of differentiation. It is as finished and perfect as the other parts of this calculus; and its great importance is in its increasing our resources by permitting us to choose, to facilitate the formation of differential equations, that system of independent variables which may appear to be most advantageous, though it may afterward be relinquished, as an intermediate step, by which, through this theory, we may pass to the final system, which sometimes could not have been considered directly.

Analytical applications. Though we can not here consider the concrete applications of this calculus, we must glance at those which are analytical, because they are of the same nature with the theory, and should be looked at in connection with it. These questions are reducible to three essential ones. First, the development into series of functions of one or more variables; or, more generally, the transformation of functions, which constitutes the most beautiful and the most important application of the differential calculus to general analysis, and which comprises, besides the fundamental series discovered by Taylor, the remarkable series discovered by Maclauring, John Bernouilli, Lagrange and others. Secondly, the general theory of *maxima* and *minima* values for any functions whatever of one or more variables; one of the most interesting problems that analysis can present, however elementary it has become. The third is the least important of the three:—it is the determination of the true value of functions which present themselves under an indeterminate appearance, for certain hypotheses made on the values of the corresponding variables. In every view, the first question is the most eminent; it is also the most susceptible of future extension, especially by conceiving, in a larger manner than hitherto, of the employment of the differential calculus for the transformation of functions, about which Lagrange left some valuable suggestions which have been neither generalized nor followed up.

It is with regret that I confine myself to the generalities which are the proper subjects of this work; so extensive and so interesting are the developments which might otherwise be offered. Insufficient and summary as are the views of the Differential Calculus just offered, we must be no less rapid in our survey of the Integral Calculus, properly so called; that is, the abstract subject of integration.

The Integral Calculus.

The division of the Integral Calculus, like that of the Differential, proceeds on the principle of distinguishing the integration of *explicit* differential formulas from the integration of *implicit* differentials, or of differential equations. The separation of these two cases is even more radical in the case of integration than in the other. In the differential calculus this distinction rests, as we have seen, only on the extreme imperfection of ordinary analysis. But, on the other hand, it is clear that even if all equations could be algebraically resolved, differential equations would nevertheless constitute a case of integration altogether distinct from that presented by explicit differential formulas. Their integration is necessarily more complicated than that of explicit differentials, by the elaboration of which the integral calculus was originated, and on which the others have been made to depend, as far as possible. All the various analytical processes hitherto proposed for the integration of differential equations, whether by the separation of variables, or the method of multipliers, or other means, have been designed to reduce these integrations to those of differential formulas, the only object which

The Integral Calculus.

Its divisions.

can be directly undertaken. Unhappily, imperfect as is this necessary basis of the whole integral calculus, the art of reducing to it the integration of differential equations is even much less advanced.

Subdivisions. As in the case of the differential calculus, and for analogous reasons, each of these two branches of the integral calculus is divided again, according as we consider functions with a single variable or functions with several independent variables. This distinction is, like the preceding, even more important for integration than for differentiation. This is especially remarkable with respect to differential equations. In fact, those which relate to several independent variables may evidently present this characteristic and higher difficulty—that the function sought may be differentially defined by a simple relation between its various special derivatives with regard to the different variables taken separately. Thence results the most difficult, and also the most extended branch of the integral calculus, which is commonly called the Integral Calculus of partial differences, created by D'Alembert, in which, as Lagrange truly perceived, geometers should have recognised a new calculus, the philosophical character of which has not yet been precisely decided. This higher branch of transcendental analysis is still entirely in its infancy. In the very simplest case, we can not completely reduce the integration to that of the ordinary differential equations.

Orders of differentiation. A new distinction, highly important here, though not in the differential calculus, where it is a mistake to insist upon it, is drawn from the higher or lower order of the differentials. We may regard this distinction as a subdivision in the integration of explicit or implicit differentials. With regard to explicit differentials, whether of one variable or of several, the necessity of distinguishing their different orders is occasioned merely by the extreme imperfection of the integral calculus; and, with reference to implicit differentials, the distinction of orders is more important still. In the first case, we know so little of integration of even the first order of differential formulas, that differential formulas of a high order produce new difficulties in arriving at the primitive function which is our object. And in the second case, there is the additional difficulty that the higher order of the differential equations necessarily gives rise to questions of a new kind. The higher the order of differential equations, the more implicit are the cases which they present; and they can be made to depend on each other only by special methods, the investigation of which, in consequence, forms a new class of questions with regard to the simplest cases of which we as yet know next to nothing.

The necessary basis of all other integrations is, as we see from the foregoing considerations, that of explicit differential formulas of the first order and of a single variable; and we can not succeed in effecting other integrations but by reducing them to this elementary case, which is the only one capable of being treated directly. This simple fundamental integration, often conveniently called *quadratures*, corresponds in the differential calculus to the elementary case of the differentiation of explicit functions of a single variable. But the integral question is, by its nature, quite otherwise complicated, and much more extensive than the differential question. We have seen that the latter is reduced to the differentiation of ten simple functions, which furnish the elements of analysis; but the integration of compound functions does not necessarily follow from that of the simple functions, each combination of which may present special difficulties with respect to the integral calculus. Hence the indefinite extent and varied complication of the question of quadratures, of which we know scarcely anything completely, after all the efforts of analysts.

<small>Quadratures.</small>

The question is divided into the two cases of *algebraic functions* and *transcendental functions*. The algebraic class is the more advanced of the two. In relation to irrational functions, it is true, we know scarcely anything, the integrals of them having been obtained only in very restricted cases, and particularly by rendering them rational. The integration of rational functions is thus far the only theory of this calculus which has admitted of complete treatment; and thus it forms, in a logical point of view, its most satisfactory part, though it is perhaps the least important. Even here, the imperfection of ordinary analysis usually comes in to stop the working of the theory, by which the integration finally depends on the algebraic solution of equations; and thus it is only in what concerns integration viewed in an abstract manner that even this limited case is resolved. And this gives us an idea of the extreme imperfection of the integral calculus. The case of the integration of transcendental functions is quite in its infancy as yet, as regards either exponential, logarithmic, or circular functions. Very few cases of these kinds have been treated; and though the simplest have been chosen, the necessary calculations are extremely laborious.

<small>Algebraic functions.</small>

<small>Transcendental functions.</small>

The theory of *Singular Solutions* (sometimes called Particular Solutions), fully developed by Lagrange in his Calculus of Functions, but not yet duly appreciated by geometers, must be noticed here, on account of its logical perfection and the extent of its applications. This theory forms implicitly a portion of the general theory of the integration of differential equations; but I have left it till now, because it is, as it were, outside of the integral calculus, and I wished to preserve the sequence of its parts. Clairaut first observed the existence of these solutions, and he saw in them a paradox of the integral calculus, since they have the property of satisfying the differential equations without being comprehended in the corresponding general integrals. Lagrange explained this paradox by showing how such solutions are always derived from the general integral by the variation of the

<small>Singular solutions.</small>

arbitrary constants. This theory has a character of perfect generality; for Lagrange has given invariable and very simple processes for finding the singular solution of any differential equation which admits of it; and, what is very remarkable, these processes require no integration, consisting only of differentiations, and being therefore always applicable. Thus has differentiation become, by a happy artifice, a means of compensating, in certain circumstances, for the imperfection of the integral calculus.

Definite integrals. One more theory remains to be noticed, to complete our review of that collection of analytical researches which constitutes the integral calculus. It takes its place outside of the system, because, instead of being destined for true integration, it proposes to supply the defect of our ignorance of really analytical integrals. I refer to the determination of *definite integrals*. These definite integrals are the values of the required functions for certain determinate values of the corresponding variables. The use of these in transcendental analysis corresponds to the numerical resolution of equations in ordinary analysis. Analysts being usually unable to obtain the real integral (called in opposition the *general* or *indefinite* integral), that is, the function which, differentiated, has produced the proposed differential formula, have been driven to determining, at least, without knowing this function, the particular numerical values which it would take on assigning certain declared values to the variables. This is evidently resolving the arithmetical question without having first resolved the corresponding algebraic one, which is generally the most important; and such an analysis is, by its nature, as imperfect as that of the numerical resolution of equations. Inconveniences, logical and practical, result from such a confusion of arithmetical and algebraic considerations. But, under our inability to obtain the true integrals, it is of the utmost importance to have been able to obtain this solution, incomplete and insufficient as it is. This has now been attained for all cases, the determination of the value of definite integrals having been reduced to entirely general methods, which leave nothing to be desired, in many cases, but less complexity in the calculations; an object to which analysts are now directing all their special transformations. This kind of *transcendental arithmetic* being considered perfect, the difficulty in its applications is reduced to making the proposed inquiry finally depend only on a simple determination of definite integrals; a thing which evidently can not be always possible, whatever analytical skill may be employed in effecting so forced a transformation.

Prospects of the Integral Calculus. We have now seen that while the differential calculus constitutes by its nature a limited and perfect system, the integral calculus, or the simple subject of integration, offers inexhaustible scope for the activity of the human mind, independently of the indefinite applications of which transcendental analysis is evidently capable. The reasons which convince us of the impossibility of ever achieving the general resolution of algebraic equations of any degree whatever, are yet more decisive against our attainment of a single method of integration applicable to all cases. "It is," said Lagrange, "one of those problems whose general solution we can not hope for." The more we meditate on the subject, the more convinced we shall be that such a research is wholly chimerical, as transcending the scope of our understanding, though the labors of geometers must certainly add in time to our knowledge of integration, and create procedures of a wider generality. The transcendental analysis is yet too near its origin, it has too recently been regarded in a truly rational manner, for us to have any idea what it may hereafter become. But, whatever may be our legitimate hopes, we must ever, in the first place, consider the limits imposed by our intellectual constitution, which are not the less real because we can not precisely assign them.

I have hinted that a future augmentation of our resources may probably arise from a change in the mode of derivation of the auxiliary quantities introduced to facilitate the establishment of equations. Their formation might follow a multitude of other laws besides the very simple relation which has been selected. I discern here far greater resources than in urging further our present calculus of indirect functions; and I am persuaded that when geometers have exhausted the most important applications of our present transcendental analysis, they will turn their attention in this direction, instead of straining after perfection where it can not be found. I submit this view to geometers whose meditations are fixed on the general philosophy of analysis.

As for the rest, though I was bound to exhibit in my summary exposition the state of extreme imperfection in which the integral calculus still remains, it would be entertaining a false idea of the general resources of the transcendental analysis to attach too much importance to this consideration. As in ordinary analysis, we find here that a very small amount of fundamental knowledge respecting the resolution of equations is of inestimable use. However little advanced geometers are as yet in the science of integrations, they have nevertheless derived from their few abstract notions the solution of a multitude of questions of the highest importance in geometry, mechanics, thermology, etc. The philosophical explanation of this double general fact is found in the preponderating importance and scope of abstract science, the smallest portion of which naturally corresponds to a multitude of concrete researches, Man having no other resource for the successive extension of his intellectual means than in the contemplation of ideas more and more abstract, and nevertheless positive.

Calculus of Variations.

By his *Calculus* or *Method of Variations*, Lagrange improved the capacity of the transcendental analysis for the establishment of equations in the most difficult problems, by considering a class of equations still more *indirect* than differential equations properly so called. It is still too near its origin, and its applications have been too few, to admit of its being understood by a purely abstract account of its theory; and it is therefore necessary to indicate briefly the special nature of the problems which have given rise to this hyper-transcendental analysis.

Problems giving rise to this Calculus. These problems are those which were long known by the name of *Isoperimetrical Problems;* a name which is truly applicable to only a very small number of them. They consist in the investigation of the *maxima* and *minima* of certain indeterminate integral formulas which express the analytical law of such or such a geometrical or mechanical phenomenon, considered independently of any particular subject.

In the ordinary theory of *maxima* and *minima*, we seek, with regard to a given function of one or more variables, what particular values must be assigned to these variables, in order that the corresponding value of the proposed function may be a maximum or a minimum with respect to those values which immediately precede and follow it:—that is, we inquire, properly speaking, at what instant the function ceases to increase in order to begin to decrease, or the reverse. The differential calculus fully suffices, as we know, for the general resolution of this class of questions, by showing that the values of the different variables which suit either the maximum or minimum must always render null the different derivatives of the first order of the given function, taken separately with relation to each independent variable; and by indicating moreover a character suitable for distinguishing the maximum from the minimum, which consists, in the case of a function of a single variable, for example, in the derived function of the second order taking a negative value for the maximum and a positive for the minimum. Such are the fundamental conditions belonging to the majority of cases; and where modifications take place, they are equally subject to invariable, though more complicated abstract rules.

The construction of this general theory having destroyed the chief interest of geometers in this kind of questions, they rose almost immediately to the consideration of a new order of problems, at once more important and more difficult,—those of *isoperimeters*. It was then no longer the values of the variables proper to the maximum or the minimum of a given function that had to be determined. It was the form of the function itself that had to be discovered, according to the condition of the maximum or minimum of a certain definite integral, merely indicated, which depended on that function. We can not here follow the history of these problems, the oldest of which is that of *the solid of least resistance*, treated by Newton in the second book of the "Principia," in which he determines what must be the meridian curve of a solid of revolution, in order that the resistance experienced by that body in the direction of its axis may be the least possible. Mechanics first furnished this new class of problems; but it was from geometry that the subjects of the principal investigations were afterward derived. They were varied and complicated almost infinitely by the labors of the best geometers, when Lagrange reduced their solution to an abstract and entirely general method, the discovery of which has checked the eagerness of geometers about such an order of researches.

It is evident that these problems, considered analytically, consist in determining what ought to be the form of a certain unknown function of one or more variables, in order that such or such an integral, dependent on that function, may have, within assigned limits, a value which may be a maximum or a minimum, with regard to all those which it would take if the required function had any other form whatever. In treating these problems, the predecessors of Lagrange proposed, in substance, to reduce them to the ordinary theory of maxima and minima. But they proceeded by applying special simple artifices to each case, not reducible to certain rules; so that every new question reproduced analogous difficulties, without the solutions previously obtained being of any essential aid. The part common to all questions of this class had not been discovered; and no abstract and general treatment was therefore provided. In his endeavors to bring all isoperimetrical problems to depend on a common analysis, Lagrange was led to the conception of a new kind of differentiation; and to these new Differentials he gave the name of Variations. They consist of the infinitely small increments which the integrals receive, not in virtue of analogous increments on the part of the corresponding variables, as in the common transcendental analysis, but by supposing that the *form* of the function placed under the sign of integration undergoes an infinitely small change. This abstract conception once formed, Lagrange was able to reduce with ease, and in the most general manner, all the problems of isoperimeters to the simple common theory of maxima and minima.

Important as is this great and happy transformation, and though the Method of Variations had at first no other object than the rational and general resolution of isoperimetrical problems, we should form a very inadequate estimate of this beautiful analysis if we supposed it restricted to this application. In fact, the abstract conception of two distinct natures of differentiation is evidently applicable, not only to the cases for which it was created, but for all which present, for any reason whatever, two different ways of making the same magnitudes vary. Lagrange himself made an immense and all-important application of his Calculus of Variations, in his 'Analytical Mechanics,' by employing it to distinguish the two sorts of changes, naturally presented by questions of rational Mechanics for the different points we have to consider, according as we compare the successive positions occupied, in virtue of its motion, by the same point of each body in two consecutive instants, or as we pass from one point of the body to another in the same instant. One of these comparisons produces the common differentials; the other occasions variations which are, there as elsewhere, only differentials taken from a new point of view. It is in such a general acceptation as this that we must conceive of the Calculus of Variations, to appreciate fitly the importance of this admirable logical instrument; the most powerful as yet constructed by the human mind.

This Method being only an immense extension of the general transcendental analysis, there is no need of proof that it admits of being considered under the different primary points of view allowed by the calculus of indirect functions, as a whole. Lagrange invented the calculus of variations in accordance with the infinitesimal conception, properly so called, and even some time before he undertook the general reconstruction of the transcendental analysis. When he had effected that important reform, he easily showed how applicable it was to the calculus of variations, which he exhibited with all suitable development, according to his theory of derived functions. But the more difficult in the use the method of variations is found to be, on account of the higher degree of abstraction of the ideas considered, the more important it is to husband the powers of our minds in its application, by adopting the most direct and rapid analytical conception, which is, as we know, that of Leibnitz. Lagrange himself, therefore, constantly preferred it in the important use which he made of the calculus of variations in his "Analytical Mechanics." There is not, in fact, the slightest hesitation about this among geometers.

Relation to the ordinary Calculus. In the section on the Integral Calculus, I noticed D'Alembert's creation of the *Calculus of partial differences*, in which Lagrange recognised a new calculus. This new elementary idea in transcendental analysis,—the notion of two kinds of increments, distinct and independent of each other, which a function of two variables may receive in virtue of the change of each variable separately,—seems to me to establish a natural and necessary transition between the common infinitesimal calculus and the calculus of variations. D'Alembert's view appears to me to approximate, by its nature, very nearly to that which serves as a general basis for the Method of Variations. This last has, in fact, done nothing more than transfer to the independent variables themselves the view already adopted for the functions of those variables; a process which has remarkably extended its use. A recognition of such a derivation as this for the method of variations may exhibit its philosophical character more clearly and simply; and this is my reason for the reference.

The Method of Variations presents itself to us as the highest degree of perfection which the analysis of indirect functions has yet attained. We had before, in that analysis, a powerful instrument for the mathematical study of natural phenomena, inasmuch as it introduced the consideration of auxiliary magnitudes, so chosen as that their relations were necessarily more simple and easy to obtain than those of the direct magnitudes. But we had not any general and abstract rules for the formation of these differential equations; nor were such supposed to be possible. Now, the Analysis of Variations brings the actual establishment of the differential equations within the reach of the Calculus; for such is the general effect, in a great number of important and difficult questions, of the *varied* equations, which, still more *indirect* than the simple differential equations, as regards the special objects of the inquiry, are more easy to form: and, by invariable and complete analytical methods, employed to eliminate the new order of auxiliary infinitesimals introduced, we may deduce those ordinary differential equations which we might not have been able to establish directly. The Method of Variations forms, then, the most sublime part of that vast system of mathematical analysis, which, setting out from the simplest elements of algebra, organizes, by an uninterrupted succession of ideas, general methods more and more potent for the investigation of natural philosophy. This is incomparably the noblest and most unquestionable testimony to the scope of the human intellect. If, at the same time, we bear in mind that the employment of this method exacts the highest known degree of intellectual exertion, in order never to lose sight of the precise object of the investigation in following reasonings which offer to the mind such uncertain resting-places, and in which signs are of scarcely any assistance, we shall understand how it may be that so little use has been made of such a conception by any philosophers but Lagrange.

We have now reviewed Mathematical analysis, in its bases and in its divisions, very briefly, but from a philosophical point of view, neglecting those conceptions only which are not organized with the great whole, or which, if urged to their limit, would be found to merge in some which have been examined. I must next offer a similar outline of Concrete Mathematics. My particular task will be to show how, supposing the general science of the Calculus to be in a perfect state,—it has been possible to reduce, by invariable procedures, to pure questions of analysis, all the problems of GEOMETRY and MECHANICS; and thus to invest these two great bases of natural philosophy with that precision and unity which can only thus be attained, and which constitute high perfection.

CHAPTER III.

GENERAL VIEW OF GEOMETRY.

Its nature.   WE have seen that GEOMETRY is a true natural science;—only more simple, and therefore more perfect, than any other. We must not suppose that, because it admits the application of mathematical analysis, it is therefore a purely logical science, independent of observation. Every body studied by geometers presents some primitive phenomena which, not being discoverable by reasoning, must be due to observation alone.

The scientific eminence of Geometry arises from the extreme generality and simplicity of its phenomena. If all the parts of the universe were regarded as immovable, geometry would still exist; whereas, for the phenomena of Mechanics, motion is required. Thus Geometry is the more general of the two. It is also the more simple, for its phenomena are independent of those of Mechanics, while mechanical phenomena are always complicated with those of geometry. The same is true in the comparison of abstract thermology with geometry. For these reasons, geometry holds the first place under the head of Concrete Mathematics.

Definition.   Instead of adopting the inadequate ordinary account of Geometry, that it is the *science of extension*, I am disposed to give as a general description of it, that it is the science of the *measurement of extension*. Even this does not include all the operations of geometry, for there are many investigations which do not appear to have for their object the measurement of extension. But regarding the science in its leading questions as a whole, we may accurately say that the measurement of lines, of surfaces, and of volumes, is the invariable aim—sometimes direct, though oftener indirect—of geometrical labors.

Idea of Space.   The rational study of geometry could never have begun if we must have regarded at once and together all the physical properties of bodies, together with their magnitude and form. By the character of our minds we are able to think of the dimensions and figure of a body in an abstract way. After observation has shown us, for instance, the impression left by a body on a fluid in which it has been placed, we are able to retain an image of the impression, which becomes a ground of geometrical reasoning. We thus obtain, apart from all metaphysical fancies, an idea of Space. This abstraction, now so familiar to us that we can not perceive the state we should be in without it, is perhaps the earliest philosophical creation of the human mind.

GEOMETRICAL MEASUREMENT.

There is another abstraction which must be made before we can enter on geometrical science. We must conceive of three kinds of extension, and learn to conceive of them separately. We can not conceive of any space, filled by any object, which has not at once *volume, surface*, and *line*. Yet geometrical questions often relate to only two of these; frequently only to one. Even when all three are to be finally considered, it is often necessary, in order to avoid complication, to take only one at a time. This is the second abstraction which it is indispensable for us to practise—to think of surface and line apart from volume; and again, of line apart from surface. We effect this by thinking of volume as becoming thinner and thinner, till surface appears as the thinnest possible layer or film: and again, we think of this surface becoming narrower and narrower till it is reduced to the finest imaginable thread; and then we have the idea of a line. Though we can not speak of a point as a dimension, we must have the abstract idea of that too; and it is obtained by reducing the line from one end or both, till the smallest conceivable portion of it is left. This point indicates, not extension of course, but position, or the place of extension. Surfaces have clearly the property of circumscribing volumes; lines, again, circumscribe surfaces; and lines, once more, are limited by points.

Kinds of extension.

The Mathematical meaning of *measurement* is simply the finding the value of the *ratios* between any homogeneous magnitudes: but geometrically, the measurement is always indirect. The comparison of two lines is direct; that of two surfaces or two volumes can never be direct. One line may be conceived to be laid upon another: but one volume can not be conceived of as laid upon another, nor one surface upon another, with any convenience or exactness. The question is, then, how to measure surfaces and volumes.

Geometrical measurement.

Whatever be the form of a body, there must always be lines, the length of which will define the magnitude of the surface or volume. It is the business of geometry to use these lines, directly measurable as they are, for the ascertainment of the *ratio* of the surface to the unity of surface, or of the volume to the unity of volume, as either may be sought. In brief, the object is to reduce all comparisons of surfaces or of volumes to simple comparison of lines. Extending the process, we find the possibility of reducing to questions of lines all questions relating to surfaces and volumes, regarded in relation to their magnitude. It is true that when the rational method becomes too complicated and difficult, direct comparisons of surfaces and volumes are employed: but the procedure is not geometrical. In the same way, the consideration of weight is sometimes brought in, to determine volume, or even surface; but this device is derived from mechanics, and has nothing to do with rational geometry.

Measurement of surfaces and volumes.

In speaking of the direct measurement of lines, it is clear that right lines are meant. When we consider

Of curved lines.

curve lines, it is evident that their measurement must be indirect, since we can not conceive of curved lines being laid upon each other with any precision or certainty. The procedure is first to reduce the measurement of curved to that of right lines; and consequently to reduce to simple questions of right lines all questions relating to the magnitude of any curves whatever. In every curve, there always exist certain right lines, the length of which must determine that of the curve; as the length of the radius of a circle gives us that of the circumference; and again, as the length of an ellipse depends on that of its two axes.

Thus, the science of Geometry has for its object the final reduction of the comparisons of all kinds of extent to comparisons of right lines, which alone are capable of direct comparison, and are, moreover, eminently easy to manage.

I must just notice that there is a primary distinct branch of Geometry, exclusively devoted to the right line, on account of occasional insurmountable difficulties in making the direct comparison; its object is to determine certain right lines from others by means of the relations proper to the figures resulting from their assemblage. The importance of this is clear, as no question could be solved if the measurement of right lines, on which every other depends, were left, in any case, uncertain. The natural order of the parts of rational geometry is therefore, first the geometry of *line*, beginning with the right line; then the geometry of *surfaces;* and, finally, that of *volumes*.

Its illimitable field. The field of geometrical science is absolutely unbounded. There may be as many questions as there are conceivable figures; and the variety of conceivable figures is infinite. As to curved Lines, if we regard them as generated by the motion of a point governed by a certain law, we can not limit their number, as the variety of distinct conditions is nothing short of infinite; each generating new ones, and those again others. Surfaces again, are conceived of as motions of lines; and they not only partake of the variety of lines, but have another of their own, arising from the possible change of nature in the line. There can be nothing like this in lines, as points can not describe a figure. Thus, there is a double set of conditions under which the figures of surfaces may vary: and we may say that if lines have one infinity of possible change, surfaces have two. As for volumes, they are distinguished from one another only by the surfaces which bound them; so that they partake of the variety of surfaces, and need no special consideration under this head. If we add the one further remark, that surfaces themselves furnish a new means of conceiving of new curves, as every curve may be regarded as produced by the intersection of two surfaces, we shall perceive that, starting from a narrow ground of observation, we can obtain an absolutely infinite variety of forms, and therefore an illimitable field for geometrical science.

The connection between abstract and concrete geometry is established by the study of the *properties* of lines and surfaces. Without multiplying in this way our means of recognition, we should not know, except by accident, how to find in nature the figure we desire to verify. Astronomy was recreated by Kepler's discovery that the ellipse was the curve which the planets describe about the sun, and the satellites about their planet. This discovery could never have been made if geometers had known no more of the ellipse than as the oblique section of a circular cone by a plane. All the properties of the conic sections brought out by the speculative labors of the Greek geometers, were needed as preparation for this discovery, that Kepler might select from them the characteristic which was the true key to the planetary orbit. In the same way, the spherical figure of the earth could not have been discovered if the primitive character of the sphere had been the only one known;—viz., the equidistance of all its points from an interior point. Certain properties of surfaces were the means used for connecting the abstract reasoning with the concrete fact. And others, again, were required to prove that the earth is not absolutely spherical, and how much otherwise. The pursuit of these labors does not interfere with the definition of Geometry given above, as they tend indirectly to the measurement of extension. The great body of geometrical researches relates to the properties of lines and surfaces; and the study of the properties of the same figure is so extensive, that the labors of geometers for twenty centuries have not exhausted the study of conic sections. Since the time of Descartes, it has become less important; but it appears as far as ever from being finished. And here opens another infinity. We had before the infinite scope of lines, and the double infinity of surfaces: and now we see that not only is the variety of figures inexhaustible, but also the diversity of the points of view from which each figure may be regarded.

There are two general Methods of treating geometrical questions. These are commonly called Synthetical Geometry and Analytical Geometry. I shall prefer the historical titles of Geometry of the Ancients and Geometry of the Moderns. But it is, in my view, better still to call them Special Geometry and General Geometry, by which their nature is most accurately conveyed.

The Calculus was not, as some suppose, unknown to the ancients, as we perceive by their applications of the theory of proportions. The difference between them and us is not so much in the instrument of deduction as in the nature of the questions considered. The ancients studied geometry with reference to the *bodies* under notice, or specially; the moderns study it with reference to the *phenomena* to be considered, or generally. The ancients extracted all they could out of one line or surface, before passing to another; and each inquiry gave little or no assistance in the next. The moderns, since Descartes, employ themselves on questions which relate to any figure whatever. They abstract, to treat by itself, every question relating to the same geometrical phenomenon, in whatever bodies it may be considered. Geometers can thus rise to the study of new geometrical conceptions, which, applied to the curves investigated by the ancients, have brought out new properties never suspected by them. The superiority of the modern method is obvious at a glance. The time formerly spent, and the sagacity and effort employed, in the path of detail, are inconceivably economized by the general method used since the great revolution under Descartes. The benefit to Concrete Geometry is no less than to the Abstract; for the recognition of geometrical figures in nature was merely embarrassed by the study of lines in detail; and the application of the contemplated figure to the existing body could be only accidental, and within a limited or doubtful range: whereas, by the general method, no existing figure can escape application to its true theory, as soon as its geometrical features are ascertained. Still, the ancient method was natural; and it was necessary that it should precede the modern. The experience of the ancients, and the materials they accumulated by their special method, were indispensable to suggest the conception of Descartes, and to furnish a basis for the general procedure. It is evident that the Calculus can not originate any science. Equations must exist as a starting-point for analytical operations. No other beginning can be made than the direct study of the object, pursued up to the point of the discovery of precise relations.

Geometry of the ancients. We must briefly survey the geometry of the ancients, in its character of an indispensable introduction to that of the moderns. The one, special and preliminary, must have its relation made clear to the other,—the general and definitive geometry, which now constitutes the science that goes by that name.

We have seen that Geometry is a science founded upon observation, though the materials furnished by observation are few and simple, and the structure of reasoning erected upon them vast and complex. The only elementary materials, obtainable by direct study alone, are those which relate to the right line for the geometry of *lines;* to the *quadrature* of rectilinear plane areas; and to the *cubature* of bodies terminated by plane faces. The beginning of geometry must be from the observation of lines, of flat surfaces angularly bounded, and of bodies which have more or less bulk, also angularly bounded. These are all; for all other figures, even the circle, and the figures belonging to it, now come under the head of analytical geometry. The three elements just mentioned allow a sufficiency of equations for the calculus to proceed upon. More are not needed; and we can not do with less. Some have endeavored to extend analysis so as to dispense with a portion of these facts: but to do so is merely to return to metaphysical practices, in presenting actual facts as logical abstractions. The more we perceive Geometry to be, in our day, essentially analytical, the more careful we must be not to lose sight of the basis of observation on which all geometrical science is founded. When we observe people attempting to demonstrate axioms and the like, we may avow that it is better to admit more than may be quite necessary of materials derived from observation, than to carry logical demonstration into a region where direct observation will serve us better.

There are two ways of studying the right line—the graphic and the algebraic. The thing to be done is to ascertain, by means of one another, the different elements of any right line whatever, so as to understand, indirectly, a right line, under any circumstances whatever. The way to do this is, first, to study the figure, by constructing it, or otherwise directly investigating it; and then, to reason from that observation. {Geometry of the right line. Graphical solutions.} The ancients, in the early days of the science, made great use of the graphic method, even in the form of Construction; as when Aristarchus of Samos estimated the distance of the sun and moon from the earth on a triangle constructed as nearly as possible in resemblance to the rightangled triangle formed by the three bodies at the instant when the moon is in quadrature, and when therefore an observation of the angle at the earth would define the triangle. Archimedes himself, though he was the first to introduce calculated determinations into geometry, frequently used the same means. The introduction of trigonometry lessened the practice; but did not abolish it. The Greeks and Arabians employed it still for a great number of investigations for which we now consider the use of the Calculus indispensable.

While the graphic or constructive method answers well when all the parts of the proposed figure lie in the same plane, it must receive additions before it can be applied to figures whose parts lie in different planes. Hence arises a new series of considerations, and different systems of *Projections*. Where we now employ spherical trigonometry, especially for problems relating to the celestial sphere, the ancients had to consider how they could replace constructions in relief by plane constructions. This was the object of their *analemmas*, and of the other plane figures which long supplied the place of the Calculus. They were acquainted with the elements of what we call Descriptive Geometry, though they did not conceive of it in a distinct and general manner.

Digressing here for a moment into the region of application, I may observe that Descriptive Geometry, formed into a distinct system by Monge, practically meets the difficulty just stated, but does not warrant the expectations of its first admirers, that it would enlarge the domain of rational geometry. Its grand use is in its application to the industrial arts;—its few abstract problems, capable of invariable solution, relating essentially to the contacts and intersections of surfaces; so that all the geometrical questions which may arise in any of the various arts of construction,—as stone-cutting, carpentry, perspective, dialling, fortification, etc.,—can always be treated as simple individual cases of a single theory, the solution being certainly obtainable through the particular circumstances of each case. This creation must be very important in the eyes of philosophers who think that all human achievement, thus far, is only a first step toward a philosophical renovation of the labors of mankind; toward that precision and logical character which can alone insure the future progression of all arts. Such a revolution must inevitably begin with that class of arts which bears a relation to the simplest, the most perfect, and the most ancient of the sciences. It must extend, in time, though less readily, to all other industrial operations. Monge, who understood the philosophy of the arts better than any one else, himself indeed endeavored to sketch out a philosophical system of mechanical arts, and at least succeeded in pointing out the direction in which the object must be pursued. Of Descriptive Geometry, it may further be said that it usefully exercises the students' faculty of Imagination,—of conceiving of complicated geometrical combinations in space; and that, while it belongs to the geometry of the ancients by the character of its solutions, it approaches to the geometry of the moderns by the nature of the questions which compose it. Consisting, as we have said, of a few abstract problems, obtained through Projections, and relating to the contacts and intersections of surfaces, the invariable solutions of these problems are at once graphical, like those of the ancients, and general, like those of the moderns. Yet, as destined to an industrial application, Descriptive Geometry has here been treated of only in the way of digression. Leaving the subject of *graphic* solution, we have to notice the other branch,—the *algebraic*.

Algebraic solutions. Some may wonder that this branch is not treated as belonging to General Geometry. But not only were the ancients, in fact, the inventors of trigonometry,—spherical as well as rectilinear,—though it necessarily remained imperfect in their hands; but algebraic solutions are also no part of analytical geometry, but only a complement of elementary geometry.

Since all right-lined figures can be decomposed into triangles, all that we want is to be able to determine the different elements of a triangle by means of one another. This reduces polygonometry to simple trigonometry.

Trigonometry. The difficulty lies in forming three distinct equations between the angles and the sides of a triangle. These equations being obtained, all trigonometrical problems are reduced to mere questions of analysis.—There are two methods of introducing the angles into the calculation. They are either introduced *directly*, by themselves or by the circular arcs which are proportional to them: or they are introduced *indirectly*, by the chords of these arcs, which are hence called their *trigonometrical lines*. The second of these methods was the first adopted, because the early state of knowledge admitted of its working, while it did not admit the establishment of equations between the sides of the triangles and the angles themselves, but only between the sides and the trigonometrical lines.—The method which employs the trigonometrical lines is till preferred, as the more simple, the equations existing only between right lines, instead of between right lines and arcs of circles.

To meet the probable objection that it is rather a complication than a simplification to introduce these lines, which have at least to be eliminated, we must explain a little.

Their introduction divides trigonometry into two parts. In one, we pass from the angles to their trigonometrical lines, or the converse: in the other we have to determine the sides of the triangles by the trigonometrical lines of their angles, or the converse. Now, the first process is done for us, once for all, by the formation of numerical tables, capable of use in all conceivable questions. It is only the second, which is by far the least laborious, that has to be undertaken in each individual case. The first is always done in advance. The process may be compared with the theory of logarithms, by which all imaginable arithmetical operations are decomposed into two parts—the first and most difficult of which is done in advance.

We must remember, too, in considering the position of the ancients, the remarkable fact that the determination of angles by their trigonometrical lines, and the converse, admits of an arithmetical solution, without the previous resolution of the corresponding algebraic question. But for this, the ancients could not have obtained trigonometry. When Archimedes was at work upon the rectification of the circle, tables of chords were prepared: from his labors resulted the determination of a certain series of chords: and, when Hipparchus afterward invented trigonometry, he had only to complete that operation by suitable intercalations. The connection of ideas is here easily recognised.

For the same reasons which lead us to the employment of these lines, we must employ several at once, instead of confining ourselves to one, as the ancients did. The Arabians, and after them the moderns, attained to only four or five direct trigonometrical lines altogether; whereas it is clear that the number is not limited. Instead, however, of plunging into deep complications, in obtaining new direct lines, we create indirect ones. Instead, for instance, of directly and necessarily determining the sine of an angle, we may determine the sine of its half, or of its double,—taking any line relating to an arc which is a very simple function of the first. Thus, we may say that the number of trigonometrical lines actually employed by modern geometers is unlimited through the augmentations we may obtain by analysis. Special names have, however, been given to those indirect lines only which refer to the complement of the primitive arc—others being in much less frequent use.

Out of this device arises a third section of trigonometrical knowledge. Having introduced a new set of lines—of auxiliary magnitudes—we have to determine their relation to the first. And this study, though preparatory, is indefinite in its scope, while the two other departments are strictly limited.

The three must, of course, be studied in just the reverse order to that in which it has been necessary to exhibit them. First, the student must know the relations between the indirect and direct trigonometrical lines: and the resolution of triangles, properly so called, is the last process.

Spherical trigonometry requires no special notice here (all-important as it is by its uses)—since it is, in our day, simply an application of rectilinear trigonometry, through the substitution of the corresponding trihedral angle for the spherical triangle.

This view of the philosophy of trigonometry has been given chiefly to show how the most simple questions of elementary geometry exhibit a close dependence and regular ramification.

Thus have we seen what is the peculiar character of Special Geometry, strictly considered. We see that it constitutes an indispensable basis to General Geometry. Next, we have to study the philosophical character of the true science of Geometry, beginning with the great original idea of Descartes, on which it is wholly founded.

Modern, or Analytical Geometry.

General or Analytical Geometry is founded upon the transformation of geometrical considerations into equivalent analytical considerations. Descartes established the constant possibility of doing this in a uniform manner: and his beautiful conception is interesting, not only from its carrying on geometrical science to a logical perfection, but from its showing us how to organize the relations of the abstract to the concrete in Mathematics by the analytical representation of natural phenomena.

Analytical representation of figures. The first thing to be done is evidently to find and fix a method for expressing analytically the subjects which afford the phenomena. If we can regard lines and surfaces analytically, we can so regard, henceforth, the *accidents* of these *subjects*.

Here occurs the difficulty of reducing all geometrical ideas to those of number: of substituting considerations of quantity for all considerations of quality. In dealing with this difficulty, we must *Position.* observe that all geometrical ideas come under three heads:—the *magnitude*, the *figure*, and the *position* of the extensions in question. The relation of the first, magnitude, to numbers is immediate and evident: and the other two are easily brought into one; for the figure of a body is nothing else than the natural position of the points of which it is composed: and its position can not be conceived of irrespective of its figure. We have therefore only to establish the one relation between ideas of position and ideas of magnitude. It is upon this that Descartes has established the system of General Geometry.

The method is simply a carrying out of an operation which is natural to all minds. If we wish to indicate the situation of an object which we can not point out, we say how it is related to objects that are known, by assigning the magnitude of the different geometrical elements which connect it with known objects.

Those elements are what Descartes, and all other geometers after him, have called the co-ordinates of the point considered. If we know in what plane the point is situated, the co-ordinates are two. If the point may be anywhere in space, the co-ordinates can not be less than three. They may be multiplied without limit: but whether few or many, the ideas of position will have been reduced to those of magnitude, so that we shall represent the displacement of a point as produced by pure numerical variations in the values of its co-ordinates. The simplest case of all, that of *plane geometry*, is when we determine the position of a point on a plane by considering its distances from two fixed right lines, supposed to be known, and generally concluded to be perpendicular to each other. These are called axes. Next, there may be the less simple process of determining the position by the distances from two fixed points; and so on to greater and greater complications. But, from some system or other of co-ordinates being always employed, the question of position is always reduced to that of magnitude.

It is clear that our only way of marking the position of a point is by the intersection of two lines. When the point is determined by the intersection of two right lines, each parallel to a fixed axis, that is the system of rectilinear co-ordinates—the most common of all. The *polar* system of co-ordinates exhibits the point by the travelling of a right line round a fixed centre of a circle of variable radius. Again, two circles may intersect, or any other two lines: so that to assign the value of a co-ordinate is the same thing as to determine the line on which the point must be situated. The ancient geometers, of course, were like ourselves in this necessary method of conceiving of position: and their geometrical *loci* were founded upon it. It was in endeavoring to form the process into a general system that Descartes created Analytical Geometry. Seeing, as we now do, how ideas of position—and, through them, all elementary geometrical ideas—can be reduced to ideas of number, we learn what it was that he effected.

Descartes treated only geometry of two dimensions in his analytical method: and we will at first consider only this kind, beginning with Plane Curves.

Lines must be expressed by equations: and again, equations must be expressed by lines, when the relation of geometrical conceptions to numbers is established. It comes to the same thing whether we define a line by any one of its properties, or supply the corresponding equation between the two variable co-ordinates of the point which describes the line. If a point describes a certain line on a plane, we know that its co-ordinates bear a fixed relation to each other, which may be expressed by an appropriate equation. If the point describes no certain line, its co-ordinates must be two variables independent of each other. Its situation in the latter case can be determined only by giving at once its two co-ordinates, independently of each other: whereas, in the former case, a single co-ordinate suffices to fix its position. The second co-ordinate is then a determinate function of the first;—that is, there exists between them a certain equation of a nature corresponding to that of the line on which the point is to be found. The co-ordinates of the point each require it to be on a certain line: and again, its being on a certain line is the same thing as assigning the value of one of the two co-ordinates; which is then found to be entirely dependent on the other. Thus are lines analytically expressed by equations.

Expression of equations by lines. By a converse argument may be seen the geometrical necessity of representing by a certain line every equation of two variables, in a determinate system of co-ordinates. In the absence of any other known property, such a relation would be a very characteristic definition; and its scientific effect would be to fix the attention immediately upon the general course of the solutions of the equation, which will thus be noted in the most striking and simple manner. There is an evident and vast advantage in this picturing of equations, which reacts strongly upon the perfecting of analysis itself. The geometrical *locus* stands before our minds as the representation of all the details that have gone to its preparation, and thus renders comparatively easy our conception of new general analytical views. This method has become entirely elementary in our day; and it is employed when we want to get a clear idea of the general character of the law which runs through a series of particular observation of any kind whatever.

Change in the line changes the equation. Recurring to the representation of lines by equations, which is our chief object, we see that this representation is, by its nature, so faithful, that the line could not undergo any modification, even the slightest, without causing a corresponding change in the equation. Some special difficulties arise out of this perfect exactness; for since, in our system of analytical geometry, mere displacements of lines affect equations as much as real variations of magnitude or form, we might be in danger of confounding the one with the other, if geometers had not discovered an ingenious method expressly intended to distinguish them always. It must be observed that general inconveniences of this nature appear to be strictly inevitable in analytical geometry; since, ideas of position being the only geometrical ideas immediately reducible to numerical considerations, and conceptions of form not being referrible to them but by seeing in them relations of situation, it is impossible that analysis should not at first confound phenomena of form with simple phenomena of position; which are the only ones that equations express directly.

To complete our description of the basis of analytical geometry, it is necessary to point out that not only must every defined line give rise to a certain equation between the two co-ordinates of any one of its points, but every definition of a line is itself an equation of that line in a suitable system of co-ordinates.

Every definition of a line is an equation.

Considering, first, what a definition is, we say it must distinguish the defined object from all others, by assigning to it a property which belongs to it alone. But this property may not disclose the mode of generation of the object, in which case the definition is merely *characteristic;* or it may express one of its modes of generation, and in that case the definition is *explanatory.* For instance, if we say that the circle is the line which in the same form contains the largest area, we offer a characteristic definition; whereas if we choose its property of having all its points equally distant from a fixed point, we have an explanatory definition. It is clear moreover that the characteristic definition always leaves room for an explanatory one, which further study must disclose.

It is to explanatory definitions only that what has been said of the definition of a line being an equation of that line can apply. We can not define the generation of a line without specifying a certain relation between the two simple motions, of translation or of rotation, into which the motion of the point which describes it will be decomposed at each moment. Now, if we form the most general conception of what a *system of co-ordinates* is, and if we admit all possible systems, it is clear that such a relation can be nothing else than the equation of the proposed line, in a system of co-ordinates of a corresponding nature to that of the mode of generation considered; as in the case of the *circle*, the common definition of which may be regarded as being the polar equation of that curve, taking the centre of the circle for the pole.

This view not only exhibits the necessary representation of every line by an equation, but it indicates the general difficulty which occurs in the establishment of these equations, and therefore shows us how to proceed in inquiries of this kind which, by their nature, do not admit of invariable rules. Since every explanatory definition of a line constitutes the equation of that line, it is clear that when we find difficulty in discovering the equation of a curve by means of some of its characteristic properties, the difficulty must proceed from our taking up a designated system of co-ordinates, instead of admitting indifferently all possible systems. These systems are not all equally suitable; and, in regard to curves, geometers think that they should almost always be referred, as far as possible, to rectilinear co-ordinates. Now, these particular co-ordinates are often not those with reference to which the equation of the curve will be found to be established by the proposed definition. It is in a certain transformation of co-ordinates then that the chief difficulty in the formation of the equation of a line really consists. The view I have given does not furnish us with a complete and certain general method for the establishment of these equations; but it may cast a useful light on the course which it is best to pursue to attain the end proposed.

Choice of co-ordinates. The choice of co-ordinates—the preference of that system which may be most suitable to the case—is the remaining point which we have to notice.

First, we must distinguish very carefully the two views, the converse of each other, which belong to analytical geometry, viz. the relation of algebra to geometry, founded on the representation of lines by equations, and, reciprocally, the relation of geometry to algebra, founded on the picturing of equations by lines. Though the two are necessarily combined in every investigation of general geometry, and we have to pass from the one to the other alternately, and almost insensibly, we must be able to separate them here, for the answer to the question of method which we are considering is far from being the same under the two relations; so that without this distinction we could not form any clear idea of it.

In the case of the representation of lines by equations, the first object is to choose those co-ordinates which afford the greatest simplicity in the equation of each line, and the greatest facility in arriving at it. There can be no constant preference here of one system of co-ordinates. The rectilinear system itself, though often advantageous, can not be always so, and may be, in turn, less so than any other. But it is far otherwise in the converse case of the representation of equations by lines. Here the rectilinear system is always to be preferred, as the most simple and trustworthy. If we seek to determine a point by the intersection of two lines, it must be best that those lines should be the simplest possible; and this confines our choice to the rectilinear system. In constructing geometrical *loci*, that system of co-ordinates must be the best in which it is easiest to conceive the change of place of a point resulting from the change in the value of its co-ordinates; and this is the case with the rectilinear system. Again, there is great advantage in the common usage of taking the two axes perpendicular to each other, when possible, rather than with any other inclination. In representing lines by equations, we must take any inclination of the axes which may best suit the particular question; but, in the converse case, it is easy to see that rectangular axes permit us to represent equations in a more simple, and even in a more faithful manner. For if we extend the geometrical *locus* of the equation into the several unequal regions marked out by oblique axes, we shall have differences of figure which do not correspond to any analytical diversity; and the accuracy of the representation will be lost.

On the whole then, taking together the two points of view of analytical geometry, the ordinary system of rectilinear co-ordinates is superior to any other. Its high aptitude for the representation of equations must make it generally preferred, though a less perfect system may answer better in particular cases. The most essential theories in modern geometry are generally expressed by the rectilinear system. The polar system is preferred next to it, both because its opposite character enables it to solve in the simplest way the equations which are too complicated for management under the first; and because polar-co-ordinates have often the advantage of admitting of a more direct and natural concrete signification. This is the case in Mechanics, in the geometrical questions arising out of the theory of circular movement, and in almost all questions of celestial geometry.

Such was the field of the labors of Descartes, his conception of analytical geometry being designed only for the study of Plane Curves. It was Clairaut who, about a century later, extended it to the study of Surfaces and Curves of double curvature. The conception having been explained, a very brief notice will suffice for the rest.

With regard to Surfaces, the determination of a point in space requires that the values of three co-ordinates should be assigned. *Determination of a point in space.* The system generally adopted, which corresponds with the rectilinear system of plane geometry, is that of distances from the point to three fixed planes, usually perpendicular to each other, whereby the point is presented as the intersection of three planes whose direction is invariable. Beyond this, there is the same infinite variety among possible systems of co-ordinates, that there is in geometry of two dimensions. Instead of the intersection of two lines, it must be that of three surfaces which determines the point; and each of the three surfaces has, in the same way, all its conditions constant, except one, which gives rise to the corresponding co-ordinates, whose peculiar geometrical effect is thus to compel the point to be situated upon that surface. Again, if the three co-ordinates of a point are mutually independent, that point can take successively all possible positions in space; but, if its position on any surface is defined, two co-ordinates suffice for determining its situation at any moment; as the proposed surface will take the place of the condition imposed by the third co-ordinate. This last co-ordinate then becomes a determinate function of the two others, they remaining independent of each other. Thus, there will be a certain equation between the three variable co-ordinates which will be permanent, and which will be the only one, in order to correspond to the precise degree of indetermination in the position of the point.

In the expression of Surfaces by Equations, and again in the expression of Equations by Surfaces, the same conception is pursued as in the analytical geometry of two dimensions. *Determination of Surfaces by Equations, and of Equations by Surfaces.* In the first case, the equation will be the analytical definition of the proposed surface, since it must be verified for all the points of this surface, and for them only. If the surface undergoes any change, the equation must, as in the case of changing lines, be modified accordingly. All geometrical phenomena relating to surfaces may be translated by certain equivalent analytical conditions, proper to equations of three variables: and it is in the establishment and interpretation of this harmony that the science of analytical geometry of three dimensions essentially consists. In the second and converse case, every equation of three variables may, in general, be represented geometrically by a determinate surface, defined by the characteristic property that the co-ordinates of all its points always preserve the mutual relation exhibited in this equation.

Thus we see in this application the complement of the original idea of Descartes: and it is enough to say this, as every one can extend to surfaces the other considerations which have been indicated with regard to lines. I will only add that the superiority of the rectilinear system of co-ordinates becomes more evident in analytical geometry of three dimensions than in that of two, on account of the geometrical complication which would follow the choice of any other.

Curves of double curvature. In determining Curves of double curvature,—which is the last elementary point of view of analytical geometry of three dimensions,—the same principle is employed. According to it, it is clear that when a point is required to be situated upon some certain curve, a single co-ordinate is enough to determine its position completely, by the intersection of this curve with the surface resulting from this co-ordinate. The two other co-ordinates of the point must thus be regarded as functions necessarily determinate, and distinct from the first. Consequently, every line, considered in space, is represented analytically no longer by a single equation, but by a system of two equations between the three co-ordinates of any one of its points. It is evident, indeed, from another point of view, that the equations which, considered separately, express a certain surface, must in combination present the line sought as the intersection of two determinate surfaces. As for the difficulty occasioned by the infinity of the number of couples of equations, through the infinity of couples of surfaces which can enter the same system of co-ordinates, and by which the line sought may be hidden under endless algebraical disguises, it must be got rid of by giving up the facilities resulting from such a variety of geometrical constructions. It is sufficient, in fact, to obtain from the analytical system established for a certain line, the system corresponding to a single couple of surfaces uniformly generated, and which will not vary except when the line itself shall change. Such is a natural use of this kind of geometrical combination, which thus affords us a certain means of recognising the identity of lines in spite of the extensive diversity of their equations.

Imperfections of Analytical Geometry. Analytical Geometry still presents some imperfections on the side both of geometry and analysis.

In regard to Geometry, the equations can as yet represent only entire geometrical *loci*, and not determinate portions of those *loci*. Yet it is necessary, occasionally, to be able to express analytically a part of a line or surface, or even a discontinuous line or surface, composed of a series of sections belonging to distinct geometrical figures. Some progress has been made in supplying means for this purpose, to which our analytical geometry is inapplicable; but the method introduced by M. Fourier, in his labors on discontinuous functions, is too complicated to be at present introduced into our established system.

In regard to analysis, we are so far from having a complete command of analytical geometry, that we can Imperfections of Analysis. not furnish anything like an adequate geometrical representation of analytical processes. This is not an imperfection in science, but inherent in the very nature of the subject. As Analysis is much more general than geometry, it is of course impossible to find among geometrical phenomena a concrete representation of all the laws expressed by analysis: but there is another evil which is due to our own imperfect conceptions; that, in our representations of equations of two or of three variables by lines or surfaces, we regard only the real solutions of equations, without noticing any imaginary ones. Yet these last should, in their general course, be as capable of representation as the first. Hence the graphic representation of the equation is always imperfect; and it fails altogether when the equation admits of only imaginary solutions. This brings after it, in analytical geometry of two or three dimensions, many inconveniences of less consequence, arising from the want of correspondence between various analytical modifications and any geometrical phenomena.

We have now seen what Analytical Geometry is. By this science we determine what is the analytical expression of such or such a geometrical phenomenon belonging to lines or surfaces: and, reciprocally, we ascertain the geometrical interpretation of such or such an analytical consideration. It would be interesting now to consider the most important general questions which would exemplify the manner in which geometers have actually established this beautiful harmony: but such a review is not necessary to the purpose of this Work, and would occupy too much space. We have seen what is the character of generality and simplicity inherent in the science of GEOMETRY. We must now proceed to ascertain what is the true philosophical character of the immense and more complex science of RATIONAL MECHANICS.

CHAPTER IV.

RATIONAL MECHANICS.

Its nature. MECHANICAL phenomena are by their nature more particular, more complicated, and more concrete, than geometrical phenomena. Therefore they come after geometry in our survey; and therefore must they be pronounced to be more difficult to study, and, as yet, more imperfect. Geometrical questions are always completely independent of Mechanics, while mechanical questions are closely involved with geometrical considerations,—the form of bodies necessarily influencing the phenomena of motion and equilibrium. The simplest change in the form of a body may enhance immeasurably the difficulties of the mechanical problem relating to it, as we see in the question of the mutual gravitation of two bodies, as a result of that of all their molecules; a question which can be completely resolved only by supposing the bodies to be spherical; and thus, the chief difficulty arises out of the geometrical part of the circumstances.

Our tendency to look for the essences of things, instead of studying concrete facts, enters disastrously into the study of Mechanics. We found something of it in geometry; but it appears in an aggravated form in Mechanics, from the greater complexity of the science. We encounter a perpetual confusion between the abstract and the concrete points of view; between the logical and the physical; between the artificial conception necessary to help us to general laws of equilibrium and motion, and the natural facts furnished by observation, which must form the basis of the science. Great as is the gain of applying Mathematical analysis to Mechanics, it has set us back in some respects. The tendency to *à priori* suppositions, drawn by us from analysis where Newton wisely had recourse to observation, has made our expositions of the science less clear than those of Newton's days. Inestimable as mathematical analysis is for carrying the science on and upward, there must first be a basis of facts to employ it upon; and Laplace and others were therefore wrong in attempting to prove the elementary law of the composition of forces by analytical demonstration. Even if the science of Mechanics could be constructed on an analytical basis, it is not easy to see how such a science could ever be applied to the actual study of nature. In fact, that which constitutes the reality of Mechanics is that the science is founded on some general facts, furnished by observation, of which we can give no explanation whatever. Our business now is to point out exactly the philosophical character of the science, distinguishing the abstract from the concrete point of view, and separating the experimental department from the logical.

We have nothing to do here with the causes or modes of production of motions, but only with the motions itself. *Its character.* Thus, as we are not treating of Physics, but of Mechanics, *forces* are only motions produced or tending to be produced; and two forces which move a body with the same velocity in the same direction are regarded as identical, whether they proceed from muscular contractions in an animal, or from a gravitation toward a centre, or from collision with another body, or from the elasticity of a fluid. This is now practically understood; but we hear too much still of the old metaphysical language about *forces*, and the like; and it would be wise to suit our terms to our positive philosophy.

The business of Rational Mechanics is to determine how a given body will be affected by any different *Its object.* forces whatever, acting together, when we know what motion would be produced by any one of them acting alone: or, taking it the other way, what are the simple motions whose combination would occasion a known compound motion. This statement shows precisely what are the data and what the unknown parts of every mechanical question. The science has nothing to do with the action of a single force; for this is, by the terms of the statement, supposed to be known. It is concerned solely with the combination of forces, whether there results from that combination a motion to be studied, or a state of equilibrium, whose conditions have to be described.

The two general questions, the one direct, the other inverse, which constitute the science, are equivalent in importance, as regards their application. Simple motions are a matter of observation, and their combined operation can be understood only through a theory: and again, the compound result being a matter of observation, the simple constituent motions can be ascertained only by reasoning. When we see a heavy body falling obliquely, we know what would be its two simple movements if acted upon separately by the force to which it is subject,—the direction and uniform velocity which would be caused by the impulsion alone; and again, the acceleration of the vertical motion by its weight alone. The problem is to discover thence the different circumstances of the compound movement produced by the combination of the two,—to determine the path of the body, its velocity at each movement, the time occupied in falling; and we might add to the two given forces the resistance of the medium, if its law was known. The best example of the inverse problem is found in celestial mechanics, where we have to determine the forces which carry the planets round the sun, and the satellites round the planets. We know immediately only the compound movement: Kepler's laws give us the characteristics of the movement; and then we have to go back to the elementary forces by which the heavenly bodies are supposed to be impelled, to correspond with the observed result: and these forces once understood, the converse of the question can be managed by geometers, who could never have mastered it in any other way.

Such being the destination of Mechanics, we must now notice its fundamental principles, after clearing the ground by a preparatory observation.

Matter not inert in Physics. In ancient times, men conceived of matter as being passive or inert,—all activity being produced by some external agency,—either of supernatural beings or some metaphysical entities. Now that science enables us to view things more truly, we are aware that there is some movement or activity, more or less, in all bodies whatever. The difference is merely of degree between what men call *brute matter* and animated beings. Moreover, science shows us that there are not different kinds of matter, but that the elements are the same in the most primitive and the most highly organized. If we knew of any substance which had nothing but weight, we could not deny activity even to that; for in falling it is as active as the globe itself,—attracting the earth's particles precisely as much as its own particles are attracted by the earth. Looking through the whole range of substances, up to those of the highest organization, we find everywhere a spontaneous activity, very various, and at most, in some cases, peculiar; though physiologists are more and more disposed to regard the most peculiar as a modification of antecedent kinds. However this may be, it would be purely absurd now to regard any portion of matter whatever as inert, as a matter of fact, or under the head of Physics.

Supposed inert in Mechanics. But in Mechanics it must be so regarded, because we can not establish any general proposition upon the abstract laws of equilibrium or motion without putting out of the question all interference with them by other and inherent forces. What we have to beware of is mixing up this logical supposition with the old notion of actual *inertia*.

Field of Rational Mechanics. As for how this is to be done—we must remember what has been just said—that in Mechanics, we have nothing to do with the origin or different nature of forces; and they are all one while their mechanical operation is uniform. It is impossible to conceive of any substance as devoid of weight, for instance; yet geometers have logically to treat of bodies as without an inherent power of attraction. They treat of this power as an external force; that is, it is to them simply a force; and it does not matter to them whether it is inherent or external—whether it is attraction or impulsion—while it is the fall of the body that they have to study. And so on, through the whole range of properties of bodies. When we have so abstracted natural properties, in our logical view, as to have before us an unmixed case of the action of certain forces, and have ascertained their laws—then we can pass from abstract to concrete Mechanics, and restore to bodies their natural active properties, and interpret their action by what we have learned of the laws of motion and equilibrium. This restoration is so difficult to effect—the transition from the abstract to the concrete in Mechanics is so difficult—that, while its theoretical domain is unbounded, its practical application is singularly limited. In fact, the application of rational mechanics is limited (accurately speaking) not only to celestial phenomena, but to those of our own solar system. One would suppose that the single property of weight was manageable enough; and that of a given form intelligible enough: but there are such complications of physical circumstances—as the resistance of media, friction, etc.—even if bodies are conceived of as in a fluid state, that their mechanical phenomena can not be estimated with any accuracy. And when we proceed to electrical and chemical, and especially to physiological phenomena, we are yet more baffled. General gravitation affords us the only simple and determinate law; and even there we are perplexed, when we come to regard certain secondary actions. It may be doubted whether questions of terrestrial mechanics will ever admit—restricted as our means are—of a study at once purely rational and precisely accordant with the general laws of abstract mechanics—though the knowledge of these laws, primarily indispensable, may often lead us to frequent and valuable indications and suggestions.

Bodies being supposed inert, the general facts, or *laws of motion* to which they are subject, are three; all results of observations. *Three laws of motion.*

The first is that law discovered by Kepler, which is inaptly called the *law of inertia*. According to it, all motion is rectilinear and uniform; that is, any body impelled by a single force will move in a right line, and with an invariable velocity. Instead of resorting to the old ways of pronouncing or imagining *why* it must be so, the Positive Philosophy instructs us to recognise the simple fact that it *is* so; that, through the whole range of nature, bodies move in a right line, and with a uniform velocity, when impelled by a single force. *Law of inertia.*

The second law we owe to Newton. It is that of the constant *equality of action and reaction;* that is, whenever one body is moved by another, the reaction is such that the second loses precisely as much motion, in proportion to its masses, as the first gains. Whether the movement proceeds from impulsion or attraction, is, of course, of no consequence. Newton treated this general fact as a matter of observation, and most geometers have done the same; so that there has been less fruitless search into the *why* with regard to this second law than to the first. *Law of equality of action and reaction.*

The third fundamental law of motion involves the principle of the independence or *co-existence of motions,* which leads immediately to what is commonly called the *composition of force.* Galileo is, strictly speaking, the true discoverer of this law, though he did not regard it precisely under the form in which it is presented here:—that any motion common to all the bodies of any system whatever does not affect the particular mo- *Law of co-existence of motions.*

tions of these bodies with regard to each other; which motions proceed as if the system were motionless. Speaking strictly, we must conceive that all the points of the system describe at the same time parallel and equal straight lines, and consider that this general motion, whatever may be its velocity and direction, will not affect relative motions. No *à-priori* considerations can enter here. There is no seeing why the fact should be so, and therefore no anticipating that it would be so. On the contrary, when Galileo stated this law, he was assailed by a host of objections that his fact was logically impossible. Philosophers were ready with plenty of *à-priori* reasons that it could not be true: and the fact was not unanimously admitted till men had quitted the logical for the physical point of view. We now find, however, that no proposition in the whole range of natural philosophy is founded on observations so simple, so various, so multiplied, so easy of verification. In fact, the whole economy of the universe would be overthrown, from end to end, but for this law. A ship impelled smoothly, without rolling and pitching, has everything going on within it just the same as if it were at rest; and, in the same way, but on the grandest scale, the great globe itself rushes through space, without its motion at all affecting the movements going on on its surface. As we all know, it was ignorance of this third law of motion which was the main obstacle to the establishment of the Copernican theory. The Copernicans struggled to get rid of the insurmountable objections to which their doctrine was liable by vain metaphysical subtleties, till Galileo cleared up the difficulty. Since his time, the movement of the globe has been considered an all-sufficient confirmation of the law. Laplace points out to us that if the motion of the globe affected the movements on it, the effect could not be uniform, but must vary with the diversities of their direction, and of the angle that each direction would make with that of the earth: whereas, we know how invariable is, for instance, the oscillating movement of the pendulum, whatever may be its direction in comparison with that of the travelling globe.

It may be as well to point out that rotary motion does not enter into this case at all, but only translation, because the latter is the only motion which can be, in degree and direction, absolutely common to all the parts of a system. In a rotating system, for instance, all the parts are not at an equal distance from the centre of rotation. When the interior of a ship is affected, it is by the rolling and pitching, which are rotary movements. We may carry a watch any distance without affecting its interior movements; but it will not bear whirling.

And, again, the forward motion of the globe could be discovered by no other means than astronomical observation; whereas, the changes which occur on the surface of the earth, produced by the inequality of the centrifugal force at its different points, are sufficient evidence of its rotation, independently of all astronomical considerations whatever.

The law or rule of the composition of forces, which is involved in the general fact just stated, is, in fact, identical with it. It is only another way of expressing the same law. If a single impulsion describes a parallelogram of forces, as the scientific term is, the effect of a second will be to describe the diagonal of the parallelogram. This is nothing more than an application of the law of the independence of forces; since the motion of any body along a straight line is in no way disturbed by a general motion which carries away, parallel with itself, the whole of this right line along any other right line whatever. This consideration leads immediately to the geometrical construction expressed by the rule of the parallelogram of forces. And thus it appears that this fundamental theorem of Rational Mechanics is a true natural law; or, at least, a direct application of one of the greatest natural laws. And this is the best account to give of it, instead of looking to logic for a fallacious *à priori* deduction of it. Any analytical demonstration, too, must suppose certain portions of the case to be evident; and to talk of a thing being evident is to refer back to nature, and to depend on observation of nature.

It is worthy of remark that those who wish to make a separate law of the composition of forces, in order to avoid introducing the third law into the prolegomena of Mechanics, and to dispense with it in the exposition of Statics, are brought back to it when entering upon the study of Dynamics. Upon this alone can be based the important law of the proportion of forces to velocities. The relations of forces may be determined either by a statical or dynamical procedure. No purpose is answered by the transposition of the general fact of the independence of forces to the dynamical department of the science: it is equally necessary for the statical; and a world of metaphysical confusion is saved by laying it down as the broad basis that in fact it is.

These three laws are the experimental basis of the science of Mechanics. From them the mind may proceed to the logical construction of the science, without further reference to the external world. At least, so it appears to me; though I am far from assigning any *à-priori* reasons why more laws may not be hereafter discovered, if these three should prove to be incomplete. There can not, in the nature of things, be many more; and I would rather incur the inconvenience of the introduction of one or two, than run any risk of surrendering the positive character of the science and overstraining its logical considerations. We can not however conceive of any case which is not met by these three laws of Kepler, of Newton, and of Galileo; and their expression is so precise, that they can be immediately treated in the form of analytical equations easily obtained. As for the most extensive, important, and difficult part of the science, the mechanics of varied motion or continuous forces, we can perceive the possibility of reducing it to elementary Mechanics by the application of the infinitesimal method. For each infinitely small point of time, we must substitute a uniform motion in the place of a varied one, whence will immediately result the differential equations relative to these varied motions. We may hereafter see what results have been obtained in regard to the abstract laws of equilibrium and motion. Meantime, we see that the whole science is founded on the combination of the three physical laws just established; and here lies the distinct boundary between the physical and the logical parts of the science.

Two primary divisions. Statics and Dynamics. As for its divisions, the first and most important is into Statics and Dynamics; that is, into questions relating to equilibrium and questions relating to motion. Statics are the easiest to treat, because we abstract from them the element of time, which must enter into Dynamical questions, and complicate them. The whole of Statics corresponds to the very small portion of Dynamics which relates to the theory of uniform motions. This division corresponds well with the facts of human education in this science. The fine researches of Archimedes show us that the ancients, though far from having obtained any complete system of rational Statics, had acquired much essential knowledge of equilibrium—both of solids and fluids—while as yet wholly without the most rudimentary knowledge of Dynamics. Galileo, in fact, created that department of the science.

Secondary divisions. Solids and Fluids. The next division is that of Solids from Fluids. This division is generally placed first, but it is unquestionable that the laws of statics and dynamics must enter into the study of solids and fluids, that of fluids requiring the addition of one more consideration,—variability of form. This however is a consideration which introduces the necessity of treating separately the molecules of which fluids are composed, and fluids as systems composed of an infinity of distinct forces. A new order of researches is introduced into Statics, relative to the form of the system in a state of equilibrium; but in Dynamics the questions are still more difficult to deal with. The importance and difficulty of the researches under this division can not be exaggerated. Their complication places even the easiest cases beyond our reach, except by the aid of extremely precarious hypotheses. We must admit the vast necessary difficulty of hydrostatics, and yet more of hydrodynamics, in comparison with statics and dynamics, properly so called, which are in fact far more advanced.

Much of the difficulty arises from the mathematical statement of the question differing from the natural facts. Mathematical fluids have no adhesion between their particles; whereas natural fluids have, more or less; and many natural phenomena are due to this adherence, small though it be in comparison with that of solids. Thus, the result of an observation of the quantity of a given fluid which will run out of a given orifice will differ widely from the result of the mathematical calculation of what it should be. Though the case of solids is easier, yet there perplexity may be introduced by the disrupting action of forces, of which abstraction must be made in the mathematical question. The theory of the rupture of solids, initiated by Galileo, Huyghens, and Leibnitz, is still in a very imperfect and precarious state, great as are the pains which have been taken with it, and much needed as it is. Not so much needed however as the mechanics of fluids, because it does not affect questions of celestial mechanics; and in this highest department alone can we, as I said before, see the complete application of rational mechanics.

There is a gap left between these two studies, which should be pointed out, though it is of secondary importance. We want a Mechanics of semi-fluids, or semi-solids,—as of sand, in relation to solids, and gelatinous conditions of fluids. Some considerations have been offered with regard to these "imperfect fluids," as they are called; but their true theory has never been established in any direct and general manner.

Such is the general view of the philosophical character of Rational Mechanics. We must now take a philosophical view of the composition of the science, in order to see how this great second department of Concrete Mathematics has attained the theoretical perfection in which it appears in the works of Lagrange, who has rendered all its possible abstract questions capable of an analytical solution, like those of geometry. We must first take a view of Statics, and then proceed to Dynamics.

SECTION I.

STATICS.

THERE are two ways of treating Rational Mechanics, according as Statics are regarded directly, or as a particular case of Dynamics. By the first method we have to discover a principle of equilibrium so general as to be applicable to the conditions of equilibrium of all systems of possible forces. By the second method, we reverse the process,—ascertaining what motion would result from the simultaneous action of any proposed differing forces, and then determining what relations of these forces would render motion null. *Converse methods of treatment.*

The first method was the only one possible in the early days of science; for, as I have said before, Galileo was the creator of the science of Dynamics. Archimedes, the founder of Statics, established the condition of equilibrium of two weights suspended at the ends of a straight lever; that is, he showed that the weights must be in an inverse ratio to their distances from the fulcrum of the lever. He endeavored to refer to this principle the relations of equilibrium proper to other systems of forces; but the principle of the lever is not in itself general enough for such application. The various devices by which it was attempted to extend the process, and to supply the remaining deficiencies, were relinquished when the establishment of Dynamics permitted the use of the second method,—of *First method. Statics by itself.* *Second method.*

Statics through Dynamics. seeking the conditions of equilibrium through the laws of the composition of forces. It is by this last method that Varignon discovered the theory of the equilibrium of a system of forces applied upon a single point; and that D'Alembert afterward established, for the first time, the equations of equilibrium of any system of forces applied to the different points of a solid body of an invariable form. At this day, this is the method universally employed. At the first glance, it does not appear the most rational,—Dynamics being more complicated than Statics, and precedence being natural to the simpler. It would, in fact, be more philosophical to refer dynamics to statics, as has since been done; but we may observe that it is only the most elementary part of dynamics, the theory of uniform motions, that we are concerned with in treating statics as a particular case of dynamics. The complicated considerations of varied motions do not enter into the process at all.

The easiest method of applying the theory of uniform motions to statical questions is through the view that, when forces are *in equilibrio*, each of them, taken singly, may be regarded as destroying the effect of all the others together. Thus, the thing to be done is to show that any one of the forces of the system is equal, and directly opposed, to the resulting force of all the rest. The only difficulty here is in determining the resulting force; that is, in mutually compounding the given forces. Here comes in the aid of the third great law of motion, and having compounded the two first forces, we can deduce the composition of any number of forces.

After having established the elementary laws of the composition of forces, geometers, before applying them to the investigation of the conditions of equilibrium, usually subject them to an important transformation, which, without being indispensable, is of eminent utility, in an analytical view, from the extreme simplification which it introduces into the algebraical expression of the conditions of equilibrium. The transformation consists in what is called the theory of *Moments*, the essential property of which is to reduce, analytically, all the laws of the composition of forces to simple additions and subtractions. Without going into an examination of this theory, it is necessary simply to say that it considers statics as a particular case of elementary dynamics, and that its value is in the simplicity which it gives to the analytical part of the process of investigation into the conditions of equilibrium. Simple, however, as may be the operation, and great as may be the practical advantage gained through the treatment of statics as a particular case of elementary Dynamics, it would be satisfactory to return, if we could, to the method of the ancients,—to leave Dynamics on one side, and proceed directly to the investigation of the laws of equilibrium regarded by itself, by means of a direct general principle of equilibrium. Geometers strove after this as soon as the general equations of equilibrium

Moments.

Want of unity in the method.

were discovered by the dynamic method. But a higher motive than even the desire to place statics in a more philosophical position impelled them to establish a direct Statical method: and this it was which caused Lagrange to carry up the whole science of Rational Mechanics to the philosophical perfection which it now enjoys.

D'Alembert made a discovery (to be treated of hereafter), by the help of which all investigation of the motion of any body or system might be converted at once into a question of equilibrium. This amounts, in fact, to a vast generalization of the second fundamental law of motion; and it has served for a century past as a permanent basis for the solution of all great dynamical questions; and it must be so applied more and more, from its high merits of simplification in the most difficult investigations. Still, it is clear that this method compels a return into statics; and Statics as independent of Dynamics, which are altogether derived from Statics. A science must be imperfectly laid down, as long as it is necessary thus to pass backward and forward between its two departments. In order to establish the necessary unity, and to provide scope for D'Alembert's principle, a complete reconstitution of Rational Mechanics was indispensable. Lagrange effected this in his admirable treatise on "Analytical Mechanics," the leading conception of which must be the basis of all future labors of geometers upon the laws of equilibrium and motion, as we have seen that the great idea of Descartes is with regard to geometrical speculations.

The principle of Virtual Velocities,—the one which Lagrange selected from among the properties of equilibrium,—had been discovered by Galileo in the case of two forces, as a general property manifested by the equilibrium of all machines. John Bernouilli extended it to any number of forces, composing any system. Varignon afterward expressly pointed out the universal use that might be made of it in Statics. The combination of it with D'Alembert's principle led Lagrange to conceive of the whole of Rational Mechanics as deduced from a single fundamental theorem, and to give it that rigorous unity which is the highest philosophical perfection of a science.

Virtual Velocities.

The clearest idea of the system of virtual velocities may be obtained by considering the simple case of two forces, which was that presented by Galileo. We suppose two forces balancing each other by the aid of any instrument whatever. If we suppose that the system should assume an infinitely small motion, the forces are, with regard to each other, in an inverse ratio to the spaces traversed by their points of application in the path of their directions. These spaces are called virtual velocities, in distinction from the real velocities which would take place if the equilibrium did not exist. In this primitive state, the principle, easily verified with regard to all known machines, offers great practical utility; for it permits us to obtain with ease the mathematical condition of equilibrium of any machine whatever, whether its constitution is known or not. If we give the name of virtual momentum (or simply of momentum in its primitive sense) to the product of each force by its virtual velocity,—a product which in fact then measures the effort of the force to move the machine,—we may greatly simplify the statement of the principle in merely saying that, in this case, the momentum of the two forces must be equal and of opposite signs, that there may be equilibrium, and that the positive or negative sign of each momentum is determined according to that of the virtual velocity, which will be considered positive or negative according as, by the supposed motion, the projection of the point of application would be found to fall upon the direction of the force or upon its prolongation. This abridged expression of the principle of virtual velocities is especially useful for the statement of this principle in a general manner, with regard to any system of forces whatever. It is simply this: that the algebraic sum of the virtual moments of all forces, estimated according to the preceding rule, must be null to cause equilibrium: and this condition must exist distinctly with regard to all the elementary motions which the system might assume in virtue of the forces by which it is animated. In the equation, containing this principle, furnished by Lagrange, the whole of Rational Mechanics may be considered to be implicitly comprehended.

While the theorem of virtual velocities was conceived of only as a general property of equilibrium, it could be verified by observing its constant conformity with the ordinary laws of equilibrium, otherwise obtained, of which it was a summary, useful by its simplicity and uniformity. But, if it was to be a fundamental principle, a basis of the whole science, it must be underived, or at least capable of being presented in its preliminary propositions as a matter of observation. This was done by Lagrange, by his ingenious demonstration through a system of pulleys. He exhibited the theorem of virtual velocities very easily by imagining a single weight which, by means of pulleys suitably constructed, replaces simultaneously all the forces of the system. Many other demonstrations have been furnished; but, while more complicated, they are not logically superior. From the philosophical point of view it is clear that this general theorem, being a necessary consequence of the fundamental laws of motion, can be deduced in various ways, and becomes practically the point of departure of the whole of Rational Mechanics. A perfect unity having been established by this principle, we need not look for any others; and we may rest assured that Lagrange has carried the co-ordination of the science as far as it can go. The only possible object would be to simplify the analytical researches to which the science is now reduced; and nothing can be conceived more admirable for this purpose than Lagrange's adaptation of the principle of virtual velocities to the uniform application of mathematical analysis.

Striking as is the philosophical eminence of this principle, there are difficulties enough in its use to prevent its being considered elementary, so far as to preclude the consideration of any other in a course of dogmatic teaching. It is for this reason that I have referred to the dynamic method, properly so called, which is the only one in general use at present. All other considerations must however be only provisional. Lagrange's method is at present too new; but it is impossible that it should for ever remain in the hands of a small number of geometers, who alone shall be able to make use of its admirable properties. It must become as popular in the mathematical world as the great geometrical conception of Descartes: and this general progress would be almost accomplished if the fundamental ideas of transcendental analysis were as widely spread as they ought to be.

Theory of Couples. The greatest acquisition, since the regeneration of the science by Lagrange, is the conception of M. Poinsot,—the theory of Couples, which appears to me to be far from being sufficiently valued by the greater number of geometers. These Couples, or systems of parallel forces, equal and contrary, had been merely remarked before the time of M. Poinsot, as a sort of paradox in Statics. He seized upon this idea, and made it the subject of an extended and original theory relating to the transformation, composition, and use of these singular groups, which he has shown to be endowed with properties remarkable for their generality and simplicity. He used the dynamic method in his study of the conditions of equilibrium: but he presented it, by the aid of his theory of couples, in a new and simplified aspect. But his conception will do more for dynamics than for statics; and it has hardly yet entered upon its chief office. Its value will be appreciated when it is found to render the notion of the movements of rotation as natural, as familiar, and almost as simple, as that of forward movement or translation.

Share of equations in producing equilibrium. One more consideration should, I think, be adverted to before we quit the subject of statics as a whole. When we study the nature of the equations which express the conditions of equilibrium of any system of forces, it seems to me not enough to establish that the sum of these equations is indispensable for equilibrium. I think the further statement is necessary,—in what degree each contributes to the result. It is clear that each equation must destroy some one of the possible motions that the body would make in virtue of existing forces; so that the whole of the equation must produce equilibrium by leaving an impossibility for the body to move in any way whatever. Now the natural state of things is for movement to consist of rotation and translation. Either of these may exist without the other; but the cases are so extremely rare of their being found apart, that the verification of either is regarded by geometers as the strongest presumption of the existence of the other. Thus, when the rotation of the sun upon its axis was established, every geometer concluded that it had also a progressive motion, carrying all its planets with it, before astronomers had produced any evidence that such was actually the case. In the same way we conclude that certain planets, travelling in their orbits, rotate round their axes, though the fact has not yet been verified. Some equations must therefore tend to destroy all progressive motion, and others all motion of rotation. How many equations of each kind must there be?

It is clear that, to keep a body motionless, it must be hindered from moving according to three axes in different planes—commonly supposed to be perpendicular to each other. If a body can not move from north to south, nor the reverse; nor from east to west, nor the reverse; nor up, nor down, it is clear that it can not move at all. Movement in any intermediate direction might be conceived of as partial progression in one of these, and is therefore impossible. On the other hand, we can not reckon fewer than three independent elementary motions; for the body might move in the direction of one of the axes, without having any translation in the direction of either of the others. Thus we see that, in a general way, three equations are necessary, and three are sufficient to establish the absence of translation; each being specially adapted to destroy one of the three progressive motions of which the body is capable. The same view presents itself with regard to the other motion—of rotation. The mechanical conception is more complicated; but it is true, as in the simpler case, that motion is possible in only three directions—in three co-ordinated planes, or round three axes. Three equations are necessary and sufficient here also; and thus we have six which are indispensable and sufficient to stop all motion whatever.

When, instead of supposing any system of forces whatever as the subject of the question, we particularize any, we get rid of more or fewer possible movements. Having excluded these, we may exclude also their corresponding equations, retaining only those which relate to the possible motions that remain. Thus, instead of having to deal with six equations necessary to equilibrium, there may be only three, or two, or even one, which it will be easy enough to obtain in each case. These remarks may be extended to any restrictions upon motion, whether resulting from the special constitution of the system of forces, or from any other kind of control, affecting the body under notice. If, for instance, the body were fastened to a point, so that it could freely rotate but not advance, three equations would suffice: and again, if it is fastened to two fixed points, two equations are enough; and even one, if these two fixed points are so placed as to prevent the body from moving on the axis between them. Finally, its being attached to three fixed points, not in a right line, will prevent its moving at all, and establish equilibrium without any condition, whatever may be the forces of the system. The spirit of this analysis is entirely independent of any method by which the equations of equilibriums will have been obtained: but the different general methods are far from being equally suitable to the application of this rule. The one which is best adapted to it is, undoubtedly, the Statical one, properly so called, founded, as has been shown, on the principle of virtual velocities. In fact, one of the characteristic properties of this principle is the perfect precision with which it analyses the phenomena of equilibrium, by distinctly considering each of the elementary motions permitted by the forces of the system, and furnishing immediately an equation of equilibrium specially relating to this motion.

When we come to the inquiry how geometers apply the principles of abstract Mechanics to the properties of real bodies, we must state that the only complete application yet accomplished is in the question of terrestrial gravity. *Connection of the concrete with the abstract question.* Now, this is a subject which can not, logically, be treated under the head of Mechanics, as it belongs to Physics. It is sufficient to explain that the statical study of terrestrial gravity becomes convertible into that of centres of gravity; and that all confusion between the two departments of research would be avoided if we accustomed ourselves to class the theory of centres of gravity among the questions of pure geometry. In seeking the centre of gravity as (according to the logical denomination of the ancient geometers) the centre of mean distances, we remove all traces of the mechanical origin of the question, and convert it into this problem of general geometry:—Given, any system of points disposed in a determinate way with regard to each other, to find a point whose distance to any plane shall be a mean between the distances of all the given points to the same plane.—The abstraction of all consideration of gravity is an assistance in every way. The simple geometrical idea is precisely what we want in most of the principal theories of Rational Mechanics; and especially when we contemplate the great dynamic properties of the centre of mean distances; in which study the idea of gravity becomes a mere encumbrance and perplexity. It is true that, by proceeding thus, we exclude the question from the domain of Mechanics, to place it in that of Geometry. I should have so classed it but for an unwillingness to break in upon established customs. However it may be as to the matter of arrangement, it is highly important for us not to misapprehend the true nature of the question.—The integral calculus offers the means of surmounting those difficulties in determining the centre of gravity which are imposed by the conditions of the question. But, the integrations in this case being more complicated than those to which they are analogous—those of quadratures and cubatures—their precise solution is, owing to the extreme imperfection of the integral Calculus, much more rarely obtained. It is a matter of high importance, however, to be able to introduce the consideration of the centre of gravity into general theories of analytical mechanics.

Such is, then, the relation of terrestrial gravitation to the science of abstract Statics. As for universal gravitation, no complete study has yet been made of it, except in regard to spherical bodies. What we know of the law of gravitation would easily enable us to compute the mutual attraction of all known bodies, if the conditions of each body were understood by us: but this is not the case. For instance, we know nothing of the law of density in the interior of the heavenly bodies. It is still true that the primitive theorems of Newton on the attraction of spherical bodies are the most useful part of our knowledge in this direction.

Gravity is the only natural force that we are practically concerned with in Rational Statics: and we see, by this, how backward this science is in regard to universal gravitation. As for the exterior general circumstances, such as friction, resistance of media, and the like, which are altogether excluded in the establishment of the rational laws of Mechanics, we can only say that we are absolutely ignorant of the way to introduce them into the fundamental relations afforded by analytical Mechanics, because we have nothing to rely on, in working them, but precarious and inaccurate hypotheses, unfit for scientific use.

Equilibrium of fluids. As for the theory of equilibrium in regard to fluid bodies—the application which it remains for us to notice—those bodies must be regarded as either liquid or gaseous.

Hydrostatics. Hydrostatics may be treated in two ways. We may seek the laws of the equilibrium of fluids, according to statical considerations proper to that class of bodies: or we may look for them among the laws which relate to solids, allowing for the new characteristics resulting from fluidity.

The first method, being the easiest, was in early times the only one employed. Till a rather recent time, all geometers employed themselves in proposing statical principles peculiar to fluids; and especially with regard to the grand question of the figure of the earth, on the supposition that it was once fluid. Huyghens first endeavored to resolve it, taking for his principle of equilibrium the necessary perpendicularity of weight at the free surface of the fluid. Newton's principle was the necessary equality of weight between the two fluid columns going from the centre—the one to the pole, the other to some point of the equator. Bouguer showed that both methods were bad, because, though each was incontestable, the two failed, in many cases, to give the same form to the fluid mass in equilibrium. But he, in his turn, was wrong in believing that the union of the two principles, when they agreed in indicating the same form, was sufficient for equilibrium. It was Clairaut who, in his treatise on the form of the earth, first discovered the true laws of the case, setting out from the evident consideration of the isolated equilibrium of any infinitely small canal; and, tried by this criterion, he showed that the combination required by Bouguer might take place without equilibrium happening. Several great geometers, proceeding on Clairaut's foundation, have carried on the theory of the equilibrium of fluids a great way. Maclaurin was one of those to whom we owe much; but it was Euler who brought up the subject to its present point, by founding the theory on the principle of equal pressure in all directions. Observation of the statical constitution of fluids indicates this as a general law; and it furnishes the requisite equations with extreme facility.

It was inevitable that the mathematical theory of the equilibrium of fluids should, in the first place, be founded, as we have seen that it was, on statical principles peculiar to this kind of bodies: for, in early days, the characteristic differences between solids and fluids must have appeared too great for any geometer to think of applying to the one the general principles appropriated to the other. But, when the fundamental laws of hydrostatics were at length obtained, and men's minds were at leisure to estimate the real diversity between the theories of fluids and of solids, they could not but endeavor to attach them to the same general principles, and perceive the necessary applicability of the fundamental rules of Statics to the equilibrium of fluids, making allowance for the attendant variability of form. But, before hydrostatics could be comprehended under Statics, it was necessary that the abstract theory of equilibrium should be made so general as to apply directly to fluids as well as solids. This was accomplished when Lagrange supplied, as the basis of the whole of Rational Mechanics, the single principle of Virtual Velocities. One of its most valuable properties is its being as directly applicable to fluids as to solids. From that time, Hydrostatics, ceasing to be a natural branch of science, has taken its place as a secondary division of Statics. This arrangement has not yet been familiarly admitted; but it must soon become so.

Liquids.

Gases.

To see how the principle of Virtual Velocities may lead to the fundamental equations of the equilibrium of fluids, we have to consider that all that such an application requires, is to introduce among the forces of the system under notice one new force,—the pressure exerted upon each molecule, which will introduce one term more into the general equation. Proceeding thus, the three general equations of the equilibrium of fluids, employed when hydrostatics was treated as a separate branch, will be immediately reached. If the fluid be a liquid, we must have regard to the condition of incompressibility,—of change of form without change of volume. If the fluid be gaseous, we must substitute for the incompressibility that condition which subjects the volume of the fluid to vary according to a determinate function of the pressure; for instance, in the inverse ratio of the pressure, according to the physical law on which Mariotte has founded the whole Mechanics of the gases. We know but too little yet of these gaseous conditions; for Mariotte's law can at present be regarded only as an approximation,—sufficiently exact for average circumstances, but not to be rigorously applied in any case whatever.

Some confirmation of the philosophical character of this method of treating hydrostatics arises from its enabling us to pass, almost insensibly, from the order of bodies of invariable form to that of the most variable of all, through intermediate classes,—as flexible and elastic bodies,—whereby we obtain, in an analytical view, a natural filiation of subjects.

We have seen how the department of Statics has been raised to that high degree of speculative perfection which transforms its questions into simple problems of Mathematical Analysis. We must now take a similar review of the other department of general Mechanics,—that more extended and more complicated study which relates to the laws of Motion.

SECTION II.

DYNAMICS.

Object. The object of Dynamics is the study of the varied motions produced by *continuous forces*. The Dynamics of varied motions or continuous forces includes two departments,—the motion of a point, and that of a body. From the positive point of view, this means that, in certain cases, all the parts of the body in question have the same motion, so that the determination of one particle serves for the whole; while in the more general case, each particle of the body, or each body of the system, assuming a distinct motion, it is necessary to examine these different effects, and the action upon them of the relations belonging to the system under notice. The second theory being more complicated than the first, the first is the one to begin with, even if both are deduced from the same principles.

With regard to the motion of a point, the question is to determine the circumstances of the compound curvilinear motion, resulting from the simultaneous action of different continuous forces, it being known what would be the rectilinear motion of the body if influenced by any one of these forces. Like every other, this problem admits of a converse solution.

Theory of rectilinear motion. But here intervenes a preliminary theory, which must be noticed before either of the two departments can be entered upon. This theory is popularly called the theory of rectilinear motion, produced by a single continuous force acting indefinitely in the same direction. It may be asked why we want this, after having said that the effect of each separate force is supposed to be known, and the effect of their union the thing to be sought. The answer to this is, that the varied motion produced by each continuous force may be defined in several ways, which depend on each other, and which could never be given simultaneously, though each may be separately the most suitable; whence results the necessity of being able to pass from any one of them to all the rest. The preliminary theory of varied motion relates to these transformations, and is therefore inaptly termed the study of the action of a single force. These different equivalent definitions of the same varied motions result from the simultaneous consideration of the three distinct but co-related functions which are presented by it,—space, velocity, and force, conceived as dependent on time elapsed. Taking the most extended view, we may say that the definition of a varied motion may be given by any equation containing at once these four variables, of which only one is independent,—time, space, velocity, and force. The problem will consist in deducing from this equation the distinct determination of the three characteristic laws relating to space, velocity, and force, as a function of time, and, consequently, in mutual correlation. This general problem is always reducible to a purely analytical research, by the help of the two dynamical formulas which express, as a function of time, velocity, and force, when the law of space is supposed to be known. The infinitesimal method leads to these formulas with the utmost ease, the motion being considered uniform during an infinitely small interval of time, and as uniformly accelerated during two consecutive intervals. Thence the velocity, supposed to be constant at the instant, according to the first consideration, will be naturally expressed by the differential of the space, divided by that of the time; and, in the same way, the continuous force, according to the second consideration, will evidently be measured by the relation between the infinitely small increment of the velocity, and the time employed in producing this increment.

Lagrange's conception of transcendental analysis excluding him from this use of the infinitesimal method for the establishment of the two foregoing dynamic formulas, he was led to present this theory under another point of view, more important than seems to be generally supposed. In his Theory of Analytical Functions, he has shown that this dynamic consideration really consists in conceiving any varied motion as compounded, each moment, of a certain uniform motion and another motion uniformly varied,—likening it to the vertical motion of a heavy body under a first impulsion. Lagrange has not given its due advantage to this conception, by developing it as he might have done. In fact, it supplies a complete theory of the assimilation of motions, exactly like the theory of the contracts of curves and surfaces, in the department of geometry. Like that theory, it removes the limits within which we supposed ourselves to be confined, by disclosing to us, in an abstract way, a much more perfect measure of all varied motion than we obtain by the ordinary theory, though reasons of convenience compel us to abide by the method originally adopted.

The first case or department of rational dynamics,— that of the motion of a point, or of a body which has all its points or portions affected by the same force,—relates to the study of the curvilinear motion produced by the simultaneous action of any different continuous forces. This case divides itself again into two,—according as the mobile point is free, or as it is compelled to move in a single curve, or on a given surface. The fundamental theory of curvilinear motion may be established in either case, in a different way; each being susceptible of direct treatment, and of being connected with the other. In the first case, in order to deduce the second, we have only to regard the active or passive resistance of the prescribed curve or surface as a new force to be added to the others proposed. In the other way, we have only to consider the moving point as compelled to describe

Motion of a point.

the curve which it must traverse; and this is enough to afford the fundamental equations, though this curve may then be primitively unknown.

Motions of a system. The other, more real and more difficult case, is that of the motion of a system of bodies in any way connected, whose proper motions are altered by the conditions of their connection. There is a new elementary conception about the measurement of forces which some geometers declare to be logically deducible from antecedent considerations, and to which they would assign the place and title of a fourth law of motion. For the sake of convenience, we may make it into a fourth law of motion; but such is not its philosophical character. The idea is, that forces which impress the same velocity on different masses are to each other exactly as those masses; or, in other words, that the forces are proportional to the masses, as we have seen them, under the third law of motion, to be proportional to the velocities. All phenomena, such as the communication of motion by collision, or in any other way, have tended to confirm the supposition of this new kind of proportion. It evidently results from this, that when we have to compare forces which impress different velocities on unequal masses, each must be measured according to the product of the mass upon which it acts by the corresponding velocity. This product is called by geometers *quantity of motion;* and it determines the *percussion* of a body, and also the *pressure* that a body may exercise against any fixed obstacle to its motion.

Proceeding to the second dynamical case, we see that the characteristic difficulty of this order of questions consists in the way of estimating the connection of the different bodies of the system, in virtue of which their mutual reactions will necessarily affect the motions which each would take if alone; and we can have no *à-priori* knowledge of what the alterations will be. In the case of the pendulum, for instance, the particles nearest the point of suspension, and those furthest from it, must react on each other by their connection—the one moving faster and the other slower than if they had been free; and no established dynamic principle exists revealing the law which determines these reactions. Geometers naturally began by laying down a principle for each particular case; and many were the principles thus offered, which turned out to be only remarkable theorems furnished simultaneously by fundamental dynamic equations. Lagrange has given us, in his "Analytical Mechanics," the general history of this series of labors; and very interesting it is, as a study of the progressive march of the human intellect. This method of proceeding continued till the time of D'Alembert, who put an end to all these isolated researches by seeing how to compute the reactions of the bodies of a system in virtue of their connection, and establishing the fundamental equations of the motion of any system. By the aid of the great principle which bears his name, he made questions of motion merge in simple questions of equilibrium. The principle is simply this. In the case supposed, the natural motion clearly divides itself into two—the one which subsists and the one which has been destroyed. By D'Alembert's view, all these last, or in other words, all the motions that have been lost or gained by the different bodies of the system by their reaction, necessarily balance each other, under the conditions of the connection which characterizes the proposed system. James Bernouilli saw this with regard to the particular case of the pendulum; and he was led by it to form an equation adapted to determine the centre of oscillation of the most simple system of weight. But he extended the resource no further; and what he did detracts nothing from the credit of D'Alembert's conception, the excellence of which consists in its entire generality.

In D'Alembert's hands the principle seemed to have a purely logical character. But its germ may be recognised in the second law of motion, established by Newton, under the name of the equality of reaction and action. They are, in fact, the same, with regard to two bodies only acting upon each other in the line which connects them. The one is the greatest possible generalization of the other; and this way of regarding it brings out its true nature, by giving it the physical character which D'Alembert did not impress upon it. Henceforth therefore we recognise in it the second law of motion, extended to any number of bodies connected in any manner.

We see how every dynamical question is thus convertible into one of Statics, by forming, in each case, equations of equilibrium between the destroyed motions. But then comes the difficulty of making out what the destroyed motions are. In endeavoring to get rid of the embarrassing consideration of the quantities of motion lost or gained, Euler, above others, has supplied us with the method most suitable for use—that of attributing to each body a quantity of motion equal and contrary to that which it exhibits, it being evident that if such equal and contrary motion could be imposed upon it, equilibrium would be the result. This method contemplates only the primitive and the actual motions which are the true elements of the dynamic problem—the given and the unknown; and it is under this method that D'Alembert's principle is habitually conceived of. Questions of motion being thus reduced to questions of equilibrium, the next step is to combine D'Alembert's principle with that of virtual velocities. This is the combination proposed by Lagrange, and developed in his "Analytical Mechanics," which has carried up the science of abstract Mechanics to the highest degree of logical perfection—that is, to a rigorous unity. All questions that it can comprehend are brought under a single principle, through which the solution of any problem whatever offers only analytical difficulties.

D'Alembert immediately applied his principle to the case of fluids—liquid and gaseous, which evidently admit of its use as well as solids, their peculiar conditions being considered. The result was our obtaining general equations of the motion of fluids, wholly unknown before. The principle of virtual velocities rendered this perfectly easy, and again left nothing to be desired, in regard to concrete considerations, and presented none but analytical difficulties. We must admit, however, that our actual knowledge obtained under this theory is extremely imperfect owing to insurmountable difficulties in the integrations required. If it was so in questions of pure Statics, much more must it be so in the more complex dynamical questions. The problem of the flow of a gravitating liquid through a given orifice, simple as it appears, has never yet been resolved. To simplify as far as they could, geometers have had recourse to Daniel Bernouilli's hypothesis of the parallelism of sections, which admits of our considering motion in regard to horizontal laminæ instead of particle by particle. But this method of considering each horizontal lamina of a liquid as moving altogether, and taking the place of the following, is evidently contrary to the fact in almost all cases. The lateral motions are wholly abstracted, and their sensible existence imposes on us the necessity of studying the motion of each particle. We must then consider the science of hydrodynamics as being still in its infancy, even with regard to liquids, and much more with regard to gases. Yet, as the fundamental equations of the motions of fluids are irreversibly established, it is clear that what remains to be accomplished is in the direction of mathematical analysis alone.

Results. Such is the Method of Rational Mechanics. As for the great theoretical results of the science—the principal general properties of equilibrium and motion thus far discovered—they were at first taken for real principles, each being destined to furnish the solution of a certain order of new problems in Mechanics. As the systematic character of the science has come out, however, these supposed principles have shown themselves to be mere theorems—necessary results of the fundamental theories of abstract Statics and Dynamics.

Statical theorems.

Law of repose. Of these theorems, two belong to Statics. The most remarkable is that discovered by Torricelli with regard to the equilibrium of heavy bodies. It consists in this; that when any system of heavy bodies is in a situation of equilibrium, its centre of gravity is necessarily placed at the lowest or highest possible point, in comparison with all the positions it might take under any other situation of the system. Maupertuis afterward, by his working out of his *Law of repose*, gave a large generalization to this theorem of Torricelli's, which at once became a mere particular case under that law; Torricelli's applying merely to cases of terrestrial gravitation, while that of Maupertuis extends throughout the whole sphere of the great natural attractive forces.

Stability and instability of equilibrium. The other general property relating to equilibrium may be regarded as a necessary complement of the former. It consists in the fundamental distinction between the cases of stability and instability of equilibrium. There being no such thing in nature as abstract repose, the term is applied here to that state of stable equilibrium which exists where the centre of gravity is placed as low as possible; while unstable equilibrium is that which is popularly called equilibrium; and it exists when the centre of gravity is placed as high as possible. Maupertuis's theorem consisted in this—that the situation of equilibrium of any system is always that in which the sum of *vires vivæ* (active forces) is a maximum or a minimum; and the one under notice, developed by Lagrange, consists in this—that in any system equilibrium is stable or unstable according as the sum of *vires vivæ* is a maximum or a minimum. Observation teaches the facts in the most simple cases; but it requires a large theory to exhibit to geometers that the distinction is equally applicable to the most compound systems.

Proceeding to the theorems relative to dynamics, the most direct way of establishing them is that used by Lagrange—exhibiting them as immediate consequences of the general equation of dynamics, deduced from the combination of D'Alembert's principle with the principle of virtual velocities. The first theorem is that of the conservation of the motion of the centre of gravity, discovered by Newton. Newton showed that the mutual action of the bodies of any system, whether of attraction, impulsion, or any other—regard being had to the constant equality between action and reaction—can not in any way affect the state of the centre of gravity; so that if there were no accelerating forces besides, and if the exterior forces of the system were reduced to instantaneous forces, the centre of gravity would remain immovable, or would move uniformly in a right line. D'Alembert generalized this property, and exhibited it in such a form that every case in which the motion of the centre of gravity has to be considered may be treated as that of a single molecule. It is seldom that we form an idea of the entire theoretical generality of such great results as those of rational Mechanics. We think of them as relating to inorganic bodies, or as otherwise circumscribed; but we can not too carefully remember that they apply to all phenomena whatever; and in virtue of this universality alone are the basis of all real science.

The second general theorem of dynamics is the principle of areas, the first perception of which is attributable to Kepler. In its simplest form it is this: that if the accelerating force of any molecule tends constantly toward a fixed point, the vector radius of the moving body describes equal areas in equal times round the fixed point; so that the area described at the end of any time increases in proportion to the time: and the reciprocal fact is clear,—that the evidence of the areas and the times proves the action upon the body of a force directed toward the fixed point. This discovery of Kepler's is the more remarkable for having been made before dynamics had been really created by Galileo. Its importance in astronomy we shall see hereafter. But though, in its simplest form, it is one of the bases of celestial

Mechanics, it is, in fact, only the simplest particular case of the great general theorem of areas, exhibited in the middle of the last century by D'Arcy, Daniel Bernouilli, and Euler. Kepler's discovery related only to the motion of a point, while the later one refers to the motion of any system of bodies, acting on each other in any manner whatever; which constitutes a case, not only more complex, but different, on account of the mutual actions involved. It yields proof, however, that though the area described by the vector radius of each molecule may be altered by reciprocal actions, the sum of the areas described will remain invariable in a given time, and will increase therefore in proportion to the time. As the theorem of the centre of gravity determines all that relates to motions of translation, this determines all that relates to motions of rotation: and the two together are sufficient for the complete study of the motion of any system of bodies, in either direction. And here comes in the facility afforded by M. Poinsot's conception—referred to under the head of Statics. By substituting for the areas or momentum of the geometers, the couples engendered by the proposed forces, a philosophical completeness is given to the theory, and a concrete value, and proper dynamic direction, to what was before a simple geometrical expression of a part of the fundamental equations of motion.

The invariable plane. Laplace elicited from the theory of areas that dynamic property which he called the *invariable plane*, the consideration of which is highly important in celestial mechanics. It is in the study of astronomy that the importance fully appears of the determination of a plane, whose direction is unaffected by the mutual action of different bodies in our own solar system; for we thus obtain a point of reference, a necessarily fixed term of comparison, by which to estimate the variations of the heavenly bodies. We are far from having yet attained precision in the determination of the situation of this plane; but this does not impair the character of the theorem in its relation to rational Mechanics. Again, we are indebted to Poinsot, who, by simplifying, has once more extended the process to which his method is applied: and he has repaired an important omission made by Laplace, in taking into the account the smaller areas described by satellites, and their rotation, and that of the sun itself; whereas Laplace has attended only to the larger areas described by the planets in their course round the sun.

Moment of inertia. Finally, there are Euler's theorems of the *moment of inertia*, and the *principal axes*, which are among Principal axes. the most important general results of rational Mechanics. By means of these, we are able to arrive at a complete analysis of the motion of rotation. By means of all the theorems just touched upon, we are put in the direct way to determine the entire motion of any body, or system of bodies whatever. Besides them, geometers have discovered some which are less general, but, though by no means indispensable, yet very important from the simplification they introduce into special researches. Students will recognise their functions, when their mere names are presented; which is all that our space allows: I refer to the theorem of the conservation of active forces,—singularly important in its applications to industrial Mechanics: the theorem, improperly called *the principle of the least action*, as old as Ptolemy, who observed that reflected light takes the shortest way from one point to another,—an observation which was the basis of Maupertuis's discovery of this property: and lastly, a theorem not usually classed with the foregoing, yet worthy of no less esteem,—the theorem of the *co-existence of small oscillations*, of Daniel Bernouilli. This discovery is as important in its physical as its logical bearings; and it explains a multitude of facts which, clearly known, could not be referred to their principles. It consists in showing that the infinitely small oscillations caused by the return of any system of forces to a state of stable equilibrium coexist without interference, and can be treated separately.

This reference to the principal general theorems hitherto discovered in Rational Mechanics concludes our review of the second branch of Concrete Mathematics.

As for our review of the whole science, I wish I could better have communicated my own profound sense of the nature of this immense and admirable science, which the necessary basis of the whole of Positive Philosophy, constitutes the most unquestionable proof of the compass of the human intellect. But I hope that those who have not the misfortune to be wholly ignorant of this fundamental science may, according to the process of thought which I have indicated, attain some clear idea of its philosophical character. *Conclusion.*

To preserve complete the philosophical arrangement of Mathematics in its present state, I ought to consider here a third branch of Concrete Mathematics; the application of analysis to thermological phenomena, according to the discoveries of Fourier. But, to avoid too great a breach of customary arrangement, I have reserved the subject, and shall place Thermology among the departments of Physics.

Mathematical philosophy being now completely characterized, we shall proceed to examine its application to the study of Natural Phenomena, in their various orders, ranked according to their degree of simplicity. By this character alone can they cast light back again upon the science which explains them; and under this character alone can they be suitably estimated. According to the natural order laid down at the beginning, we now proceed to that class of phenomena with which Mathematics is most concerned,— the phenomena of ASTRONOMY.

BOOK II.

ASTRONOMY.

CHAPTER I.

GENERAL VIEW.

Its nature. It is easy to describe clearly the character of astronomical science, from its being thoroughly separated, in our time, from all theological and metaphysical influence. Looking at the simple facts of the case, it is evident that though three of our senses take cognizance of distant objects, only one of the three perceives the stars. The blind could know nothing of them; and we who see, after all our preparation, know nothing of stars hidden by distance, except by induction. Of all objects, the planets are those which appear to us under the least varied aspect. We see how we may determine their forms, their distances, their bulk, and their motions, but we can never know anything of their chemical or mineralogical structure; and, much less, that of organized beings living on their surface. We may obtain positive knowledge of their geometrical and mechanical phenomena; but all physical, chemical, physiological, and social researches, for which our powers fit us on our own earth, are out of the question in regard to the planets. Whatever knowledge is obtainable by means of the sense of Sight, we may hope to attain with regard to the stars, whether we at present see the method or not; and whatever knowledge requires the aid of other senses, we must at once exclude from our expectations, in spite of any appearances to the contrary. As to questions about which we are uncertain whether they finally depend on Sight or not,—we must patiently wait, for an ascertainment of their character, before we can settle whether they are applicable to the stars or not. The only case in which this rule will be pronounced too severe is that of questions of temperatures. The mathematical thermology created by Fourier may tempt us to hope that, as he has estimated the temperature of the space in which we move, we may in time ascertain the mean temperature of the heavenly bodies: but I regard this order of facts as for ever excluded from our recognition. We can never learn their internal constitution, nor, in regard to some of them, how heat is absorbed by their atmosphere. Newton's attempt to estimate the temperature of the comet of 1680 at its perihelion could accomplish nothing more, even with the science of our day, than show what would be the temperature of our globe in the circumstances of that comet. We may therefore define Astronomy as the science by which we discover the laws of the geometrical and mechanical phenomena presented by the heavenly bodies. *Definition.*

It is desirable to add a limitation which is important, though not of primary necessity. The part of the science which we command from what we may call the Solar point of view is distinct, and evidently capable of being made complete and satisfactory; while that which is regarded from the Universal point of view is in its infancy to us now, and must ever be illimitable to our successors of the remotest generations. Men will never compass in their conceptions the whole of the stars. The difference is very striking now to us who find a perfect knowledge of the solar system at our command, while we have not obtained the first and most simple element in sidereal astronomy—the determination of the stellar intervals. Whatever may be the ultimate progress of our knowledge in certain portions of the larger field, it will leave us always at an immeasurable distance from understanding the universe. *Restriction.*

Throughout the whole range of science, there exists a constant and necessary harmony between our needs and our knowledge. We shall find this to be true everywhere. The fact is, we need to know only what, in some way or other, acts upon us; and the influence which acts upon us becomes, in turn, our means of knowledge. This is evidently and remarkably true in regard to Astronomy. It is of the highest importance to us to know the laws of the solar system: and we have attained great precision with regard to them; but, if the knowledge of the starry universe is forbidden to us, it is clear that it is of no real consequence to us, except as a gratification of our curiosity. The interior mechanism of each solar system is essentially independent of the mutual action of distant suns; as it may well be, considering the distance of these suns from each other, in comparison with the distance of planets from their suns. Our tables of astronomical events, constructed in advance, proceed on the supposition of there being no other system than our own; and they agree with our direct observations, precisely and necessarily. This is our proper field; and we must remember that it is so. We must keep carefully apart the idea of the solar system and that of the universe, and be always assured that our only true interest is in the former. Within this boundary alone is astronomy the supreme and positive science that we have determined it to be; and, in fact, the innumerable stars that are scattered through space serve us scientifically only as providing positions which may be called fixed, with which we may compare the interior movements of our system.

Means of exploration. We shall find, as we proceed through the whole gradation of the science, that the more complex the science, the more various are the means of exploration; whereas, it does not at all follow, as we shall see, that the completeness of the knowledge obtained is in any proportion to the abundance of our means. Our knowledge of astronomy is more perfect than that of any of the sciences which follow it; yet in none are our means of exploration so few.

The means of exploration are three:—direct observation; observation by experiment; and observation by comparison. In the first case, we look at the phenomenon before our eyes; in the second, we see how it is modified by artificial circumstances to which we have subjected it; and in the third, we contemplate a series of analogous cases, in which the phenomenon is more and more simplified. It is only in the case of organized bodies, whose phenomena are extremely difficult of access, that all the three methods can be employed; and it is evident that in astronomy we can use only the first. Experiment is, of course, impossible; and comparison could take place only if we were familiar with abundance of solar systems, which is equally out of the question. Even simple observation is reduced to the use of one sense,—that of sight alone. And again, even this sense is very little used. Reasoning bears a greater proportion to observation here, than in any science that follows it; and hence its high intellectual dignity. To measure angles and compute times are the only methods by which we can discover the laws of the heavenly bodies; and they are enough. The few incoherent sensations concerned would be, of themselves, very insignificant; they could not teach us the figure of the earth, nor the path of a planet. They are combined and rendered serviceable by long-drawn and complex reasonings; so that we might truly say that the phenomena, however real, are constructed by our understanding.

Its rank. The simplicity of the phenomena to be studied, and the difficulty of getting at them, constitute, by their combination, the eminently mathematical character of the science of astronomy. On the one hand, the perpetual necessity of deducing from a small number of direct measures, whether angular or horary, quantities which are not themselves immediately observable, renders the use of abstract mathematics indispensable; while, on the other hand, astronomical questions being always, in themselves, problems of geometry, or else of mechanics, must fall into the department of concrete mathematics. Again, the regularity of astronomical forms admits of geometrical treatment; and the simplicity of astronomical movements admits of mechanical treatment with a very high degree of precision. There is perhaps no analytical process, no geometrical or mechanical doctrine, which is not employed in astronomical researches, and many of them have as yet had no other aim. Considering the simple nature of astronomical investigations, and the easy application to them of mathematical means, it is evident why astronomy is, by common consent, placed at the head of the natural sciences. It deserves this place, first, by the perfection of its scientific character; and, next, by the preponderant importance of the laws which it discloses.

Passing over, for the present, its utility in the measurement of time, the exact description of the globe, and the perfecting of navigation, which are not circumstances that could determine its rank, we may just observe that its affords an instance of the necessity of the loftiest scientific speculations to the satisfaction of the most ordinary wants. Hipparchus began to apply astronomical theory to the finding the longitude at sea. A prodigious amount of geometrical science has gone to improve our tables of longitude up to their present point; and if we can not now get within half-a-dozen miles of a true estimate in the seas under the line, it is for want of more science still.

Those who say that science consists in an accumulation of observed facts may here see how imperfect is their account of the matter. The Chaldeans and Egyptians collected facts from observation of the heavens; but there was no astronomical science till the early Greek philosophers referred the diurnal movement to geometrical laws. *When it became a science.* The aim of astronomical researches was to establish what would be the state of the sky at some future time; and no accumulation of facts could effect this, till the facts were made the basis of reasonings. Till the rising of the sun, or of some star, could be accurately predicted, as to time and place, there was no astronomical science. Its whole progress since has been by introducing more and more certainty and precision into its predictions, and in using smaller and smaller data from direct observation for a more and more distant prevision. No part of natural philosophy manifests more strikingly the truth of the axiom that *all science has prevision for its end:* an axiom which separates science from erudition, which relates the events of the past, without any regard to the future.

However impossible may be the aim to reduce the phenomena of the respective sciences to a single law, *Reduction to a single law.* supreme in each, this should be the aim of philosophers, as it is only the imperfection of our knowledge which prevents its accomplishment. The perfection of a science is in exact proportion to its approach to this consummation; and, according to this test, astronomy distances all other sciences. Supposing it to relate to our solar system alone, the point is attained; for the single general law of gravitation comprehends the whole of its phenomena. It is to this that we must recur when we wish to show what we mean by the *explanation* of a phenomenon, without any inquiry into its first or final cause; and it is here that we learn the true character and conditions of scientific *hypothesis,*—no other science having applied this powerful instrument so extensively or so usefully. After having exhibited these great general properties of astronomical philosophy, I shall apply them to perfect the philosophical character of the other principal sciences.

Regarding astronomical science, apart from its method, and with a view to the natural laws which it discloses, its pre-eminence is no less incontestable. I have always admired, as a stroke of philosophic genius, Newton's title of his treatise on Celestial Mechanics, "The Mathematical Principles of Natural Philosophy;" for it would be impossible to indicate with a more energetic conciseness that the general laws of astronomical phenomena are the basis of all our real knowledge.

Relation to other sciences. We may see at a glance that astronomy is independent of all the natural sciences, depending on Mathematics alone: and though, philosophically speaking, we put Mathematics at the head of the whole series, we practically regard it less as a natural science of itself (from the paucity of phenomena which it presents to observation) than as the repository of principles by which the natural sciences are interpreted and investigated. Philosophically speaking, astronomy depends on Mathematics alone, owing nothing to Physics, Chemistry, or Physiology, which were either undiscovered, or lost in theological and metaphysical confusion, while astronomy was a true science in the hands of the ancient geometers. But the phenomena of the other sciences are dependent naturally as well as systematically, on astronomical facts, and can be perfectly studied only through astronomy. We can not thoroughly understand any terrestrial phenomenon without considering what our globe is, and what part it bears in the solar system, as its situation and motions affect the conditions of everything upon it; and what would become of our physical, chemical, and physiological ideas, without consideration of the law of gravitation? In the remotest case of all, that of Social phenomena, it is certain that changes in the distance of the earth from the sun, and consequently in the duration of the year, in the obliquity of the ecliptic, etc., which in astronomy would merely modify some coefficients, would largely affect or completely destroy our social development. It is no exaggeration to say that Social physics would be an impossible science, if geometers had not shown us that the perturbations of our solar system can never be more than gradual and restricted oscillations round a mean condition which is invariable. If astronomical conditions were liable to indefinite variations, the human existence which depends upon them could never be reduced to laws.

Not less important is the influence of astronomical science on our own intelligence. It has done much more than relieve us from superstitious terrors and absurd notions about comets and eclipses, —notions which, as Laplace observed, would spring up again immediately if our astronomy were forgotten. This science has done much more for our understandings than that. It has done more than any other pursuit—simply because it is the most scientific of all—to expose and destroy the doctrine of final causes, which is generally regarded by the moderns as the basis of every religious system, though it is in fact a consequence and not a cause. The knowledge of the motion of the earth has overthrown the very foundation of the doctrine, which supposed the universe to be subordinated to our globe, and therefore to Man. Since Newton's time, the development of celestial Mechanics has deprived theological philosophy of its principal intellectual office, by proving that the order maintained throughout our system and the whole universe is by the simple gravitation of its parts. If we took an *à-priori* view, we should say that, as we exist, our system must be such as to admit of our existence; and one necessary condition of this is such a degree of stability in our system as we actually find. This stability we scientifically perceive to be a simple consequence of mechanical laws working among the incidents of our system,—the extremely small planetary bodies in their relation to the larger sun; the small eccentricity of their orbits, and moderate inclination of their planes; which incidents, again, are necessary consequences of the mode of formation of the entire system. The stability by virtue of which we hold our existence is not found in the case of comets, whose perturbations are not only great, but liable to indefinite increase; and their being inhabited is inconceivable. Thus, the doctrine of final causes would be reduced to the truism that there are no inhabited bodies in our system but those which are habitable. This brings us back to the principle of the conditions of existence, which is the true positive transformation of the doctrine of final causes, and of far superior scope and profit in every way.

We have next to consider the divisions of the science. These arise immediately out of the fact, now familiar to us, that astronomical phenomena are either geometrical or mechanical. They are *Celestial Geometry*, which is still called Astronomy, from its having possessed a scientific character before the other; and *Celestial Mechanics*, of which Newton was the immortal founder. Though our business is with our own system, the same division extends to Sidereal astronomy, supposing that kind of exploration to be within our power. As before, we see geometry to be more simple in its phenomena than mechanics, and that mechanics is dependent on geometry, without reciprocity. In fact, men were successfully inquiring into the forms and sizes of the heavenly bodies, and studying their geometrical laws, before anything was known of the forces which changed their positions. Whereas, the province of Celestial Mechanics is to analyze the motions of the stars, in order to refer them, by the rules of Rational Mechanics, to the elementary motions regulated by a universal and invariable mathematical law; —thence, again, departing to perfect the knowledge of real motions by scientifically determining them *à priori*, taking from observation the necessary data—the fewest possible—for the calculations of general mechanics. This is the link by which astronomy and physics are connected, and connected so closely that some great phenomena render the transition almost insensible; as in the theory of the Tides. But it is evident that the whole reality of celestial mechanics consists in its having issued from the exact knowledge of true movements, furnished by celestial geometry. It was for want of this point of departure that all attempts before the time of Newton, even Descartes', however valuable in other ways, failed to establish systems of celestial mechanics. This division of the science into two parts has therefore nothing arbitrary in it, nor even scholastic; it is derived from the nature of the science, and is at once historical and dogmatic. As for the subdivisions, we need not trouble ourselves with them now.

In regard to the point of view from which the science should be regarded, Lacaille thought it would simplify matters extremely to place his observer on the surface of the sun. And so it would, if the thing could be done in accordance with positive knowledge; but undoubtedly the solar station should be the ultimate and not the original one, under a rational system of astronomical study. And when, as in the case of this work, the object is the analysis of the scientific method, and the observation of the logical filiation of the leading scientific ideas, it matters less to obtain a clearer exposition of general results than to adhere to the positive method.

I suppose my readers to be well acquainted with the two fundamental facts of the diurnal and annual rotation of our globe, as data without which nothing could be clearly understood of the essential methods and general results of astronomical science. I am not giving a treatise on astronomy, nor even a summary; but a series of philosophical considerations upon the different parts of the science, in which any extended special exposition would be misplaced.

We must first see what methods of observation astronomers need, and are possessed of.

CHAPTER II.

METHODS OF STUDY OF ASTRONOMY.

SECTION I.

INSTRUMENTS.

Observation. ALL astronomical observation is, as we have seen, comprehended in the measurement of times and of angles. The two considerations concerned in attaining the great perfection we have reached are the perfecting the instruments, and the application by theory of certain corrections, without which their precision would be misleading.

The observation of shadows was the first resource of astronomers, when the rectilinear propagation of light was established. Solar shadows, and also lunar, were very valuable in the beginning; and much was obtained from the simple device of a style, so fixed as to cast a shadow corresponding with the diurnal rotation to be observed: but the alterations rendered necessary by the annual motion, and impossible to make on that apparatus, rendered the instrument unfit for precise observations. Again, by comparing the length of the shadow cast by a vertical style with the height of the style, the corresponding angular distance of the sun from the zenith was computed: and a valuable method this was: but the penumbra rendered the accurate measurement of the shadow impossible. The difficulty, aggravated by its unequal amount at different distances from the zenith, was partly removed by the use of very large gnomons; but not completely. These imperfections determined astronomers to get rid as soon as possible of the process of gnomonic measurement. Shadows will always be at hand to measure by when better means are wanting: and one application of this instrument remains in our observatories—as the basis of the meridian line, regarded as dividing into two equal parts the angle formed by the horizontal shadows of the same length which correspond to the two equivalent parts of the same day. In this case, the penumbra is harmless, as it affects the two parts equally; and as for the obliquity of the sun's motion, that may be mainly got rid of by choosing the period of either solstice—especially the summer one. It is easy, too, to rectify the observation by the stars.

Proceeding to more exact methods, and, first, with regard to measurement of time, it is clear that the most perfect of all chronometers is the sky. It seems as if it would be enough, after knowing precisely the latitude of one's observatory, to measure the distance of any star from the zenith, and learn its horary angle, and, as an immediate consequence, the time that has elapsed, by resolving the spherical triangle formed by the pole, the zenith, and the star. If a sufficiently wide observation of this kind had been made, and numerical tables formed for certain selected stars, great results might have been obtained from this natural method; but it is insufficient; and it has the defect of making the measure of time depend on that of angles, which is the least perfect of the two, in our day. This method is therefore used only in the absence of a better, as in nautical astronomy; and its commonest service is, in regulating other chronometers, by a comparison with that of the heavens themselves. Artificial methods of measuring time are therefore indispensable in astronomy.

Every phenomenon which exhibits continuous change might serve, in a rough way, to mark time: various chemical processes, or even the beating of our own pulses, might afford a measure, more or less inaccurate: astronomical phenomena are excluded, because they are what we want to measure: and we therefore have recourse to physical means, and find weight the best. The ancients tried it in the form of the flow of liquids; and the water clocks succeeded the hour-glass; but the uncertainty of these led to solids being preferred; and in the form of weight having a vertical descent. By no care, however, could the disturbances caused by natural forces be remedied, till Galileo, by his creation of rational dynamics, suggested the pendulum. Whether it is or is not correct to assign to Galileo the idea of using the pendulum as a measurer of time, it is certain that his discoveries suggested it, and that Huyghens enabled us to use it. He had recourse to the highest principles of science to render this service, and discovered the principle of vires vivæ, which, besides being scientifically indispensable, afforded to art new means of modifying oscillations without changing the dimensions of the apparatus. Considered as a collection of discoveries for a single aim, Huyghens's treatise *De Horologio oscillatorio*, is perhaps the most remarkable example of special researches that the history of the human mind has yet exhibited. From that time, the perfecting of astronomical clocks became merely a matter of art. In regard to fixed clocks, two things have to be attended to;—the diminution of friction, by improved methods of suspension, and the correction by a compensating apparatus of irregularities caused by variations of temperature. As for portable chronometers, worked by a spiral spring, they are a marvellous invention; but they belong to the province of art, and not science.

In regard to the measurement of angles, it is clear that an instrument which would admit of an allowance for minutes and seconds, must be of a size incompatible with minute precision. It must always be that large apparatus must be so affected in its weight and temperature as to be impaired in its accuracy. The large telescopes of modern times are intended to show us stars otherwise invisible: and no one thinks of using them for purposes of precise measurement. It is generally agreed now that instruments for measuring angles should not be more than ten feet in diameter when we are dealing with an entire circle; and they are usually not more than six or seven. The wonder then is how we are to estimate angles to a second, as we do every day, with circles whose size would scarcely indicate minutes. It is done by the concurrent use of three methods—the eye-piece, the use of the vernier (so called after its inventor), and the repetition of angles.

It was long before it occurred to astronomers to use their lenses for any other purpose than the discovery of new objects: but at last it occurred to them to replace the ancient transoms and modern sights by an eye-piece which should secure the advantages without the inaccuracies of a large instrument. Morin first made this use of a lens. Auzout followed with his invention of the reticle; and, a century after, Dollond gave us a power of absolute precision by his invention of the achromatic object glass. Vernier proposed in 1631 to divide intervals into parts much more minute than could be marked. He enabled us to ascertan angles, within half a minute, or circles divided only into sixth-parts of a degree. The precision obtainable by his simple apparatus is indefinite, being limited only by our difficulty in detecting the coincidence of the line of the vernier with that of the limb. The union of the third method with these two, gives us the perfection we have attained. It is strange that we should have been so long in perceiving that, the imperfections of angular instruments having nothing to do with the dimensions of the angle to be measured, we should gain much by increasing, in fixed proportions, the magnitude of the angles, which is equivalent to diminishing the imperfection of the instrument. The repetition of angles served every purpose immediately, with regard to terrestrial objects, on account of the steadiness of the point of view; but there was the difficulty, with regard to the heavenly bodies, of their perpetual change of place. Borda applied himself to measure the distance from the zenith of the stars when they crossed the meridian; and the star then remains sensibly at the same distance from the zenith long enough to allow the operation of the multiplication of the angle. By these means, angular instruments are matched with horary in regard to precision. They require from the observer a diligent patience in applying all the minute precautions and rectifications which experience has proved to be indispensable to the fullest use of these instruments.

Then, we have Roemer's meridional eye-glass, which fixes the instant of the passage of a star over the meridian. The plane of the meridian is made in this case purely geometrical, by being described by the optic axis of a simple eye-glass, properly disposed; which is enough when all we want to know is the precise moment of the star's passage. Then there are the micrometrical instruments, by which we measure the diameters of stars, and generally all small angular intervals. These are the material instruments of observation,—horary and angular. We must now advert to the intellectual means,—that is, to the corrections which astronomers must apply to the results exhibited by their instruments. There would be little use in perfecting our instruments, if refraction and parallax introduced as much error into our observation as we had got rid of by the improvement of our apparatus.

The corrections required are of two kinds. The first relate to the errors caused by the position of the observer,—the ordinary refractions and parallax. No deep astronomical knowledge is required for the correction of these. The second class, arising from the same cause, since they proceed from the observer being on a moving planet, are founded on primary astronomical theories: they are the annual parallax, the precession of the equinoxes, aberration, and nutation. Our business now is with the first and most important class.

Requisite corrections.

SECTION II.

REFRACTION.

Refraction. The light which comes to us from any star must be more or less turned aside by the action of the terrestrial atmosphere. We must estimate the amount of this deviation before our observations can answer any theoretical purpose. The star is, by this refraction, made to appear too near the zenith, while left in the same vertical plane. Only at the zenith is the error absent, while it increases as the star descends to the horizon. This error, primarily affecting distances from the zenith, must affect, indirectly, all other astronomical measurements, except azimuths: but it would be easy to calculate them, if we once knew the law of diminution and increase of refraction at different distances from the zenith. Philosophers have tried the logical way and the empirical, and have ended by combining the two.

If our atmosphere were homogeneous, the refraction of light would be uniform and calculable. But our atmosphere is composed of strata; and the consequent refractions are excessively unequal, and increasing as the light penetrates a denser stratum, so that its passage constitutes a curve of the last degree of complication. Even this would be calculable, with more or less pains, if we knew the law of variation of these atmospheric densities: but we do not and can not know that law. We have no exact knowledge of the laws of temperature, and can not estimate atmospheric changes, either as to number or degree: and all mathematical processes founded on laws of pressure, etc., may be good as exercises, but are of no value in estimating refraction. As to the empirical method, if the refraction remained always constant at the same height, we might construct tables; and, by extending our observations and instituting various comparisons, we might hope to obtain such a mass of materials as would afford us some certain results. This is what astronomers have, in fact, patiently and laboriously done, by the help of the improved instruments we have spoken of. They have used whatever geometrical help they could make applicable: but the results are discouraging enough. There is nothing like uniformity in the results: for the changes in the atmosphere are beyond our calculation and measurement. We study the barometer, the thermometer, and the hygrometer, at the right moment; we can learn from them only the changes taking place on the spot in which we are; and our tables of refraction vary as our observatories, and even in one observatory at different times. Delambre found differences of four or five minutes between one day and another, after taking all imaginable pains. All that we can do is to confine our observations to the nearest possible approach to the zenith, and to place no reliance on what we attempt near the horizon. By doing this, we shall find our astronomical observations less affected by the unmanageable difficulties of refraction than might be anticipated.

SECTION III.

PARALLAX.

THE difficulty of the parallaxes can be dealt with much more easily and satisfactorily than that of the refractions. Observations of the heavenly bodies made in different places could not be exactly compared without a reference, in idea, to those which would be made from an imaginary observatory, situated in the middle of the earth, which is besides the true centre of apparent diurnal motions. This correction, which is called the parallax, is analogous to that which is constantly made in measurements of the earth's surface, under the more logical name of reduction to the centre of the station.

Parallax.

The effect of the parallax, like that of refraction, is upon the distance of stars from the zenith alone, leaving the star in the same vertical plane, and placing it too far from the zenith, instead of too near, as in the case of refraction. In this instance too, as in the other, though not according to the same law, the deviation increases as the star descends to the horizon. In like manner, too, there must be secondary modifications for all the other astronomical quantities, except with regard to the azimuths. The rectification is easy in comparison with the other case, from the absence of the hopeless difficulties caused by our ill-understood atmosphere. The similar course of the two difficulties, producing counteracting effects, has, we may observe, relaxed the attention of astronomers to the facts of refraction and parallax, by partly concealing their influence on actual observations.

The parallax does not, like refraction, affect all the stars alike, but, on the contrary, affects all unequally, and each according to its position. It is insensible with regard to all which lie outside the limits of our system, on account of their immense distance: and it varies extremely within our system, from the horizontal parallax of Uranus, which can never reach a half-second, to that of the moon, which may at times exceed a degree. Here lies the radical distinction, in astronomical calculations, between the theory of parallaxes and that of refractions. The determination of questions of parallaxes does not wholly depend, like that of refraction, on methods of observation in astronomy, but is truly a portion of science. Depending as it does, ultimately, on the estimate of the distances of the stars from the earth, it pertains to celestial geometry, through the necessity of knowing the law of motion of each star. Thus, it constitutes a part of the science itself; though, in the absence of direct knowledge of the distances of stars, an empirical method of determining the coefficients, analogous to that employed in the case of refraction, may be adopted. The method which will suffice is to choose a place and time which will show the proposed star passing the meridian very near the zenith: then to measure, for several consecutive days, its polar distance, so as to know pretty nearly the amount of this distance at any moment of the process: and this being laid down, then to calculate, for this instant, according to the horary angle and its two sides, the true distance from the star to the zenith, when it is considerably remote from it, without being too near the horizon (say from 75° to 80°): and then, the comparison of this distance with that which is actually observed at the moment, will evidently disclose the corresponding parallax, and therefore the horizontal parallax, provided the due correction for the refraction has been made. This is the method by which it is most easily established that the parallax of all the stars is absolutely insensible.

It is a serious inconvenience in this method, that all the uncertainty of the case of refraction is introduced into that of parallax. In regard to a body whose parallax is very great, as the moon, the uncertainty is of small consequence; but in regard to the sun, or other distant body, an error of one third, or even one half, in the value of its horizontal parallax, might be occasioned. The method is absolutely inapplicable to the remotest of our planets; and not only to Uranus, but to Saturn and Jupiter. The rational method must be resorted to, in the case of these. The empirical method has been mentioned here from the philosophical interest which attaches to the fact that, up to a certain point, the true distances of stars from the earth, at least in proportion to its radius, may be ascertained by observations made in one place; a thing which appears, at first sight, goemetrically impossible.

SECTION IV.

CATALOGUE OF STARS.

Catalogue of stars. I AM disposed to give a place here, contrary to custom, to the Catalogue of stars, which I think should be reckoned among our necessary means of observation in astronomy. This catalogue is a mathematical table of directions by which we find the different stars. Such a determination is a basis of direct knowledge in regard to Sidereal astronomy: while, in regard to our own system, it is simply a valuable means of observation, which supplies us with terms of comparison indispensable for the study of the interior movements of the system. Such has been the essential use of catalogues of stars, from Hipparchus, who began them, to this day. In order to fulfil their purposes, these catalogues should contain the greatest possible number of stars, spread over every region of the sky. Astronomers have done their duty well; for it is a settled habit with them to determine, as far as they can, the co-ordinates of every new star which they observe; and thus our catalogues are very voluminous, and for ever augmenting. Our business here is not with the system of classification and nomenclature adopted in these catalogues. The nomenclature, bearing as it does the marks of the primitive theological state of astronomy, might be easily replaced by one of a methodical character—the objects to be classified being of the simplest nature, and the distinctions being, in fact, only those of position. But it is this very simplicity which prevents the need from being felt as it would among more complex elements—useful as a rational system would no doubt be in finding and assigning the places of stars. The change will be made in time, no doubt, and the need is not urgent. Stars are not known by their names, for astronomical purposes, but by their descriptions; and the classification and nomenclature in the catalogue, resulting from the fundamental division of the circle, are as perfect as possible; and all else is of little importance. I would only ask that we should cease to speak of the *magnitudes* of stars, as marking their rank, and substitute the word *brightness*, in order to avoid all risk of supposing stars to be large or small in proportion to their brightness or dimness. The word *brightness* would be a simple declaration of the fact, without judging the causes, which we are far from understanding.

By viewing these methods as I have brought them together, we may trace the progress of the science from its earliest days. With regard to angular measurement, for instance, the ancients observed with exactness a degree at the utmost; Tycho Brahe carried up the precision to a minute, and the moderns to a second;—a perfection so recent, that observations which lie more than a century behind our time are considered, from their want of precision, inadmissible in the formation of astronomical theories.

My object has been, chiefly, to show the harmony which exists among these different methods of observation; a harmony which, while it tends to perfect them all, up to a certain point, still restricts them all, by making each a limit to the rest. No improvement in horary or angular instruments, for instance, could carry us far, while our knowledge of refraction remains as imperfect as it is. But there is no reason to suppose that we have approached the limits imposed by the conditions of the subject.

CHAPTER III.

GEOMETRICAL PHENOMENA OF THE HEAVENLY BODIES.

SECTION I.

STATICAL PHENOMENA.

The phenomena of our solar system divide themselves into two classes—the *Statical* and the *Dynamical*. The first class comprehends the circumstances of the star itself, independent of its motions; as its distance, magnitude, form, atmosphere, etc.: the other comprehends the facts of its displacements, and the mathematical considerations belonging to its different positions. According to the usual analogy, the first is independent

Two classes of phenomena.

of the second; while the second could have no existence without the first. The Statical phenomena would exist if the system was immovable: while the dynamical are wholly determined by the statical conditions.

Planetary distances. The first thing necessary to be known about any heavenly body is its distance from the earth: and the difficulty of obtaining this ground for further observations is extremely great—the smallness of the base of our triangle, and the immensity of the distance of the planet, rendering all accuracy hopeless in very many cases. Toward the middle of the last century, when it was desired to determine the horizontal parallax of the moon—the most manageable of the heavenly bodies—Lacaille went to the Cape of Good Hope, and Lalande to Berlin, to observe its distance at the same moment from the zenith—that moment being appointed—as the middle of an anticipated eclipse. The stations were so chosen as to afford a pretty accurate knowledge of the extent of the line of the base—which was about as long a one as our globe could afford. The observations of the two distances of the moon from the zenith must thus afford the necessary data for the resolution of the triangle which must give the distance sought: and thus we have obtained a very exact knowledge of the moon's distance, which, at its mean, is about sixty terrestrial diameters, and about which we are sure that we can not be mistaken to the extent of more than twelve miles. The same method might serve to give us, though with much less precision, the distance of Venus and even Mars, if the observation was made when they were nearest to the earth; but it becomes too uncertain with regard to the sun. It would leave an uncertainty of at least an eighth, or about twelve millions of miles. Of course, it is of no avail with regard to yet more distant bodies.

The method used by astronomers under this difficulty is to measure, first, distances for which our small terrestrial bases will serve; and on these, according to their related phenomena, to erect other calculations; thus making of the first a basis for the support of new estimates. Aristarchus of Samos conceived of an ingenious method of discovering the distance of the sun through that of the moon; but the uncertainty about seizing the exact moment of the quadrature of the moon introduced fatal inaccuracy into the calculation. Halley's method, by means of the passage of Mercury and Venus over the sun, is more circuitous, and suitable only to an advanced state of geometrical science; but it is far more accurate, and the only one now admissible, for determining the parallax of those planets and of the sun, and therefore the distance of the sun from the earth, through the differences in the transit observable at two very distant stations. By this method we can estimate, within a hundredth part, the distance of the sun from the earth. This distance being ascertained, we have it for a basis for other calculations. We have only to observe the angular distance from the sun to the proposed body, at two periods separated by six months

—that is, from opposite points of the earth's orbit. This gives us an immense triangle, the base of which is twice the length of the distance of the earth from the sun: and thus it is that our knowledge of the earth's motion has helped us to a base twenty-four thousand times longer than the longest that can be conceived on our own globe. It is true, the planet observed will have changed its place in the interval; but the remoter planets—which alone are in question here—move very slowly;—Saturn's circuit, for instance, occupying thirty years; and our times of observation being practically reducible to a shorter time than six months—even to two or one, with regard to those planets of our system which move more rapidly; while the slower ones may be considered almost stationary, during such short periods of time; and again, allowance can be made for this small change of place, according to the geometrical theory of its proper motion. It is in this way that astronomers have attained to their knowledge of the positions of the remotest bodies of our system. The numbers by which we express their relations to the distance of the earth from the sun, are now certain to the third decimal at least.

The vast increase of the basis of observation afforded us by our knowledge of the earth's movement is clearly the greatest that we can attain. If we have cleared the bounds of our globe, we certainly can not go beyond its orbit. Great as this distance appears to us, it vanishes when we want to ascertain the distances of stars outside our system. All measurement is here so out of the question that the most we can do is to fix a limit within which they certainly are not,—saying, for instance, that the nearest star is at least two hundred thousand times more remote than the sun, or ten thousand times further off than the remotest planet of our system; which is quite sufficient to establish the independence of our system.

When we have ascertained the distance of the planets from the earth, it is easy to understand how we may find their distances from each other, since, in the triangle in which each is contained, two sides are already given, and the angle to the earth can always be measured. It is only with regard to the sun and the moon that the distances to the earth are of importance. It is enough to know the distances of the planets from the sun, and of the satellites from their planets, which involve little variation. These are our means for ascertaining astronomical distances. As we might anticipate, our assurance is in proportion to the nearness; and great remoteness baffles us entirely. We see here again, as everywhere, that the most simple and elementary determination depends on the most delicate and complex scientific theories. This first case exhibits so much of the spirit of astronomical procedure, that we may go more rapidly through the other statical heads of celestial geometry.

The distances of the stars from our globe being once ascertained, we can learn whatever we desire about their form and size by observation, if it be but precise enough. Their very distance is favorable to this; for, while their motion

Form and size.

or ours displays in turn all their possible aspects, our distance enables us to see at once the whole of each aspect. With regard to the most distant and the smallest, however,—to the stars outside our system, and the satellites of Uranus, and the small planets between Mars and Jupiter,—they can appear to us only as points of vivid light, and their sphericity is concluded upon only through a bold induction. But, in observing the larger planets of our system, we have only to measure their apparent diameter in all directions, after allowing for refraction and parallax. It is much easier to us to learn the form and size of sun and moon than of our own globe, since we have had the aid of glasses. The only case of difficulty is that of Saturn's rings; as it once was with the moon, whose changing aspects greatly puzzled the ancients. The most simple geometry now solves the last difficulty, and Huyghens has helped us over the first. With these exceptions, direct observation assures us that the planets are all round, with more or less flattening at the poles and bulging at the equator, in proportion to the rapidity of their rotation.

As for the size of the heavenly bodies, it is easily calculated from the measurement of the apparent diameter combined with the determination of the distance; and the only reason why men were so long and so widely mistaken about the dimensions of the planets was that their real distances were unknown. No rule as yet appears which connects these results with the order of the distance of the planets from the sun. All we know is that the sun is larger than all the other bodies of the system put together, and in general that the satellites are much smaller than their planets, as the laws of celestial mechanics require. With regard to the bodies outside of our system, as we have no knowledge of their distances, we are, of course, ignorant of their dimensions.

Planetary atmospheres. It is by the occultation of stars, as starry eclipses are called, that we make observations on the atmospheres of the planets, by seeing what deviation their atmospheres cause in the light of the remote stars which they eclipse. As the sun's light is prolonged to us by the refraction of our atmosphere, the atmosphere of a planet defers (only in a much greater degree) the occultation of the star, and also shortens it; and the comparison of the apparent duration of the eclipse with that which it would otherwise be, gives us data for the calculation of the atmosphere which causes the deviation. It is thus that we learn that the moon has no appreciable atmosphere. The horizontal refraction which, on our globe, would reach thirty-four minutes, does not in the moon amount to a single second. And the inference that an atmosphere is wanting there is confirmed by M. Arago, who in a different path of inquiry, about the polarization of light reflected from liquid surfaces, has established the fact that there are not, on the surface of the moon, any great liquid masses, fitted to form an atmosphere. The next best-known case is that of Venus, which exhibits a horizontal refraction of thirty minutes, twenty four seconds. As for the extent of the atmospheres, it may be roughly conjectured from the cessation of the refracting power; but such conjectures must be very loose, as the refracting power may become imperceptible to us, far within the limits of an atmosphere becoming attenuated toward its verge. The strangest phenomenon is that of the telescopic planets, with the exception of Vesta; the atmosphere of Pallas, for instance, being more than twelve times the diameter of the planet. The usual condition, however, appears to be that shared by our globe,—of an atmosphere which is very shallow in proportion to the dimensions of the planet: and this is nearly all we know.

The remaining statical topic is that of the form and size of the earth, which has been left to the last, on account of its special nature. *Earth's form and size.*

No glance of the eye will aid us here, not any direct observation whatever. A long accumulation of indirect observations, serving as a basis for complex mathematical reasonings, are our only means. The geometrical aspect of the question must be taken first, though it depends on the highest mechanical theories, and arises from a mechanical beginning. In the infancy of mathematical astronomy, the variations exhibited in different places by the diurnal movement furnished the first geometrical proof of the earth being round. It was enough to establish its evidently and exclusively spherical character, that the change exhibited by the height of the pole on each horizon was always in exact proportion to the length traversed according to any meridian whatever: and this remains the source of all our geometrical knowledge of the form and dimensions of our planet. Astronomers reached their knowledge of its precise form through that of its size; for it was long before its deviation from the perfect spherical form was understood. In this, as in every case, the form of any body is appreciable only by measuring its dimensions in various directions; and here the only difficulty is in the measuring. The first principles of the discovery were given by Eratosthenes, in the early days of the school of Alexandria; but his method was never effectually employed till the middle of the seventeenth century, when Picard undertook to measure the degree between Paris and Amiens. This was the great starting point of the measuring operations, which must have revealed, as they became more perfect, the truth that the earth is not a perfect sphere; but Newton, by his theory of gravitation, and with his one fact of the shortening of the seconds pendulum at Cayenne, settled the matter, by deciding that our globe must necessarily be flattened at the poles, and bulge at the equator, in the relation of 229 to 230. The astronomers could not at once pronounce against the evidence of direct measurement, while the geometers saw the fact to be certain; and the controversy between these two orders of philosophers, for half a century, led to those scientific operations which have brought us all to one mind. The question was settled by the great expedition sent forth, above a century ago, by the French Academy, to measure, *Means of discovery.*

at the equator and the pole, the two extreme degrees of latitude which must exhibit the widest variation from each other: and the comparison of these with each other, and with Picard's degree, terminated the controversy, and established, not only the truth of Newton's discovery, but the very near accuracy of his calculation. All the experiments made since, in various countries, have united in confirming the fact of the continual lengthening of degrees in approaching the pole. It does not follow that the figure of the earth has been ascertained with absolute precision. There are slight discrepancies which may either be from imperfection in our estimates, or from the earth not being precisely an ellipsoid of revolution; but whatever may be the results of future labors, we know that we are near enough to the truth for all practical purposes, unless in questions of extreme delicacy. We have no absolute knowledge here, any more than in any other department; and we must be content to make our approximations more complex as new phenomena arise to demand it. Such is the true character of the advances that have been made in this science from the beginning. Superficial observers may call its theories arbitrary, from the incessant changes of view that have arisen: but the knowledge gained has always been positive; every scientific opinion has corresponded with the facts which gave rise to it; and such opinions remain therefore useful and sound at this day, within their own range. The science has thus always exhibited a character of stability, through all incidents of progression, from the earliest days of the Alexandrian school till now.

Such are the statical aspects of the planets of our system. We have now to look at the geometrical theory of their motions.

Planetary motions. Like all other bodies, the planets have a motion composed of translation and rotation. The connection of these two motions is so natural, that when we know of the one we look for the other. Yet they present very different degrees of difficulty, and require separate consideration.

The progression of the stars was observed long before their rotation,—the unassisted eye being enough for the first; yet the geometrical study of their rotations is easier, because the motions of the observer have no effect upon them; whereas they largely affect questions of translation. And again, the question of orbits is the chief difficulty of the study of translations; and it does not enter into that of rotations. The latter nearly approaches to the character of statical questions; and therefore it ought to be taken first in the exposition of celestial geometry.

Rotation. Galileo introduced the study of rotations by discovering that of the sun, which was sure to follow closely on the invention of the telescope. The method used is obvious enough, and the same in all cases;—to observe any marks that may exist on the surface of the body, their displacement and return. The more such points of observation are multiplied, the more accurate and complete will be the calculations of time, magnitude, uniformity of movement, etc., deducible from them. There is no more delicate task than this, except with regard to the sun and moon; and none that more absolutely requires a special training of the eye. It is a proof of this, that a careful and honest observer, Bianchini, supposed the rotation of Venus to be twenty-four times slower than it is. Some bodies, as Uranus, are too remote, and others, as the satellites and new planets, too small, to have their rotation established at all, though it is concluded from analogy and induction. We, as yet, know of no law determining the time of these rotations: they are not corrected with distances, nor with magnitudes; and they seem only to have some general, but no invariable, connection with the degree of flattening at the poles.* But if the duration is, though regular in each case, altogether irregular as regards the different bodies, the case is much otherwise with the direction; for it is always, throughout our system, from west to east, and on planes slightly inclined to that of the solar equator: and this constitutes an important general datum in the study of our globe.

The study of translations, much more complex, is also much more important, if we consider the great end of astronomical pursuit—the exact provision of the state of the heavens at some future time. Besides that the movement of the earth constitutes an important element in such a study, it must make a difference with regard to other stars, whether the observer is fixed or moving, as his own movement must affect his observation of other motions. We might indeed decide with certainty, without this introductory knowledge, that the sun and not the earth is the true centre of the motions of all the planets, as Tycho Brahe did when he denied our own motion; for it is enough, with this view, to establish that the distances from the planets to the sun scarcely vary at all, while their distance from us varies excessively; and again, that the solar distance between each inferior planet and the sun is less, and between each superior planet and the sun is greater, than our distance from the sun. But we can not go further than this,—we can not determine the form of the planetary orbits, or the mode in which they are traversed, without making a careful and exact allowance for the displacement of the observer. Deferring for the present the subject of the earth's motion, we will briefly notice some important data connected with the planetary motions, which may be obtained without reference to our own movement, and which are so simple as to rank among statical researches. I mean particularly the knowledge of the planes of orbits, and of the duration of the sidereal revolutions, which has nothing to do with the form of the orbits or the variable velocity of the planets. A plane being determined by three points, it is enough to observe three positions of a star to draw a geometrical conclusion about the situation of the plane of its orbit. Astronomers do not now

Translation.

---

* The rotations of some of the satellites are known. They all follow the law of the moon's rotation, namely, the time corresponds with the orbital periods.—J. P. N.

use, in these operations, the declinations and right ascensions, which are the only co-ordinates directly observed, but, for the sake of convenience, two other spherical co-ordinates, improperly called astronomical latitude and longitude, which are analogous, with regard to the ecliptic, to the others with regard to the equator. After having determined the latitude and longitude of the planet in the three positions, its nodes are found; that is, the points at which its orbit meets the plane of the ecliptic, and the inclination of the orbit to this plane. It is evident that confirmation may be obtained by observing other positions of the body, if they are chosen sufficiently remote from each other; and thus we may obtain a far greater precision than in the case of rotations. It is thus that we have learned that the planes of all the planetary orbits pass through the sun; and the same with regard to the satellites of any planet; and that these planes are in general slightly inclined to the ecliptic, and more slightly still to the plane of the solar equator, except the newly-discovered planets, in whose case we find the inclination much more considerable.

Sidereal revolution. The duration of the sidereal revolutions may, of course, be directly observed, in the first instance, by looking for the return of the star to the same spot in relation to the centre of its motion. If we suppose its motion to be uniform, which we may for a first approximation, we can estimate its course by observing the time required between any of the three positions, without waiting for the total revolution, which is sometimes very slow. The geometrical law of this motion permits us to determine, from this kind of observation, the exact time of the planetary revolution. The values of these periodic times are not irregularly divided among the bodies of our system, like the other data that we have noticed. The shorter the course, the more rapid the motion; and the duration increases more rapidly than the corresponding distance; so that the mean velocity diminishes in proportion as the distance increases. We owe to Kepler the discovery of the harmony between these two essential elements, and it is one of the most indispensable bases of celestial mechanics.

Such is the spirit of the methods by which celestial geometry is made to yield us the elementary data which characterize the bodies of the solar system. We have still to consider those of our own planet, before we proceed to the geometrical laws of the planetary motions.

Motion of the Earth.

We are accustomed to think of the motions of translation and rotation as inseparable; but, in the transition from supposing the earth to be motionless, to the present state of our knowledge, a *Evidences of the Earth's Motion.* theory existed that it whirled round its axis, but was stationary in space. We now perceive that, in addition to the general evidence of the double motion of the planetary bodies, we have special evidence about our own globe—that the annual motion could not exist without the diurnal, though we might logically suppose beforehand that it could.

As the rotation of the earth can not be absolutely uniform in all parts of its surface, some indications of its course must exist among terrestrial phenomena. We must therefore distinguish between the celestial and terrestrial proofs of our diurnal motion, while the annual motion admits only of the former.

Immediate appearances go for nothing in this case; *Ancient conceptions.* for it is clear that, to our eyes (as we do not feel the rotation), it must be exactly the same thing whether we move round among the heavenly bodies, or whether they, fixed in a system, move round us in a contrary direction. There was nothing absurd in the latter supposition, in the old days when men had no doubt of the stars being very near, and not much larger than they appear to the eye, while they exaggerated the size of the earth. They could not avoid supposing that such a mass must be immovable, while the small stars, with their little intervals, were seen moving every day. Even when the Greek astronomers had sketched out the true geometrical theory of the movements of the planets, they treated only of the directions, and had no idea of measuring distances; and it required the whole strength of positive evidence of dimensions and distances to uproot men's strong and natural persuasion of the stability of their globe. From the moment of our obtaining an idea of the proportions of the universe, the old conception became too revolting to reason to be sustained. When it was understood that the earth is a mere point in the midst of prodigious intervals, and that its dimensions are extremely small in comparison with that of the sun, and even of other bodies of our own system, it was absurd to suppose that such a universe could travel round us every day. What velocities would be required to enable the outlying stars to complete such a daily circuit—making allowance for their being twenty-four thousand times nearer the earth, if the earth describes no orbit—and how small the movement of the earth, while those prodigious masses were travelling at such speed! On mechanical grounds, the centrifugal force would be seen to be unmanageable. In every way, the supposition was perceived to be monstrous. Again, the passage of stars before each other, and in a contrary direction to that of the general movement of the sky, showed that they were at different distances from each other, and not bound into an unvarying fabric. Hence arose the notion of Aristotle and Ptolemy, of a system of solid and transparent firmaments. But the existence of comets alone was enough to confute this, appearing as they do in all regions of the *How they gave way.* sky in turn. As Fontenelle said, this theory put the universe in danger of being fractured. It was, curiously enough, Tycho Brahe, the most illustrious opponent of the Copernican system, who provided for the overthrow of his own arguments by first presenting the true geometrical theory of comets. Long before modern precision was attained, men had been prepared by such

Earth's rotation. considerations as the above to conclude upon the rotation of the earth. Long before Copernicus, a rough conception of the truth existed. Even Tycho Brahe felt the astronomical superiority of the true theory; but it seemed to be contradicted by what is before our eyes—the fall of heavy bodies, etc. Copernicus himself could not remove the objections which arose out of men's ignorance of the laws of Mechanics. These objections held their ground for a century, till Galileo established the great law which we have recognised as one of the three on which Rational Mechanics is based—that the relative motions of different bodies are independent of the common motion of the whole. Till this was established, the supposition of the rotation of the earth was inadmissible. It is a curious fact, casting much light upon the action of the human mind, that the opponents of Galileo taunted him with the so-called fact that a ball let down from the top of the mast of a ship in motion would not fall at the foot of the mast, but some way behind—neither they nor anybody else having tried the experiment, which would have shown them that their supposed fact was a mistake. The followers of Copernicus did worse—they admitted the so-called fact, but tried to reason away its bearings with fantastic subtleties. The matter was not settled even by the demonstrations of Galileo, nor till Gassendi compelled observation by a public experiment in the port of Marseilles.

That order of experiments has been carried on, and would be of high value if we could obtain perpendicular stations of sufficient height for the purpose. It is clear that a lofty tower must describe a larger circle in the same time at the top than at the base; and that any body dropped from it must share the higher rate of velocity, having a slight horizontal velocity in the direction of the earth's rotation—falling therefore a little to the east of the base of the tower. Omitting the consideration of the resistance of the air, this amount is calculable in the function of the height of the tower and of its latitude; but experiment would also be valuable; and it is to be hoped that it will be tried at the equator, where the deviation must be greater than anywhere else.

Influence of centrifugal force upon gravity. The most certain terrestrial proof of the earth's rotation is found by tracing the influence of the centrifugal force upon the direction and intensity of weight. This has been done by that observation of Richer, on the shortening of the seconds-pendulum at Cayenne, which has been mentioned as having emboldened Newton to declare the true figure of the earth. The deviation from the spherical form is too small to account for more than one third of the effect observed; and the other two thirds are precisely what would be required, at the equator, where the centrifugal force is greatest, on the supposition of the earth's rotation. Wherever the delicate observation can be made with sufficient precision on other points of the globe's surface, the result answers to the theory. Thus, we should have sufficient assurance, in the absence of the abundant astronomical proofs that we possess of the rotation of the earth. Probably no one fact has ever, in the history of our race, produced such consequences as that observation of Richer's—two thirds of the estimated effect having completely established the rotation of our globe, and the other third having led Newton to the ascertainment of its form.

The movement of translation is ascertainable only by astronomical proofs, for the difference in velocity of the various parts of the globe, in virtue of this motion, is too slight to be sensible to us, or to produce any effect on terrestrial phenomena. When the circuits of other planets were known, men's minds were prepared for that of the earth,—the question then being whether the earth was in analogy with Venus, Mars, Jupiter, etc., or whether, while they continued their courses round the sun, the sun made a yearly circuit round the earth. Reason must declare, in such a case, that any uncertainty must arise simply from the position of the observer, who, placed on any other planet, would have doubted whether he was not the centre of the heavenly motions. Any observation of mere appearances must evidently go for nothing in this case; as appearances must be exactly the same,—the parallelism of its axis of rotation being unaltered,—whether the earth or the sun is in the ecliptic, and the other in the centre. The proofs must be derived from better testimony than mere appearances; and they so abound that we have only to choose among those which are presented by the whole range of the heavens. *Earth's translation.*

The phenomenon called the precession of the equinoxes was observed by Hipparchus, who was struck by the difference of two degrees which he observed between the longitudes of stars in his time and those which had been recorded a century and a half before. To account for such a phenomenon, successive astronomers imagined other heavens; a process that they repeated with regard to nutations, which was a phenomenon too minute for their observation. To account for it, on the supposition of the earth being stable, a third general movement of the whole heavens must be supposed. Newton indicated, and Bradley afterward proved, that very slight alterations in the parallelism of the earth's axis,—such alterations as must result from the influence of the sun, and yet more of the moon, upon the equatorial bulge,—precisely account for the perturbations which create such confusion under the ancient view of the earth's stability. The most unquestionable proof of all, however, is in that class of phenomena called the retrogradations and stations of the planets, which are perfectly explained by the annual circuit of our globe, and are otherwise quite incomprehensible. If two boats are gliding down a river, at different rates of speed, the one must appear to the other advancing, stationary, or retrograde, according as its own speed is smaller, equal, or greater. With regard to the heavenly bodies, their velocities and other circumstances are known to us, so that we can calculate what their courses ought to be to our eyes, on the supposition of our own *Precession of the equinoxes.* *Retrogradations and stations of the planets.*

annual movement. The appearances answering to our scientific expectation, the proof is practically complete. If the earth moves, the retrogradation of the larger planets ought to happen, as it does, when they are in opposition, and that of Venus and Mercury when they are in inferior conjunction. The regular occurrence of this coincidence was not even attempted to be explained by the ancients.

We have called these proofs practically complete; and they were held to be so by Copernican philosophers before the time of Kepler and Galileo: but our age is not satisfied without a more strict mathematical evidence, amounting to demonstration.

Aberration of light. The one demonstration on which modern science rests is that derived from the various phenomena of the aberration of light, which are quite incompatible with the stability of the globe. Roemer's observations of the satellites of Jupiter suggested to him the use of light as a measure of distance. Knowing what changes must be taking place at various distances from us in the heavens, and knowing the velocity of light, the variations in time at which the changes become visible to us will be a measure of our change of place and distance. For instance, the first satellite of Jupiter is eclipsed every forty-two hours and a half. The eclipse will take place in a shorter or a longer time than this to our eyes, in proportion as we are removed to the one side or the other of our mean distance from Jupiter, on account of the smaller or greater space that the light will have to travel through. By extending our observation, not only to the other satellites of Jupiter, but to those of Saturn and Uranus, we have obtained further verifications of the relation of our orbit to theirs, and also proof of the uniformity in the passage of light,—at least within our own system. If the earth were immoveable, we might have an error of time, with regard to distant stars, but not of place: but, by compounding the velocity of the earth in its orbit with that of light, which is about ten thousand times greater, we can calculate how far any star ought to appear to deviate from its position. This deviation is found not to exceed, at its *maximum*, twenty seconds in any direction; and therefore forty seconds is the greatest deviation which can appear in the position of any star in the course of the year. It was the striking periodicity of these deviations which led Bradley to seek for the true theory in the combination of the motion of the earth with that of light, and to work it out with the mathematical exactness permitted by modern science; and there is nothing in the case to prevent the direct application of the mathematical process to the visible phenomena. The result is an unquestionable demonstration of the annual movement of the earth, with which all the phenomena of the case precisely agree, and without which they could not exist.

It is evident that this knowledge of aberrations compels us to add another correction to those of refraction and parallax, and the same is the case with regard to the precession of the equinoxes and nutation. Thus, as science advances, the preparation of a phenomenon, observed with the best instruments, for scientific use, becomes a delicate and laborious operation.

These are the considerations which have led men to the knowledge of the double motion of the planet we inhabit. No other intellectual revolution has ever so thoroughly asserted the natural rectitude of the human mind, or so well shown the action of positive demonstration upon definitive opinions; for no other has had such obstacles to surmount. A very small number of philosophers, working apart, without any other social superiority than that which attends positive genius and real science, have overthrown, within two centuries, a doctrine as old as our intelligence, directly established upon the plainest and commonest appearances, intimately connected with the whole system of existing opinions, general interests, and dominant authorities, and supported moreover by human pride, powerful in the recesses of each individual mind. The whole system of theological belief rested on the notion that the entire universe was ordained for Man—a notion which appears truly absurd the moment it is seen that our globe is only a subaltern star,—not any centre whatever, but circulating in its place and season, among others, round the sun, whose inhabitants might, with more reason, claim the monopoly of a system which is itself scarcely perceptible in the universe. The notion of final causes and providential laws undergoes dissolution at the same time; for, the once clear and reasonable idea of the subordination of all things to the advantage of Man being exploded, no assignable purpose remains for such providential action. As the admission of the motion of the earth overthrows the whole theory founded on the human destination of the universe, it is no wonder that religious minds revolted from the great disclosure, and that the sacerdotal power maintained a bitter rage against its illustrious discoverer.

<small>Influence of scientific fact upon Opinion.</small>

The Positive philosophy never destroys a doctrine without instantly substituting a conviction, adequate to the needs of our human nature. If the vanity of Man was grievously humbled when science disabused him of his notion of his supreme importance in the universe, to this vanity at once succeeded a lofty sentiment of his true intellectual dignity, when he saw what means were in his power, under such difficulties as his position imposed upon him, for the discovery of such a truth as he had attained. Laplace has pointed this out, showing how to the fantastic and enervating notion of a universe arranged for Man has succeeded the sound and vivifying conception of Man discovering, by a positive exercise of his intelligence, the general laws of the world, so as to be able to modify them, for his own good, within certain limits. Which is the nobler lot? Which is most in harmony with our highest instincts? Which is the most stimulating to our faculties? And which is the most animating to our feelings?

One more remark suggested by these discoveries is that a clear distinction is for ever established between our system and the universe at large. The old notion of the universe as a single system was founded on the error of the stability of the earth as its centre. The discovery of the earth's revolution at once transported all the external stars to distances infinitely more considerable than the greatest planetary intervals, and has left no place for the idea of system at all, beyond the limits of our sun's influence. We do not know, more or less, and men will probably never know, whether the innumerable suns that we see compose a general system, or any number, large or small, of partial systems entirely independent of each other. The idea of the universe, therefore, is excluded from positive philosophy; and that philosophy is, strictly speaking, bounded by the limits of the solar system, in regard to definite results; and this circumscription is, as elsewhere, to be regarded as real progress. This restriction is further justified by the knowledge we have obtained of all really universal phenomena being essentially independent of the interior phenomena of our system, since the astronomical tables of the state of our system, prepared without reference to any other sun than our own, invariably coincide with the minutest direct observations. The theory of the earth's revolution has not as yet exerted its due influence on our views, and especially in regard to this last consideration. This is doubtless owing to the imperfections of our education, which keep back these high philosophical truths till even the best minds have been possessed with an opposite doctrine: so that the positive knowledge which they afterward attain, commonly does little more than modify and restrain the bad tendencies of their education, instead of ruling and guiding their highest faculties.

Kepler's Laws.

The first idea that occurs to us when we are once satisfied of the revolution of the earth is that our point of view ought henceforth to be the centre of the sun. This transformation of our observation is called the *annual parallax*, and follows the same rules as the diurnal parallax, allowance being made for the much greater distance. Whether our observations of the sidereal heavens are *geocentric* or *heliocentric*—from the middle of the earth or of the sun—is of no appreciable consequence; but within our system the annual parallax is of sensible importance. When, from the central point of view, the orbits of the planets are determined, we can proceed to that great aim and end of the science—the prevision of future conditions of the heavens at appointed times.

Annual parallax.

The earliest supposition was that the motions of the planets were uniform and circular. The ancients had a superstition, as their writings abundantly show, that the circle was the most perfect of all forms, and therefore the most suitable for the motion of such divine existences as the stars. The choice of the form was wise: they had to suppose some form, while that of the circle answered best to what they saw; and we ourselves

Circles.

now take it provisionally in forming the theory of a new star. But the superstitious attachment of the ancients to this form was a serious impediment to the advance of astronomy. For every deviation and new appearance a new circle was supposed, till all the simplicity of the original hypothesis was lost in a complication of epicycles. By the end of the sixteenth century the number of circles supposed necessary for the seven stars then known amounted to seventy-four, while Tycho Brahe was discovering more and more planetary movements for which these circles could not account. Thus it is that men cleave to old ideas and methods till they are utterly worn out, and proved beyond recall to be ineffectual, under all additions that can be made to them.

Then came Kepler, the first man for twenty centuries who had the courage to go back to the beginning, as if nothing had been done in the way of theory. He took for his materials the complete system of exact observations which were the result of the life of his illustrious precursor, Tycho Brahe. Notwithstanding the natural hardihood of his genius, his works reveal to us how strenuously he had to maintain his enthusiasm, in order to support the toils of so bold and difficult an enterprise—rational as it was. He chose the planet Mars for study; and it was a happy choice; because the marked eccentricity of that planet was most apt to suggest the true law of irregularity. Mercury is more eccentric still; but it does not admit of continuous observation. He discovered three great laws, which, extended from the case of Mars to that of all the other planets in our system, constitute the foundation of Celestial Mechanics. The first law regulates the velocity; the second determines the figure of the orbit; the third establishes harmony among all the planetary motions.   *Kepler.*   *His three laws.*

It had long been remarked that the angular velocity (that is, the larger or smaller angle described, in a given time, by its vector radius) of each planet increases constantly in proportion as the body approaches the centre of its motion; but the relation between the distance and the velocity remained wholly unknown. Kepler discovered it by comparing the maximum and minimum of these quantities, by which their relation became more sensible. He found that the angular velocities of Mars at its nearest and furthest distance from the sun were in inverse proportion to the squares of the corresponding distances. Another way of expressing this law is used by himself; that the area described in a given time by the vector radius of the planet is of a constant magnitude, though its form is variable: or, again, in other words, that the areas described increase in proportion to the times. Thus he destroyed the old notion of the uniformity of the planetary motions, and showed that the uniformity was not in the arcs described, but in the areas.   *First law.*

The second law was less difficult to discover, when once Kepler had surrendered his attachment to the   *Second law.*

circle. The next figure that presented itself must naturally be the elipse, which is the simplest form of closed curve, after the circle. The Greek geometers had advanced the abstract theory of this curve some way. Kepler could not long hesitate where to place the sun in it: it must be either in the centre or in one of the two foci. No mathematical labor was needed to show him that it could not be in the centre: and thus, in constructing elliptic orbits, Kepler was necessarily led to place the sun in the focus for all the planets at once. His hypothesis once formed, it was easy to verify it by comparison with observations, the first principles of the required calculations being laid down beforehand. The second law of Kepler, then, is that the planetary orbits are elliptical, having the sun for their common focus.

Third law. These two laws determined the course of each planet; but the movements of all round their common focus seemed to be purely arbitrary, till Kepler discovered his third law. Being distinguished by the most remarkable genius for analogy ever seen in man, Kepler sought, and successfully, to establish some kind of harmony among all these various movements. He spent much time in pursuing the old metaphysical ideas of certain mystic harmonies which must exist in the universe: but, beyond the general conception of harmony, he obtained no assistance from these vague notions. The ground on which he proceeded was, in fact, the observation of astronomers that the planetary revolutions are always slow in proportion to the extent of their orbits. If he had confined himself to this ground, this discovery would certainly not have occupied seventeen years of assiduous toil. At last his labor issued in the discovery that the squares of the times of the planetary revolutions are proportional to the cubes of their mean distances from the sun: a law which all subsequent observations have verified. One important result of this law is that we may determine the periodic times and mean distances of all the planets by any one. By it, for instance, we have determined the duration of the year of Uranus, when once we knew its distance from the sun; and, conversely, if we discovered a new planet very near the sun, we need only observe its short revolution, to be able to calculate its distance, which, in that position, we could not effect by other means. Astronomers are every day using this double facility, afforded them by Kepler's third law.

These are the three laws which will for ever constitute the basis of celestial geometry, in regard to planetary motions. They answer for all the bodies in our system, regulating the satellites, by placing the origin of areas and the focus of the ellipse in the centre of the respective planets. Since Kepler's time, the number of bodies in our system has more than trebled; and all have in turn verified these laws. By them, motions of translation require for their determination nothing more than a simple geometrical problem, which demands from direct observation only a certain number of data—six for each planet. And thus is a perfectly logical character given to astronomy.

The application of these laws, restricted to our own system, is naturally divided into three problems;—the problem of the planets; that of the satellites; and that of the comets. These are the three general cases of our system; and, by the application to them of Kepler's laws, we may assign to every body within the system, its precise position, in all time past and all time to come; and thence again, we can exhibit all the secondary phenomena, past and future, which must result from such relative positions. The next striking fact of this kind to the general mind is the prediction of eclipses, absolutely conclusive as it is, with regard to the accuracy of our geometrical knowledge. This kind of prediction, quite apart from the vague prophesying of ancient times, when eclipses occurred, as they do now, necessarily from the planetary orbits being all closed curves, and which men's experience told them must return—began in the immortal school of Alexandria; and its degrees of precision, to the hour, then to the minute, then to the second, faithfully represents the great historical phases of the gradual perfecting of celestial geometry. It is this which will, apart from all other considerations, for ever make the observation of eclipses a spectacle as interesting for philosophers as for the public, and on grounds which the spread of the positive spirit will render, we may hope, more and more analogous, though unequally energetic.

Three problems.

Prediction of Eclipses.

We are learning to make more use of this class of phenomena, and to make out new uses from them, as time goes on. Independently of their practical utility in regard to the great problem of the longitudes, they have been found, within a century, very important in determining with more exactitude the distance of the sun from our earth. Whether it be an eclipse by the moon, or the transit of Venus or Mercury, the difference in duration of the phenomenon, observed in different parts of the earth, will furnish the relative parallax of that body and the sun, and consequently the distance of the sun itself. Some bodies are more fit than others for this experiment, certain conditions being necessary, which are not common to all. Of the three known bodies which can pass between us and the sun, two—the Moon and Mercury—are excluded by these conditions; and there remains only Venus. Halley taught us how to conduct and use the observation. The parallax, in such a position, offers suitable proportions, being nearly three times that of the sun; and the angular velocity is small enough to allow the phenomenon (lasting from six to eight hours) to present differences of at least twenty minutes between well-chosen observatories. I have specified this case, on account of its extreme importance to the whole system of astronomical science; but it would be quitting our object and plan to notice any other secondary cases.

Transit of Venus.

I must remark upon one very striking truth which becomes apparent during the pursuit of astronomical science;—its distinct and ever-increasing opposition as it attains a higher perfection to the theological and metaphysical spirit. Theological philosophy supposes everything to be governed by will; and that phenomena are therefore eminently variable and irregular,—at least virtually. The Positive philosophy, on the contrary, conceives of them as subjected to invariable laws, which permit us to predict with absolute precision. The radical incompatibility of these two views is nowhere more marked than in regard to the phenomena of the heavens; since, in that direction, our prevision is proved to be perfect. The punctual arrival of comets and eclipses, with all their train of minute incidents, exactly foretold, long before, by the aid of ascertained laws, must lead the common mind to feel that such events must be free from the control of any will, which could not be will if it was thus subordinated to our astronomical decisions.

Foundation of Celestial Mechanics. The three laws of Kepler form the foundation of the higher conception to which we are next to pass on; the mechanical theory of astronomical phenomena. By this ulterior study, we obtain new determinations; but a more important office of the Mechanical theory is to perfect celestial geometry itself, by giving more precision to its theories, and establishing a sublime connection among all the parts of our solar system, without exception. The laws of Kepler, inestimable as they are, have come to be regarded as a sort of approximation,—supposing, as they do, various elements to be constant, while they are subject to more or less alteration. The exact knowledge of the laws of these variations constitutes the principal astronomical result of celestial mechanics, independently of its own high philosophical importance.

SECTION II.

DYNAMICAL PHENOMENA.

Gravitation.

Character of laws of Motion. The laws of Motion, more difficult to discover than those of extension, and later in being discovered, are quite as certain, universal, and positive in character: and of course it is the same with their application. Every curvilinear displacement of any kind of body,—of a star as well as a cannon-ball,—may be studied under the two points of view which are equally mathematical: geometrically, in determining by direct observation the form of the trajectory and the law by which its velocity varies, as Kepler did with the heavenly bodies; and mechanically, by seeking the law of motion which prevents the body from pursuing its natural straight course, and which, combined with its actual velocity, makes it describe its trajectory, which may henceforth be known *à priori*. These inquiries are evidently equally positive, and in like manner founded upon phenomena. If we find still in use some terms which seem to relate to the nature and cause of motion, they are only vestiges of a mode of thinking long gone by; and they do not affect the positive character of the research.

The two motions which constitute the course of the cannon-ball are perfectly known to us beforehand; but we have not the geometrical knowledge of its trajectory. With regard to the star, our knowledge of its trajectory compensates exactly for the difficulty of our preliminary ignorance about its elementary motions. If the law of the fall of weights had not been directly established, we should have learned it, indirectly, but no less surely, from the observation of the curvilinear motions produced by weight.

Celestial Mechanics was then founded on a firm basis, when through Kepler's laws, and by the rules of rational Dynamics, discovery was made of the law of direction and intensity of the force which must act upon the planet to divert it from the tangent which it would naturally describe. This fundamental law once discovered, all astronomical researches enter into the domain of Mechanics, in which the motions of bodies are calculated from the forces which impel them. This was the course philosophically and perseveringly pursued by Newton. *Their History.*

It does not detract from Newton's merits that Kepler had some foresight of the results of his great laws. He carried their dynamic interpretation as far as the science of his day permitted; and, seeking for what could not yet be found, he wandered off among fantasies. The true precursors of Newton, as founders of dynamics, were Huyghens and Galileo,—especially the last: yet history tells of no such succession of philosophical efforts as in the case of Kepler, who, after constituting celestial geometry, strove to pursue that science of celestial mechanics which was, by its nature, reserved for a future generation. As the means were wanting, he failed; but the example is not the less remarkable.

The first of Kepler's laws proves that the accelerating force of each planet is constantly directed toward the sun. The accelerating force, however great it may be supposed, does not at all affect the magnitude of the area which would be described in a given time by the vector radius of the planet, in virtue of its velocity, if its direction passes exactly through the sun, while it would inevitably change it on any other supposition. Thus, the permanence of this area,—the first general datum of observation,—discloses the law of direction. The great difficulty of the problem, gloriously solved by Newton, lies in the discovery, by means of Kepler's other two theorems, of the law of the intensity of this action, which we speak of as exercised by the sun on the planets.

When Newton began to work on this conception, he took Kepler's third law as his basis, supposing the orbits, as he might do for such a purpose, to be circular and uniform. The solar action, equal, and opposed to the centrifugal force of the planet, thus became necessarily constant at the different points of the orbit, and could not vary but in passing from one planet to another. This variation between one planet and another was provided for by the theorems of

Huyghens relating to the centrifugal force in the circle. This force being in proportion to the relation between the radius of the orbit and the square of the periodic time, must vary from one star to another inversely to the square of its distance from the sun, in virtue of the permanence which Kepler showed to exist of the relation between the cube of this distance and this same square of the periodic time, for all the planets. It was this mathematical consideration which put Newton in the way of his great discovery, and not any metaphysical reasonings, such as prevailed before it, and which probably never entered his mind, one way or another.

There remained the difficulty of explaining how this law of the variation of the solar action agreed with the geometrical nature of the orbits, as exhibited by Kepler. The elliptical orbit presented two remarkable points,—the aphelion and the perihelion, in which the centrifugal force was directly opposed to the action of the sun, and consequently equal to it; and the change in this action there must be at the same time more marked. The curve of the orbit was evidently identical at these two points; the action then had simply to be measured, according to Huyghens' theorems, by the square of the corresponding velocity. Thence, it was easily deduced, from Kepler's first law, that the decrease of the solar action, from the perihelion to the aphelion, must be inversely to the square of the distance. Here was a full confirmation of the law which related to the different planets by an exact comparison between the two principal positions of each of them. Still, however, the elliptical motion had not been considered. Any other curve would, thus far, have served as well as the ellipse, provided its two extremities had shown an equal curvature. The remaining portion of the demonstration,—the measurement of the solar action throughout the extent of the orbit,—is to be obtained only by transcendental analysis. The process is necessary for carrying on the comparison of the solar action and the centrifugal force; and the theory of the curvature of the ellipse is required. Huyghens made a near approach to the principle of this great process; but it could not be completed without the aid of the differential analysis, of which Newton was the inventor, as well as Leibnitz. By the aid of this analysis, the force of the solar action in all parts of the orbit is easily esti-

Newton's demonstration. mated, in various ways; and it is found to vary inversely to the square of the distance, and that it is independent of the direction. Furthermore, the same method shows, in accordance with Kepler's third law, that the action varies in proportion to distance alone: so that the sun acts upon all the planets alike, whatever may be their dimensions, their distance only being the circumstance to be considered. Thus Newton completed his demonstration of the fundamental law that the solar action is, in every case, proportionate, at the same distance, to the mass of the planet; in the same way that, by the identity of the fall of all terrestrial bodies in a vacuum, or by the precise coincidence of their oscillations, proof had already been obtained of the proportion between their weight and their masses. We thus see how the three laws of Kepler have concurred in establishing, according to the rules of rational mechanics, this fundamental law of nature. The first shows the tendency of all the planets toward the sun; the second shows that this tendency, the same in every direction, changes with the distance from the sun, inversely to its square; and the third teaches that this action is always simply proportionate, the distance being equal, to the mass of each planet. In accordance with the laws of Kepler, which relate to the whole interior of our system, the same theory applies to the connection between the satellites and their planets.

Newton thought it necessary to complete his demonstration by presenting it in an inverse manner; that is, by determining *à priori* the planetary motions which must result from such a dynamic law. The process brought him back, as it must do, to Kepler's laws. Besides furnishing some means of simplifying the study of these motions, this labor proved that, whereas, by Kepler's laws, the orbit might have had more figures than one, the ellipse was the only one possible under the Newtonian law.

It was once a great perplexity to some people, which others could not satisfactorily explain, that when the planet is travelling toward its aphelion we can not say that it tends toward the sun. But the difficulty arose out of the use of inappropriate language. The question is, not whether the planet is nearer to the sun than it lately was, but whether it is nearer than it would have been without the force that sends it forward. It is always tending toward the sun to the utmost that is allowed by the other force to which it is subjected. The orbit is always concave toward the sun; and it would evidently have been insurmountable if the trajectory could have been convex. In the same way, when a bomb ascends, its weight is not suspended or reversed: it always tends toward the earth, and is, in fact, falling toward it more rapidly every moment, even if ascending, because it is every moment farther below the point at which it would have been but for the action of the earth upon it; and its trajectory is always concave to the ground.

Old difficulty explained.

I have thus far carefully avoided giving any name to the tendency of the planets toward the sun, and of the satellites toward the planets. To call it *attraction* would be misleading; and we, in truth, can know nothing of its nature. All that we know is that these bodies are connected, and that their effect upon each other is mathematically calculable. It is by quite another property of Newton's great discovery that this effect is *explained*, in the true sense of the word,—that is, comprehended from its conformity with the ordinary phenomena which gravity continually produces on the surface of our globe. Let us now see what this property of the discovery is.

Term Attraction inadmissible.

We owe a great deal to the moon. If the earth had no satellite, we might calculate the celestial motions by the rules of dynamics, but we could not connect them with those which are under our immediate observation. It is the moon which affords this connection by enabling us to establish the identity of its tendency toward the earth with weight, properly so called: and from this knowledge, we have risen to the view that the mutual action of the heavenly bodies is nothing else than weight properly generalized: or, putting it the other way, that weight is only a particular case of the general action. The case of the moon is susceptible of the most precise testing. The data are known; and by dynamical analysis, the intensity of the action of the earth upon the moon is exactly ascertainable. We have only to suppose the moon close to the earth, with the due increase of this intensity, inversely to the square of the distance, and compare it with the intensity of weight on the earth, as manifest to us by the fall of bodies, or by the pendulum. A coincidence between the two amounts to proof; and we have, in fact, mathematical demonstration of it. It was in pursuing this method of proof that Newton evinced that philosophical severity which we find so interesting in the anecdote of his long delay, because he could not establish the coincidence, while confident that he had discovered the fact. He failed for want of an accurate measurement of a degree on the earth's surface; and he put aside this important part of his great conception till Picard's measurement of the earth enabled him to establish his demonstration.

Extent of the demonstration. The identity of weight and the moon's tendency toward the earth places the whole of celestial mechanics in a new light. It shows us the motions of the stars as exactly like that of projectiles which we have under our immediate observation. If we could start our projectiles with a sufficient and continuous force, we should, except for the resistance of the air, find them the models of the planetary system: or, in other words, astronomy has become to us an artillery problem, simplified by the absence of a resisting medium, but complicated by the variety and plurality of weights.—If our observation of weight on our globe has helped us to a knowledge of planetary relations, our celestial observations have in turn taught us the law of the variation of weight, imperceptible in terrestrial phenomena. Men had always conceived weight to be an inalterable property of bodies, finding that no metamorphosis,—not even from life to death,—made any change in the weight of a body, while it remained entire. This was the one particular in which men might suppose they had found the Absolute. In a moment, the Newtonian demonstration overthrew this fast-rooted notion, and showed that weight was a relative quality,—not under the circumstances in which it had hitherto been observed, but under the new one,—the position of the observed body in the system,—its distance from the centre of the earth. The human mind could hardly have sought out this fact directly: but, once revealed in the course of astronomical study, the verification easily followed; and experiments on our own globe, in the vertical direction, and yet more in the horizontal, have established the reality of the law, by experiments too delicate, from the necessity of the case, to be appreciable, if we had not known beforehand what differences must be found to exist.

It is to express briefly the identity between weight and the accelerating force of the planets that the happy term *Gravitation* has been devised. This term has every merit. It expresses a simple fact, without any reference to the nature or cause of this universal action. It affords the only explanation which positive science admits; that is, the connection between certain less known facts and other better known facts. Since the creation of this term, there has been no excuse for the continued use of the word *attraction*. It is desirable to avoid pedantry in language; but it is of high importance to preserve pure the positive character of so fundamental a conception as this, by using a term which expresses exactly what we know, and dismissing one which assumes what is purely fanciful, and wholly incorrect. Attraction is *a drawing toward*. Now, when we draw anything toward us, the distance is of no importance: the same force draws the same body with equal ease three feet or thirty feet, which is directly contradictory to the facts of gravitation. Our business is with the fact of the action, and not at all with its nature. It was the use of this metaphysical term, it now appears, which occasioned the opposition that the Newtonian theory encountered so long, and especially in France. Descartes had, by laborious efforts, banished the notions of occult qualities, which he perceived to be so fatal to science; and in this theory of attraction, his followers saw a falling back into the old metaphysical delusions. We perceive this in the writings of John Bernouilli and Fontenelle: and it appears that the clear and positive scientific intellect of France did good service in stripping off from the sublime discovery of Newton the metaphysical appearance which obscured its reality for a time.

Term Gravitation unobjectionable.

One more consideration remains to be adverted to. We have regarded the heavenly bodies thus far as points, without reference to their forms and dimensions. But as it is proved that the intensity of the action of the sun on the planets, and of the planets on their satellites, is proportioned to the mass of the body acted upon, it is clear that the force operates directly only on molecules, which are all independently affected by it; and equally, their distance being the same. The gravitation of molecules is therefore the only real one; and that of masses is simply its mathematical result. In the mathematical study of motions, however, it is necessary to have a conception of a single force, instead of such an infinity of elementary actions: and hence arises that preliminary part of celestial mechanics which consists in compounding in one result all the mutual gravitation of the molecules of two stars. Newton founded this portion, with all the rest; and the two theorems which he established for the purpose still remain the commonest expression of this important theory.

Gravitation is that of molecules.

He proved that if the stars were truly spherical, and their strata were homogeneous, the gravitation of their particles would be so balanced that the bodies might be treated as points, in the study of their motions of translation. But the irregularity of their forms, however slight, must be considered in the theory of their rotations, to which these theorems cease to be applicable. For any other form than the sphere, the problem becomes very complicated; and the analytical difficulties can be surmounted only by approximation, notwithstanding all the perfections introduced into the theory in recent times. And unless we could also learn what is the law of density in the interior of the stars,—a kind of knowledge which seems to be for ever beyond our reach,—we can not attain a perfect solution.

Secondary gravitation. The fundamental law of Rational Mechanics, which declares the necessary equality of action and reaction, shows that gravitation must be mutual,—that the sun must tend toward the planets, and the planets toward their satellites. The extreme inequality of the masses renders the ascertainment of the inverse gravitation extremely difficult; yet its reality is established by various secondary phenomena. The gravitation of the planets toward each other is a necessary part of the whole conception; but it was not mathematically demonstrated till Newton's successors deduced from it an exact explanation of the perturbations observed in the principal motions of the planets. Their labors have established secondary gravitation as positively as the primary.

Thus has every kind of proof concurred to establish that great fundamental law which is the noblest result of our aggregate studies of nature. All the molecules of our system gravitate toward each other, in proportion to their masses, and inversely to the squares of their distances.

Domain of the law. I dare not, as many do, confidently extend the application of this law to the entire universe. There can be no objection to entertaining it analogically till we obtain some knowledge of the mechanism of the sidereal heavens; but we must remember that we have not yet that knowledge, and that we can not promise ourselves that we ever shall. Without the phenomena of our own system, the theory of its motions would be only an intellectual exercise and sport: there can be no positive science apart from phenomena, and of the phenomena of the universe beyond our own system we are not in scientific possession. It must be understood that I advocate simply a suspension of judgment where there is no ground for either affirmation or denial. I merely desire to keep in view that all our positive knowledge is relative; and, in my dread of our resting in notions of anything absolute, I would venture to say that I can conceive of such a thing

---

* M. Comte omits here all notice of such positive applications as we are able to make in Sidereal astronomy. He takes no notice of the fact that the motion of the multiple stars in elliptical orbits, and in accordance with Kepler's law of the velocities, demonstrates the existence of a law of force, according to the inverse square of the distance. —J. P. N.

as even our theory of gravitation being hereafter superseded. I do not think it probable; and the fact will ever remain that it answers completely to our present needs. It sustains us, up to the last point of precision that we can attain. If a future generation should reach a greater, and feel, in consequence, a need to construct a new law of gravitation, it will be as true as it now is that the Newtonian theory is, in the midst of inevitable variations, stable enough to give steadiness and confidence to our understandings. It will appear hereafter how inestimable this theory is in the interpretation of the phenomena of the interior of our system. We already see how much we owe to it, apart from all specific knowledge which it has given us, in the advancement of our philosophical progress, and of the general education of human reason. Descartes could not rise to a mechanical conception of general phenomena without occupying himself with a baseless hypothesis about their mode of production. This was, doubtless, a necessary process of transition from the old notions of the absolute to the positive view; but too long a continuance in this stage would have seriously impeded human progress. The Newtonian discovery set us forward in the true positive direction. It retains Descartes' fundamental idea of a Mechanism, but casts aside all inquiry into its origin and mode of production. It shows practically how, without attempting to penetrate into the essence of phenomena, we may connect and assimilate them, so as to attain, with precision and certainty, the true end of our studies,—that exact prevision of events which *à-priori* conceptions are necessarily unable to supply.

CHAPTER IV.

CELESTIAL STATICS.

Kepler's laws connected celestial phenomena to a certain degree, before Newton's theory was propounded: but they left this imperfection,—that phenomena which ranked under two of these laws had no necessary connection with each other. Newton brought under one head all the three classes of general facts, uniting them in one more general still; and since that time we have been able to perceive exactly the relation between any two of the phenomena which are all connected with the common theory. As far as we can see, there is nothing more to gain in this direction. *Consummation by Newton.*

We have seen what this great conception is in itself. We have now to observe its application to the mathematical explanation of celestial phenomena, and the perfecting of their study. For this purpose, we will recur to our former division of subjects, and contemplate the phenomena of planets as *Statical considerations.*

immoveable first, and of planets in motion afterward; the statical phenomena first, and the dynamical afterward.

To know the mutual gravitation of the heavenly bodies, we must know their masses. Such knowledge once appeared inaccessible from its very nature; but the Newtonian theory has put it within our power, and furnished us with a wholly new set of ideas about these bodies. There are three ways in which the inquiry has been prosecuted, all differing from each other, both in generality and in simplicity. The first method, the most general, the only one in fact which is applicable to all cases, is the most difficult. It consists in analyzing the special share of each body in the perturbations observed in the principal motions of another,—both of translation and rotation. Here two elements are concerned,—the distance, and the mass of the star in question. The first is well known, the other is not; and only an approximate determination is possible. It is difficult to apportion the shares in the action; and geometers place little dependence on the computation of masses obtained by this method, in comparison with that obtained by either of the others.

First method of inquiry into masses.

Nearest in generality to this first method is that which Newton employed with regard to planets that had a satellite; that of comparing the motion of the satellite round the planet with that of the planet round the sun. The law which determines the action by the distance being compared, in its results, in the two cases, gives the relation of the masses of the sun and the planet. The mass of Jupiter, determined by Newton in this way, has undergone little change of statement by methods since employed; and what difference there is is almost wholly owing to the data of the process being now better known.

Second method.

The third method is the most direct and simple of all; but it is the most restricted, as it is necessarily confined to the planet inhabited by the observer. It consists in estimating the relative masses by the comparison of the weights which they produce. If we knew the mass of any planet, we should know what would be the weight of things on its surface, or at a given distance; and reciprocally, the weight being known, we are able to estimate the mass. With the pendulum we have measured terrestrial weight with absolute precision; and, diminishing it, inversely to the square of the distance, we shall know its value at the distance of the sun. We have then only to compare it with the amount, before well known, which expresses the sun's action upon the earth, to find immediately the relation of the mass of the earth to that of the sun. With regard to every other planet, on the contrary, it must be the estimate of its mass which would yield that of its corresponding gravity. All these methods being practicable in the case of the earth, its mass, in comparison with that of the sun, must be considered the best known of all within our system. The mass of the moon, and that of Jupiter, are now estimated almost as perfectly; and those of Saturn and Uranus come next. We are

Third method.

less sure about the other three which have been calculated—Mercury, Venus, and Mars; though the uncertainty about them can not be very great. Of the telescopic planets and the comets we know scarcely anything, owing to their extreme smallness, which precludes their exerting any sensible influence on perturbations. Comets pass, during their prodigious course, near very small stars, such as the satellites of Jupiter and Saturn, without producing any perceptible derangement. As for the satellites, we have no knowledge except of the moon, and approximately, of those of Jupiter. No comparison of results has as yet exhibited any harmony whatever between them. The only essential circumstance which they present is the vast superiority of the size of the sun to the whole contents of the system. Those entire contents, if thrown together, would scarcely amount to a thousandth part of the mass of the sun. Looking abroad from the sun, we see alternating, without any visible order, here decreasing, there increasing masses. We might have supposed, *à priori*, as Kepler did, that the masses were regularly connected with the volumes (which are themselves irregular, however), so that the mean densities should be continually less in mathematical proportion to their distances from the sun. But, independently of this numerical law, which is never exactly observed, the simple fact of the decrease of density presents some exceptions, in regard to Uranus, among others. No rational ground can be assigned for this.

SECTION I.

WEIGHT OF THE EARTH.

These are the means by which the masses of the bodies of our system are ascertained. The remaining process is to bring them into relation with our estimates of weight by ascertaining the total weight of the earth. Bouguer was the first who distinctly perceived the possibility of such an estimate, during his scientific expedition to Peru, when he found that the neighborhood of vast mountains slightly affected the direction of weight. We see how, in accordance with the law of gravitation, a considerable mass, regarded as condensed in its centre of gravity, may affect the plumb-line, however slightly, if it be brought close enough, subjecting it to a secondary gravitation, which affords data for a comparison between the action of the earth and that of the mountain. By this, some estimate may be formed of the proportion of the mountain to the globe. In the time of Bouguer science was not advanced enough to admit of more than the conception of how the thing could be done. Half a century later, Maskelyne observed the mountain Schehallion in Scotland, and found that it occasioned an alteration in the natural direction of weight of from five to six seconds; and Hutton deduced from this that the weight of the earth is equal to four and a half times that of a similar volume of distilled water at its maximum of density. Anything like exactness, however, is out of the question while there must be so much uncertainty about the weight of the mountain, which can be calculated only from its volume.

When Coulumb had invented his Torsion Balance, intended to measure the smallest forces, Cavendish saw how the earth might be weighed by comparing it, by means of this balance, with artificial masses which might be computed. By his immortal experiment, he discovered the mean density of our globe to be five and a half times equal to that of water; whence we can, if we think proper, deduce the weight of the earth in cwts. and tons.—We thus obtain, among other advantages, some insight into the constitution of our globe, which by its positivity, puts to flight many fanciful notions. The density of the parts near the surface is so far below the average—water occupying much space, for instance—that the density nearer the centre must be much above the average. This is in accordance with the indications of Celestial Mechanics; and it furnishes us with one condition of the interior of the globe. There can be no void there. What there is we know not, further than that it must be something consistent with the condition of superior density.

SECTION II.

FORM OF THE PLANETS.

Form of the planets. The next great statical inquiry relates to the form of the heavenly bodies, as deduced from the theory of their equilibrium.

Geometers suppose the planetary bodies to have been originally fluid, because their equilibrium can thus consist with only one form; whereas, if they had been always solid, as our earth is now, their equilibrium might have been compatible with any form whatever. Several phenomena indicate this supposition, and it agrees remarkbly with the whole of our direct observations.

If the planets had no motion of rotation, their being perfectly spherical would accord with the equilibrium of their molecules: but the centrifugal force engendered by the rotation must necessarily modify the primitive form, by altering, more or less, the direction, or the intensity of weight, properly so called. Huyghens established this with regard to the direction, and Newton with regard to the intensity. We thus become easily assured of the general fact of the nearly spherical form of all the planets, and of their being slightly flattened at the poles: but, when we go further, and attempt to estimate their forms mathematically, and learn the precise degree of the flattening at the poles, the question becomes one of transcendental analysis, and is involved in difficulty which can never be entirely surmounted. The inquiry involves a sort of vicious circle, which does not admit of a logical issue. In order to form an equation of the surface, we ought, by the law of equilibrium of fluids, to know the weight of the molecules concerned; whereas, by the law of gravitation, this can be ascertained only through the knowledge of the form of the planet, and even of the mode of variation of its interior density. All that can be done is to discover whether the proposed form fulfils such and such conditions. Maclaurin discovered a theorem, highly valued by geometers, which has become the basis of all our inquiries on this subject, and which shows that the ellipsoid of revolution precisely fulfils the conditions of equilibrium. But this supposes the structure of the body to be homogeneous; which it is not, in any case. The labors of geometers have however brought within very narrow limits the possible variations of the polar flattening. The result with regard to the earth is that the mathematical rule perfectly agrees with direct observation.

In the case of the planets we have another resource. *Estimate from perturbations.* Their flattening affects certain phenomena of perturbation, by the study of which we obtain materials for an estimate. Altogether, the calculations and measurements agree more closely than we could have ventured to hope. The only case which seems to present a real exception is that of Mars, which, by its magnitude, its mass, and the time of its rotation, should be little more flattened than the earth; whereas, if the observations of Herschel are exact, it is almost as much so as Jupiter.—We must observe, moreover, that though, as Maclaurin has shown, equilibrium is compatible with the ellipsoid form, this form is not to be supposed the only one:—witness, in our own system, the rings of Saturn, which are a remarkable example to the contrary: and Laplace has demonstrated how these rings could, even in a fluid state, be in equilibrium.

The most useful consequence of the mathematical theory of the planetary forms is that it has established an important relation between the value of the *Indirect estimate of the earth's form.* different degrees on the earth's surface and the intensity of the corresponding gravity, measured by the length of the seconds pendulum in different latitudes. We can thus, with great ease multiply our indirect observations about the form of our globe; whereas the geometrical estimate of degrees is a long and laborious operation, which can not be often repeated with due care. But, generally speaking, the more indirect a measurement is, *cæteris paribus*, the more uncertain it is: and there remains the uncertainty arising from our ignorance of the law of interior density in our earth; so that our chief reliance should still be on mathematical measurement, conducted with due care.

An interesting question belonging to the hydrostatic theory of the planetary forms is of the conditions of stability of equilibrium of the fluids which are collected *Hydrostatic theory of planetary forms.* on a part or the whole of the surface of the planets. Laplace shows this stability to depend, under all circumstances, on the density of the fluid being less than the mean density of the planet; a view established with regard to the earth by Cavendish's fine experiment.

SECTION III.

THE TIDES.

Question of the tides. THERE remains the question of the tides—the last important inquiry under the head of celestial statics. Under the astronomical point of view, this is evidently a statical question—the earth being, in that view, regarded as motionless; and it is not less a statical question in a mathematical view, because what we are looking at is the figure of the ocean during periods of equilibrium, without thinking of the motions which produced that equilibrium. Moreover, this inquiry naturally belongs to the study of the planetary forms.

A particular interest attaches to this question, from its being the link between celestial and terrestrial physics—the celestial explanation of a great terrestrial phenomenon. Descartes did much for us in establishing this. He failed to explain the phenomenon; but he cast aside the metaphysical conceptions which had prevailed before, and showed that there was a connection between the change of the tides and the motions of the moon; and this certainly helped to put Newton in the way of the true theory. As soon as it was known that the cause of the tides was to be looked for in the sky, the theory of gravitation was certain to afford its true explanation. Newton, therefore, gave out the simple principle that the unequal gravitation of the different parts of the ocean toward any one of the bodies of our system, and particularly toward the sun and moon, was the cause of the tides; and Daniel Bernouilli afterward perfected the theory. The same theory answers for the atmosphere; but we had better study it in the case of the seas alone; on account of the uncertainty of our knowledge of the vast gaseous covering of our globe, whose diffused mass almost defies precise observation.

Theory of the tides. Suppose the earth joined to any heavenly body by a line passing through the earth's centre. It is clear that the point of the earth's surface which is nearest the other body will gravitate toward it more, and the remoter point less, than the centre, inversely to the squares of their respective distances. The first point tends away from the centre: and the centre tends away from the second point; and in each case the fluid surface must rise; and in nearly the same degree in both cases. The effect must diminish in proportion to the distance from these points in any direction: and at a distance of ninety degrees it ceases. But there the level of the waters must be lowered because of the exhaustion in that place caused by the overflow elsewhere. And here enters a new consideration, difficult to manage:—the changes in the terrestrial gravity of the waters, occasioned by their changes of level. Thus the action of any heavenly body causes the ocean to assume the form of a spheroid, elongating it in the direction of that body. Newton calculated the chief part of the phenomenon of the tides on the supposition of an ellipsoid of homogeneous structure, as he had done in estimating the effect of the centrifugal force on the earth's figure, substituting for the centrifugal force the difference between the gravitation of the centre of the globe and that of its surface next the proposed body. After that, Maclaurin's theorem served Daniel Bernouilli for a basis of an exact theory of the tides.

Thus far, we have regarded the tides only as if they were a fixed accumulation of waters under the proposed star. This is the mathematical basis of the whole question; but the most striking part has yet to be considered—the periodical rise and fall. It is the diurnal motion of our globe which causes this rise and fall, by carrying the waters successively into all the positions in which the other body can raise or depress them. Hence arise the four nearly equal periodical alternations, when the two greatest elevations take place during the two passages of the heavenly body over the meridian of the place, and the lower levels at its rising and setting; the total period being precisely fixed by combining the terrestrial rotation with the proper daily movement of the heavenly body. The last indispensable element of the question is the valuation of the powers of the different heavenly bodies. This calculation is easily made from the difference between the gravitation of the centre of our globe and that of the extreme points of its surface next the observed body. Guided by the law of gravitation, we can determine which, among all the bodies of our system, are those which can participate in the phenomenon, and what is the share taken by each. We thus find that the sun by its immense mass, *Influence of the sun.* and the moon by its proximity, are the only ones which produce any appreciable tides: that the action of the moon is from two and a half to three times more powerful than that of the sun; and that, consequently, when they act in opposite directions, that of the moon prevails; which explains the primary observation of Descartes about the coincidence of the tidal period with the lunar day. *Of the moon.*

Thus far, we have considered only the effect of a single heavenly body upon the tides; that is, the case of a simple and abstract tide. *Composite influence.* The complication is very great, when the action of two such bodies has to be considered. But the resources of science are sufficient to meet this case—even deriving from it new means of estimating the mass of the sun and moon;— and also of calculating the modifications arising out of the various distances of the earth from either body; and again, of tracing the changes of direction caused by the diurnal movement of the proposed body—whether in accordance with the earth's axis of rotation, or parallel with the equator, which makes the difference between the tides of our equinoctial and solstitial lunar months. As for the difference of the phenomenon in various climates, the consideration of latitude is the only one which affords much result. At the poles, there can of course be no other tides than such as are caused by the flux and reflux of waters elsewhere, as the earth has no rotation there. The equator must exhibit the tides at their extremes, not only on account of the diminished gravitation there, but yet more on account of the more complete diversity of the successive positions occupied by the waters during the daily rotation. Elsewhere the greatness of the tide may vary in proportion to the force of the rotation.

Requisites for exactitude. The mathematical theory of the tides accords with direct observation to a degree which is really wonderful, considering how many hypotheses geometers must have recourse to, to make the questions calculable at all, and how many inaccessible data would be required to make an estimate thoroughly logical. It would not even be enough to know the extent and form of the bed of the ocean. Something beyond that in difficulty is required—the true law of density, in the interior of the earth, as with regard to the figure of the planets. We ought to know, too, whether the interior strata are solid or fluid, in order to know whether they participate in tidal phenomena, and whether they therefore modify those at the surface or not. These considerations show the soundness of the advice given by one who was full of the true mathematical spirit, consisting above all in the relation of the concrete to the abstract, Daniel Bernouilli, who recommended geometers "not to urge too far the results of formulas, for fear of drawing conclusions contrary to truth."

The comparison between mathematical theory and direct observation has never been carried out to any advantage—all the measurements having been taken in the ports, or near the shore. The tides in such places are very indirect; and they can not properly represent the regular tides from which they issue, their force being chiefly determined by the form of the soil—at the bottom as well as on the surface—and even perhaps affected by its structure. These are incidents which can not enter into mathematical estimates; and to them we must doubtless refer the vast differences in the height of the tides at the same time, and in nearly the same place—as, for instance, the tides of Bristol and Liverpool, of Granville and Dieppe. The only way of making an effectual direct observation would be to note the phenomena of the tides in a very small island, at the equator, and thirty degrees at least from any continent, for such a course of years as would allow of repeated record of variations, as repeatedly foreseen. In this way, and in no other, might the mathematical theory of the tides be verified and perfected.

Whatever may be the uncertainty with regard to some of the data of this great theory, it has that conclusive sanction,—the fulfilment of its previsions;—a fulfilment so exact as to guide our conduct; and this, as we know, is the true end of all science. The principal local circumstances, except the winds, being calculable, it has been found practicable to assign for each port the mean height of the tides and their times; and thus have mathematical determinations been proved to be sufficiently conformable to reality, and a class of phenomena which, a century ago, were regarded as inexplicable, have been referred to invariable laws, and shown to be as little arbitrary as anything else.

Such are the philosophical characteristics of the three great questions which compose the statical department of Celestial Mechanics. We must next look into the dynamical apartment, as represented by the phenomena of our system.

CHAPTER V.

CELESTIAL DYNAMICS.

Perturbations.

THE principal motion of the planets is, as we have seen, determined by the gravitation of each of them toward the focus of its orbit. The regularity of this movement must be impaired by the mutual gravitation of the bodies of the system. The most striking of those derangements were observed by the School of Alexandria, in the first days of Mathematical Astronomy; others have been observed, in proportion as our knowledge became more precise; and now, all are explained with such completeness by the theory of gravitation, that the smallest perturbations are known before they are observed. This is the last possible test and triumph of the Newtonian system.

There are, as Lagrange pointed out, two principal kinds of perturbations, which differ as much in their mathematical theory as in the circumstances which constitute them; instantaneous changes, from shocks or explosions, and gradual changes or perturbations, properly so called, caused by secondary gravitation, requiring time. The first kind may never have taken *Instantaneous.* place in our system; but it is necessary to consider it, not only because it is of possible occurrence, but because it is a necessary preliminary to the study of the other kind,—the gradual perturbations being treated theoretically as a series of little shocks.

The first case is easy of treatment. No collision or explosion would affect Kepler's laws: and, if the form of the orbit was altered, the accelerating forces would remain the same; and thus, the new variation once understood, our calculations might proceed as before. Supposing a collision between two planets, or the breakage of one planet into several fragments by an internal explosion; there might be any variations whatever in the astronomical elements of their elliptical movement; but there are two relations which are absolutely unalterable, and which might, in my opinion, generally enable us to establish the reality of such an event at any period whatever: these are the essential properties of the continu-

12

ous motion of the centre of gravity, and the invariableness of the sum of the areas,—both resting on that great law of the quality of action and reaction to which all changes must conform. From these must result two important equations between the masses, the velocities, and the positions of the two bodies, or the two fragments of the same body, considered before and after the event. No indication at present leads us to suppose that the case of collision has ever occurred in our system; and it is evident that such an encounter, though not mathematically impossible, would be very difficult. But it is far otherwise with regard to explosions.

The little planets discovered between Mars and Jupiter have mean distances and periodic times so nearly identical, that Mr. Olbers has conjectured that they once formed a single planet, which had exploded into fragments. Lagrange added a supposition, from the irregularity of their form, that the event must have happened after the consolidation of the primitive planet. When their masses become known, I think this conjecture may be subjected to mathematical proof,—in this way. By calculating the positions and successive velocities of the centre of gravity of the system of these four planets, we might, if they had such an origin, retrace the principal motion of the primitive planet. If we should then find this centre of gravity describing an ellipse round the sun as a focus, and its vector radius tracing areas proportioned to the times, this event would be as completely established as any fact that we have not witnessed. We have not yet the materials for such a test; but it is interesting to see how celestial mechanics may establish, in a positive manner, events like these which appear to have left no evidence behind them. It is obvious that the instantaneous character of such a change must preclude our fixing any date for it, since the phenomena would be precisely the same, whether the explosion were recent or long ago. It is otherwise with regard to perturbations, properly so called.

Lagrange believed that these explosions had been frequent in our system, and that this was the true explanation of comets, judging from the greatness of their eccentricity and inclination, and the smallness of their masses. We have only to conceive that a planet may have burst into two very unequal fragments, the larger of which would proceed pretty nearly as before, while the smaller must describe a very long ellipse, much inclined to the ecliptic. Lagrange showed that the amount of impulsion necessary for this change is not great; and that it is less in proportion as the primitive planet is remote from the sun. This opinion is far from having been demonstrated; but it appears to me more satisfactory than any other that has been proposed on the subject of comets.

Gradual perturbations. The important and difficult subject of perturbations is the principal object of celestial mechanics, for the perfecting of astronomical tables. They are of two classes; the one relating to motions of translation, the other of rotation. The latter are, as before, the most difficult: but the motions of rotation are less altered than the other class, within our own system; and they are less important to be known.

In the study of motions of translation, the planets must be treated as if they were condensed in their centres of gravity. *Perturbations of translation.*

The direct method, the only rational one, of calculating the differential equations of the motion of any one planet, under the influences of all the rest, is impracticable, from the unmanageable complication of the problem. It would make an inextricable analytical enigma. Geometers have therefore been obliged to analyze directly the motion of each planet round that which is its focus, taking for modification only one at a time. This is what constitutes in general the celebrated problem of three bodies, though this denomination was at first employed only for the theory of the moon. *Problem of three bodies.* It is easy to see what circumvolutions are involved in this method, since the modifying body, being in its turn modified by others, compels a return to the study of the primitive body, to understand its perturbations. The determination of the motions of the whole of our system must, by its very nature, be a single problem. It is the imperfection of our analysis which obliges us to divide it into detached problems, and to overload our formulas with multiplied modifications. The elementary problem of two bodies,—one of these even being regarded as fixed,—is the only one that we are capable of bringing to a solution; the problem of the elliptical motion, represented by Kepler's laws; and here the calculations are extremely laborious. It is to this type that geometers have to refer the motions of the planets, by extremely complicated approximations, accumulating the perturbations separately produced by every body that can be supposed to exert any influence; and these perturbations prescribe the series required for the integration of the equations belonging to the case of the three bodies.

Then follows the task of choosing the perturbations which have to enter into the estimate. The law of gravitation enables us to compare the secondary influences involved in each case,—the masses of all within our own system being supposed to be known. It is a favorable circumstance to mathematical research that our system is constituted of bodies of very small mass in comparison with the sun (making the perturbations extremely small); moreover, very few, very far from each other, and very unequal in mass; the result of all which is that, in almost every case, the principal motion is modified by only one body. If the contrary had been the case, the perturbations must have been very great, and extremely varied, since a great number of bodies must have powerfully acted in each disturbance. Celestial Mechanics must then, we should think, have presented an inextricable complication, being incapable of reduction to the problem of three bodies.

This study of modified motions divides itself into three parts, answering, as in a former case, to the planets, satellites, and comets. Rigorously speaking, we ought to make a fourth case of the sun, which can not here be regarded as motionless, because the planets react upon it. In fact, we can not allow ourselves to consider any point within the system as motionless, except the centre of gravity of the system itself, which is the true focus of planetary motion, and round which the sun itself must oscillate, in directions which vary according to the positions of the planets. This point is always between the centre and the surface of the sun. But we can not approach nearer to the fact than this: we shall probably never be able to indicate this centre precisely; and it is enough for practical purposes, and necessary to them, to consider the sun as fixed, except as to its rotary motion. The same conclusion must be come to with regard to the planets and their satellites,—even in the case of the earth and moon, where the variations of the primary body are greatest. The centre of gravity falling within the mass of the primary body, its variations from that centre may be neglected as having no appreciable influence on the motion of translation; and thus, celestial mechanics presents, in this branch, no other problems than those treated, under another point of view, by celestial geometry.

Centre of the Solar System.

The simplest problem is here, as before, that of the planets, and for the same reasons,—the smallness of their eccentricities, and of the inclinations of their orbits. There is also a considerable uniformity of perturbations, since each planet remaining in the same regions of the sky, continues in the same mechanical relations, though their intensity varies within certain limits. The least privileged of these bodies in these matters is unhappily our own planet, on account of the heavy satellite which escorts it so closely, and to which its chief perturbations are due; though this does not save it from being sensibly troubled by others, at the period of opposition, and especially by such a mass as that of Jupiter. No other planet with satellites, not even Jupiter, is in so unfavorable a case; for Jupiter's motion could not be very much deranged by the action of his satellites, however near in position, since the mass of the largest is less than a ten-thousandth part of his, while the mass of our moon is a sixty-eighth part of that of the earth. Jupiter's circulation is sensibly affected by Saturn alone. The simplest case of all seems to be that of Uranus, from its being the last planet, and very remote from the next; and its six satellites do not appear to trouble its motion.

Problem of the Planets.

The problem of the satellites is necessarily more complicated than that of the planets, on account of the instability of the focus of the principal motion, as in celestial geometry. Besides their own perturbations, the satellites have reflected upon them all those to which their planet is liable. The founders of Celestial Mechanics were long perplexed, for instance, by the perpetual acceleration of the mean motion of the moon; it was considered inexplicable, till Laplace discovered its cause in the slight variation to which the eccentricity of the earth's orbit is subject.

Problem of the Satellites.

In regard to the direct perturbations of the satellites, there is an essential distinction between the case of one, and that of several satellites. In the first,—the single case of our moon,—the disturbing body is the sun, on account of its unequal action on the planets and its satellite. If the difficulties arising out of this position are greater than in the case of any other satellite, it is partly because the case more immediately concerns us, and because our opportunities of observation disclose more fully the imperfection of our means. For, in the mathematical point of view, there must be more complexity in the case of several satellites; all that is true in regard to one being true in regard to each one, with the addition of the mutual action of the members of the group. Their perturbations are reduced by the preponderating size of their planet; but, from there being so many of them, of such nearly equal sizes and direction, and all so close together, the difficulty of calculating their motions is so great that the only theory as yet established is that of the satellites of Jupiter. For the motions of three of them, Laplace found means completely to account. Those of Saturn and Uranus are known only geometrically, we having not even an approximate estimate of their masses. It is to be remembered, however, that we do not need so perfect a knowledge of them as of the moon; and that a much less exact theory will suffice for them than for the moon, whose slightest irregularity is very evident to us.

The comets intervene to increase our difficulties about the satellites. From the extreme prolongation of their orbits, and their inclination in all directions, comets are in a state of ever variable mechanical relations, from the number of bodies that they approach in their course; while the planets, and even the satellites, have always the same relations, the variation being only in the intensity. The perturbation which, in every other case, bears a very small proportion to the gravitation, may, in the case of comets, exceed it; so that it is conceivable that a comet might be diverted from its orbit, and become a satellite, when it passes near so considerable a body as Jupiter, Saturn, or even Uranus. Besides the eccentricities of comets, there are other circumstances, such as their small weight, and their possible loss of weight by parting with some of their atmosphere to the bodies they approach, which tend to perplex the study of their perturbations. These are the incidents which make it so difficult to foresee exactly the return of these little bodies. When we have studied them so long and so laboriously as to have, to the best of our belief, mastered their case, we find that their periods are entirely changed through one omitted circumstance. A memorable example of this was the comet of 1770, calculated by Lexell. This comet had then a revolution of less than six years: but it has never appeared since, having been entirely deranged by passing too near Jupiter. The imperfection of our knowledge about these small bodies is from the same cause that renders them of very little consequence to us. From their vast distances, their action upon any one body of the system is little more than momentary; and their lightness prevents even the satellites from being affected by their passage. The passage of the comet of 1770 among the satellites of Jupiter proved this, in a striking manner. Their tables, constructed beforehand, without any idea of such an incident, perfectly agreed with direct observations; a proof that the intrusion of the comet did not sensibly affect their motions. There is, therefore, no more occasion for the puerile fears of our day than for the religious terrors of former times, in regard to the passage of comets. Their collision with the earth is all but impossible; and they could not otherwise be felt at all. Their mere approach, however near, could have no other effect than to raise somewhat the corresponding tide. If a comet could pass two or three times nearer to us than the moon (which no known comet could do) its very small mass could produce no other effect than an imperceptible rise of the tides. We have, therefore, no immediate and practical reason to regret the imperfection of our cometary theories.

Perturbations of rotation. Passing from the perturbations proper to motions of translation, we must notice those belonging to rotation.

The ellipsoid bodies of our system must, whether they began or not, have ended, sooner or later, with turning round one of their axes,—and that one the most stable,—that of their smallest diameter: for, as we have seen, it is their rotation that has produced their deviation from a perfectly spherical form, and determined the direction favorable to stability. The regularity of this rotation is

The planets. evidently so indispensable to the existence of living bodies on the surface of a planet, that we might, *a priori*, assert this stability wherever life is possible, from the time when it became possible. But, stable as each planet is in itself, its mutual gravitation with others must introduce certain secondary modifications, the bearing of which must be upon the direction of its axis in space. It is only with regard to the earth that these modifications concern us; for, however great they might be in any other body, they could in no way affect us.

If the planets were perfect spheres, the total gravitation of their particles must pass through their centres of gravity; and thus, it is only through their slight failure in sphericity that they can act at all upon one another's rotation; that failure being caused by the rotation itself. We see here how the same necessity which secures the stability of the rotations, with regard to their duration and their poles, determines, from another point of view, the inevitable alteration of the parallelism of their axes.—In our own planet the precession of the equinoxes, modified by the nutation, results from the action of the other bodies of our system,—especially of the sun and moon,—upon our equatorial protuberance. The power of each body is, as in the case of the tides, in the direct ratio of its mass, and inversely to the cube of its distance; so that the sun and moon are the only bodies whose influence need be considered. Further, the extent of the deviation depends on the mass and magnitude of the earth, on the time of its rotation, on its degree of flattening, and on the obliquity of the ecliptic. The intensity of the influence must vary, as in the case of the tides, with the variable distance of the sun from the earth, and yet more of the moon; but the want of uniformity is too slight to be perceptible to direct observation.— These are the general causes which determine the small changes which the rotation of our globe undergoes, in regard to the direction of its axis in space.—The case of the other planets bears a general likeness to that of the earth, varied according to the different inclinations of their axes to their orbits, their position, their mass, their size, the duration of their rotation, and the degree of their flattening at the poles. On all these grounds, the perturbations of Mars are the most remarkable.

The rotation of the satellites presents one consideration of the highest interest,—that remarkable equality between the duration of this rotation and that of their circuit round their planet, by which they present always the same hemisphere, except from those very small oscillations called librations, whose law is well understood. The fact is absolutely certain only with regard to the moon; but our mechanical principles justify our erecting it into a general law of all the satellites. Lagrange has shown that it results from the preponderance that, by the action of the planet, the nearer hemisphere must acquire at the outset, whence arises a natural tendency in the satellite to return perpetually to the same position. If it is thus with the moon, there is every reason to suppose the same fact with regard to satellites belonging to heavier planets, to which they are proportionally nearer.

The satellites.

Such are the various kinds of perturbations produced in the movements of the bodies of our system, by their mutual action. This study may be simplified and rendered much more exact, by the device of referring all these movements to a plane whose position must necessarily be independent of all their variations. Among several planes which have been proposed, differing in their degrees of variableness, M. Poinsot has discovered one which is the only truly invariable one, but which is extremely difficult to determine, since it requires not only an estimate of the planetary masses, but data dependent on the mathematical law of the interior density of the heavenly bodies,—a law which is still very hypothetical. The theory is complete; but its precise application is at present impossible. Whatever may be the practical difficulties, we can not but feel a deep interest in seeing how Celestial Mechanics has accomplished the fixing of an invariable plane in the midst of all the interior perturbations of our system, as Newton had first recognised an inalterable velocity,—that of the centre of general gravity. These are the only two elements in our system which are rigorously independent of all the events that can occur in its interior;—of even the vastest commotions that our imagination can suggest. Such variations as they can be conceived to have could relate only to the most general phenomena of the universe, produced

Device of an invariable plane.

by the mutual action of different suns, of which they would afford us the clearest manifestations, if such knowledge were within our reach.

Stability of our system. We end this study of perturbations with a recognition of the stability of our own system, in regard to all its most important constituent bodies. Setting aside the comets, all the variations whatever of any perceptible value are periodical; and their period is usually very long, while their extent is very small; so that the whole of our planetary system can only oscillate with extreme slowness round a mean state, from which it deviates very little. Through all starry changes the translations of our planets present the almost rigorous invariableness of the great axes of their elliptical orbits, and of the duration of their sidereal revolutions: and their rotation shows a regularity even more perfect, in its duration, in its poles, and even, though in a somewhat smaller degree, in the inclination of its axis to the corresponding orbit. We know, for instance, that from the time of Hipparchus, the length of the day has not varied the hundredth part of a second. Amidst all this general regularity, we perceive a special and most marked stability with regard to the elements which are concerned in the continued existence of living beings.—Such are the sublime theorems of natural philosophy for which humanity is indebted to the sum of the great works executed in the last century by the successors of Newton.

The general cause of these important results lies in the small eccentricity of all the principal orbits, and the small divergence of their planes. If the planets had had cometary orbits and planes, there would have been no regularity—no periodicity,—and, we may add, no life upon their surface. No planets can be habitable but such as have their oscillations restricted within very narrow limits.

Resistance of a Medium. The Mathematical theory of celestial mechanics has taken no notice, thus far, of the resistance of any general medium, in which these motions are proceeding. The conformity of our mathematical tables with observed facts shows that the resistance is imperceptible in degree; yet, as it is manifestly impossible that it should be null, the geometers have endeavored to prepare beforehand a general analysis of it. Considered apart from its intensity, this action is of a totally different nature from that of perturbations, though gradual like them: for it can not be periodical, and must always be exercised in the same direction, so as continually to diminish all velocities, and the more the greater they are. It can not alter the positions of the orbits, but can by possibility affect only their dimensions, and periodic times, and the duration of rotations: that is, it affects the elements which are spared by the perturbations. Thus, the rotations must become slower, the orbits must grow smaller and rounder, and their periodic times shorter; because, as velocity diminishes, the solar action must become more powerful, and these effects are not only continuous, but always increasing in rapidity. So, in a future too remote to be assigned, all the bodies of our system must be united to the solar mass, from which it is probable that they proceeded: and thus the stability of the system is simply in relation to the perturbations properly so called. These are among the incontestable indications of Celestial Mechanics.

As yet, we practically fail to recognise the effect of a resisting medium. We neither trace its operations, nor should know how to calculate it if we could trace it. Whenever we do, it will be by the study of comets; for their small mass, and the great surface which they present to the action of the medium when their atmospheres are widely diffused, must render its resistance much more appreciable than in the case of planets,—their velocity being besides naturally at its *maximum* at the moment of this expansion. Some contemporary astronomers believe that they have established the effect of this resistance in regard to one or two comets. Hitherto the study of these bodies seems to be only negatively useful, to prevent the return of the absurd terrors which they formerly occasioned. We now see that there is no body in our system, however insignificant, whose theory may not offer to us a direct and positive interest, since we may owe to comets the knowledge of one of the most important general laws of the system to which we belong, and that which, in a remote future, must chiefly rule its destinies.*

In our geometrical review we saw, by the agreement of astronomical tables with direct observation, that our system is independent of all that lies outside. This incontestable truth is confirmed by the mechanical view. If our system gravitated toward any of the suns outside, the action of other suns would nearly neutralize the tendency. Again, it would be only by an unequal action of those suns upon our planets that any change could be occasioned. Again, the vast distances would, according to our law of gravitation, make the action of remote suns imperceptible. The nearest body, if a million times heavier than our system, would produce an effect incalculably smaller than the action which occasions our tides. We may therefore pronounce the independence of our system to be perfectly certain. I notice this because we seem to find here the only exception to the great encyclopedical law which is the basis of this work,—that the most general phenomena rule the most particular, without being in any degree reciprocally influenced. Thus our astronomical phenomena regulate those of our own globe,—whether physical, chemical, physiological, or social. Yet here we find that the phenomena of the universe have no influence over those of the solar system. There is no difficulty about this to persons who, like myself, admit that our researches are limited by the boundaries of our own system, and that positive knowledge can not go beyond it. The study of the universe forms no part of natural philosophy; a truth which will become more ap-

Independence of the solar system.

---

* M. Comte estimates too lightly the indications of a medium given by Encke's comet.—J. P. N.

parent, and be seen to be more important the further our studies extend.

At the close of this brief review of celestial dynamics, we see that, great as are the achievements since Newton's time, we are reminded in many directions of the imperfection which results from the insufficiency of our mathematical analysis. In the execution of astronomical tables it has to borrow from celestial geometry other aid than the estimate of indispensable data, derived from direct observation; and this in regard not only to bodies whose mechanical theory is but just initiated, but with regard to some with which we are best acquainted.

<small>Achievements of Celestial Dynamics.</small> We see, however, that, beside the sublime direct knowledge afforded to us, celestial dynamics has powerfully contributed to perfect the whole body of astronomical theories in regard to their definite aim—the exact prevision of the state of the heavens at any period whatever, past or future. Kepler's laws might suffice to determine the state of our system for a short time, proper data being chosen; but if we wish to extend the inquiry, back or forward, to any considerable period, we find the most perfect theory of perturbations absolutely necessary. It is to celestial dynamics that we owe our power of ranging up and down the centuries, to fix the precise moments of various celestial phenomena, such as eclipses, with certainty, and with a minuteness only inferior to that which is possible in the case of present events.

Though we have, according to my view, completed our consideration of astronomical science, it would be felt to be a great omission if we passed over altogether what is now called Sidereal Astronomy. We will therefore see how much there is that we can conceive to be positive in regard to cosmogony.

CHAPTER VI.

SIDEREAL ASTRONOMY AND COSMOGONY.

<small>Multiple Stars.</small> THE only branch of Sidereal Astronomy which appears to admit of exact study is that of the relative motions of the Multiple Stars, first discovered by Herschel. By multiple stars astronomers understand stars very near each other, whose angular distance never exceeds a half minute, and which, for this reason, appear to be one, not only to the naked eye, but to ordinary telescopes, only the most powerful lenses being able to separate them. The relative movements of these stars tend to deceive us as to their precise multiple character, as, for instance, by mutual occultations, which do not permit us to separate them.

Among some thousands of multiple stars registered in the catalogues, before the southern heavens had been really explored, almost all were only double, and we have found none which are more than triple—a circumstance which may be owing solely to the imperfection of our telescopes, as we knew of none but single stars before Herschel's time. However interesting the study of them is, they constitute only a particular case in the universe, as the intervals of the stars which compose them are probably much smaller than those which divide the suns of the universe, so that the study of their relative motions does not lead us up to any of the great general phenomena of the heavens, and the speciality would be more conspicuous if astronomers did what I think they ought—form their catalogues of those double stars only whose motions they have fully established. With regard to others, we can not be sure whether their duality is a real relation or an accident. Knowing nothing whatever of their interval, or of the distance of either of them from us, we can not be sure whether they form a system any more than any other two stars combined by chance in the heavens. Because a few incontestable examples are before us of a binary system, in which the smaller circulates round the larger, it is anything but philosophical to conclude the same to be the case with the whole multitude of double stars, some of which may appear so merely through an accident of position, apparent only to our own system. Analogy is not applicable here; as what looks like analogy is merely the imperfection of our investigations. No astronomer would venture to assert that if our telescopes were what they may one day become, we might not find between stars now apparently independent a multitude of clustered intermediate stars which should render the case of duality almost general. The apparent nearness would not then be a sufficient ground for presuming their mutual revolutions, because it is in virtue of their very small number that analogy now suggests that presumption. The only positive study in sidereal astronomy is that of the known relative motions of certain double stars, at present not more than seven or eight in number. We could never hope to assign with accuracy their orbits, or their periodic times, or any solid basis for dynamical conclusions.* The importance of such inquiries is much diminished by the consideration that our system, which, in such a case, means our sun, belongs to no groups of the kind—either investigated or merely pointed out. This circumstance seems to me not at all accidental; for, if our system made a part of a double star, which it is not difficult to imagine, it would probably be impossible for us ever to be aware of the other part of such a duality, because in the direction of the sun it would be so near that its light would be lost to us in that of our sun. Such a case might, however, have

---

* M. Comte quite underrates the importance of the phenomena of the multiple stars. The orbits of a very considerable number are now distinctly ascertained, and the laws of motion in their orbits. The existence of a motion of revolution is fixed, with regard to the far greater number.—J. P. N.

a scientific interest for us, not only as elucidating the displacements of our system, but as allowing such great precision as might arise from the position of the inquirer on one of the stars of the couple.

The first of the few orbits of double stars known to us was investigated by Savary. They all present a very considerable eccentricity, the smallest of which is double, and the greatest four times greater than that of the most eccentric in our system. Of their periodic times, the shortest slightly exceeds forty years, and the longest six hundred. We can not perceive that the eccentricity and the duration bear any fixed relation to each other, and neither seems to depend at all on the angular distance of the respective pairs of stars. This is the sum of what we know about the double stars; and unless we could learn something of their linear distance from our system and from each other, our conceptions can neither be accurate, nor of great importance. M. Savary has proposed a method, founded on the known velocity of light, by which these distances will, if ever, be estimated;* but the uncertainty of some of the elements which must enter into the question is so great that the most that can be hoped for is the fixing of certain limits within which the real distance may be supposed to lie; and this is all that M. Savary himself proposed. At present, we know only the nearer limit, beyond which, not only the double stars, but the whole starry host, are known to lie.

Our Cosmogony. Proceeding now to ascertain what we may rationally conceive of our own cosmogony, I need hardly say that we must put aside altogether any notion of *creation*, as unintelligible—all that we are able to conceive of being successive *transformations* in the sky; and of these, only such as have produced its present state. Here, again, we find our own system to be the only subject of knowledge. We are in possession of some facts in regard to it which may bear testimony to its immediate origin; but we can form no reasonable conjectures about the formation of the suns themselves. The phenomena necessary for such a purpose are not only not explored, but not explorable. Whatever may be the interest of Herschel's curious observations on the progressive condensation of the nebulæ, they do not warrant his conclusion of their transformation into stars;† for from such a conclusion must flow consequences about form and motion which must be in harmony with established phenomena; and of these we have absolutely none.

Origin of Positive Cosmogony. The beginning of positive cosmogony was when geometers, pursuing the mathematical theory of the figures of the planets, showed that they were originally in a state of fluidity. We can not go further back than this; and we must set out with an existing sun, turning on its axis with an indeterminate

---

* They are calculated; but by strictly geometrical methods.

† This portion of Herschel's speculation must be abandoned. What he fancied to be instances of nebulous matter turn out to be galaxies, or vast groups of stars.—J. P. N.

velocity, admitting, for the formation of the planetary system, no agencies which we do not now see at work, in the phenomena which we habitually witness, though they may have wrought formerly on a larger scale. These restrictions are indispensable to the scientific character of the inquiry; and, after all, our cosmogonic theories, however guarded, must remain essentially conjectural, if ever so plausible. No mathematical principles can enter here as into celestial mechanics, leading us up to a definite theory, and excluding every other. No abstract theory of formations is possible; and the utmost we can do is to collect such information as can be had, construct hypotheses from it, and compare them carefully and continuously with the whole of the phenomena that we explore. Such hypotheses, whatever degree of consistency they may attain, can never, like the law of gravitation, take rank among general facts: for we can never be sure that some other hypothesis may not turn up which would equally well answer the present purpose, and some others besides.

The cosmogony of Laplace seems to me to present the most plausible theory of any yet proposed. It has the eminent merit of requiring, for the formation of our system, only the simple agents, weight and heat, which meet us everywhere, and which are the only two principles of action which are absolutely general. The point in which I differ from Laplace is with regard to comets, which he regards as strangers in our system; whereas Lagrange's view of them before cited, appears to be preferable, as being consistent with the independence of our solar group. *Cosmogony of Laplace.*

The hypothesis of Laplace tends to explain the general circumstances of our system, viz., the common direction of all the planets from west to east; that of their rotations; and that of all their satellites: also, the small eccentricity of all the orbits; and finally, the small inclination of their planes, especially in comparison with that of the solar equator.

It is supposed by this theory that the solar atmosphere was originally extended to the limits of our system, in virtue of its extreme heat; that it was successively contracted by cooling; and that the planets were formed by this condensation. The theory rests on two mathematical considerations. The first involves the necessary relation between the successive expansions or contractions of any body whatever and the duration of its rotation; by which the rotation should be quickened as the dimensions lessen and becomes slower as they increase, so that the angular and linear variations sustained by the sum of the areas become exactly compensated. The other consideration relates to the connection between the angular velocity of the sun's rotation and the possible extension of its atmosphere, the mathematical limit of which is at the distance at which the centrifugal force, due to this rotation, becomes equal to the corresponding gravity: so that if any portion of the atmosphere should be outside of this limit, it would cease to belong to the sun, though it must continue to revolve with the velocity it had at the moment of separation. From that moment, it ceases to be involved in any further consequences from the cooling of the solar atmosphere. It is evident, from this, how the solar atmosphere must have diminished, as to its mathematical limit, without intermission, in regard to the parts situated at the solar equator, as the cooling was for ever accelerating the rotation. Portions of the atmosphere, thus parted with, must form gaseous zones, situated just beyond the respective limits; and this constituted the first condition of our planets. By the same process the satellites were formed out of the atmospheres of their respective planets. Once detached from the sun, our planets must become first liquid and then solid, in the course of their own cooling, without being further affected by solar changes: but the irregularity of the cooling, and the unequal density of parts of the same body must change in almost every case, the primitive annular form, which remains in the rings of Saturn alone. In most cases, the whole gaseous zone has gathered, in the way of absorption, round the preponderating portion of the zone as a nucleus: thence the body assumed its spheroidal form, with a revolving motion in the same direction as its movement of translation, on account of the excess of the velocity of the upper molecules in comparison with that of the lower.

This theory answers to all the appearances of our system, and explains the difficulty of the primitive impulsion of the planets. It shows, also, that the formation of the system has been successive, the remotest planets being the most ancient, and the satellites the most modern.*

If from points of view like these the stability of our system can scarcely be regarded as absolute, what it may lead us to suspect is that, by the continuous resistance of the general medium, our system must at length be reunited to the solar mass from which it came forth, till a new dilatation of this mass shall occur in the immensity of a future time, and organize in the same way a new system, to follow an analogous career. All these prodigious alternations of destruction and renewal must take place without affecting the most general phenomena, occasioned by the mutual action of the suns; so that these revolutions of our system, too vast to be more than barely conceived of by our minds, can be only secondary, even local events, in relation to really universal transformations. It is not less remarkable that the natural history of our system should be, in its turn, as certainly independent of the most prodigious changes that the rest of the universe can undergo: so that whole systems are, perhaps frequently, developed or condensed in other regions of space, without our attention being in any way drawn toward these immense events.

* The author subjoins a proposed mathematical verification of Laplace's cosmogony, which is not given in the text, as it does not seem to rest on adequate foundations. If an arithmetical verification be ever obtained, it will probably be in connection with the periods of the *rotations* of the different planets;—periods already in so far connected with the nebular hypothesis by the investigations of an American inquirer—Mr. Kirkwood.—J. P. N.

Recapitulation. The end I had in view in this exposition of astronomical philosophy will be attained if I have clearly exhibited, in regard both to method and to doctrine, the true general character of this admirable science, which is the immediate foundation of the whole Natural Philosophy. We have seen the human mind, by means of geometrical and mechanical researches, and with the help of constantly improving mathematical aids, attaining to a precision of logical excellence superior to any that other branches of knowledge admit of. We see the various phenomena of our system numerically estimated, as the different aspects of the same general fact, rigorously defined, and continually reproduced before our eyes in the commonest terrestrial phenomena; so that the great end of all our positive studies, the exact prevision of events, has been attained as completely as could be desired, in regard alike to the certainty and extent of the prevision. We have seen how this science must operate in liberating the human intellect for ever from all theological and metaphysical thraldom by showing that the most general phenomena are subjected to invariable relations, and that the order of the heavens is necessary and spontaneous. This last consideration belongs more particularly to a subsequent part of this work; but it has been our business to point out as we went along how the development of astronomical science has shown us that the universe is not destined for the passive satisfaction of Man; but that Man, superior in intelligence to whatever else he sees, can modify for his good, within certain determinate limits, the system of phenomena of which he forms a part —being enabled to do this by a wise exercise of his activity, disengaged from all oppressive terror, and directed by an accurate knowledge of natural laws. Lastly, we have seen that the field of positive philosophy lies wholly within the limits of our solar system, the study of the universe being inaccessible in any positive sense.*

---

* As before remarked, M. Comte speaks much too absolutely here, in oversight of what modern astronomical researches have really accomplished.—J. P. N.

BOOK III.
PHYSICS.

CHAPTER I.
GENERAL VIEW.

Imperfect condition of the science. ASTRONOMY was a positive science, in its geometrical aspect, from the earliest days of the School of Alexandria; but Physics, which we are now to consider, had no positive character at all till Galileo made his great discoveries on the fall of heavy bodies. We shall find the state of Physics far less satisfactory than that of Astronomy, not only on account of the greater complexity of its phenomena, but under its speculative aspect, from its theories being less pure and systematized, and, under its practical aspect, from its previsions being less extended and exact. The precepts of Bacon and the conceptions of Descartes have advanced it considerably in the last two centuries, in its character of a positive science; but the empire of the primitive metaphysical habits is not to be at once overthrown; and Physics could not be immediately imbued with the positive spirit, which Astronomy itself, our only completely positive science, did not assume in its mechanical aspect till the middle of that period. The further we go among the sciences, the more we shall find of the old unscientific spirit, and not only in their details, but impairing their fundamental conceptions. If we now compare the philosophy of Physics with the perfect model offered to us by astronomical philosophy, I hope we shall perceive the possibility of giving to it, and afterward to the other sciences in their turn, the same positivity as the first, though their phenomena are far from admitting of an equal perfection of simplicity and generality.

First, we must see what is the domain of Physics, properly so called.

Its domain. Taken together with Chemistry (for the present), the object of the two is the knowledge of the general laws of the Inorganic world. This study has marked characters, to be analyzed hereafter, distinguishing it from the science of Life, which follows it in our encyclopedic scale, as well as from that of astronomy which precedes it. The distinction between Physics and Chemistry is much less easy to establish; and it is one more difficult to pronounce upon from day to day, as new discoveries bring to light closer relations between them. Though the division between these sciences is less obvious than between any other two in the scale, it is not the less real and indispensable, as we shall see by three considerations which, perhaps, might be insufficient apart, but which, when united, leave no uncertainty.

Compared with Chemistry.

First, the generality which characterizes physical researches contrasts with the speciality inherent in the chemical. Every physical consideration is applicable to all bodies whatever, while chemistry studies the action appropriate to a particular substance. If we look at their classes of phenomena, we find that gravity manifests itself in the same way in all bodies; and the same with phenomena of heat, of sound, of light, and even electrical effects. The difference is only in degree. But, in the compositions and decompositions of Chemistry, we have to deal with specific properties, which vary not in elementary substances, but in their most analogous combinations. The only exception which can be alleged, in the whole domain of Physics, is that of magnetic phenomena; but modern researches tend to prove that they are a mere modification of electrical phenomena, which are unquestionably general. The general properties of Physics were, in the metaphysical days of the science, regarded as consisting of two classes—those which were necessarily, and those which were contingently universal. But the false distinction arose from the notion of that age, that the business of science was to inquire into the nature of bodies,—the study of their properties being a mere secondary affair. Now that we know our business to be with the properties alone, we see the error, and need only ask whether we can conceive of any body absolutely devoid of weight, or of temperature.

Its generality.

In the second place, Physics relates to masses, and Chemistry to molecules; insomuch that chemistry was formerly called Molecular Physics. But, real as this distinction is, we must not carry it too far, but remember that purely physical action is often as molecular as chemical action; as in the case of gravity. Physical phenomena observed in masses are usually only the sensible results of those which are going on among their particles: and, at most, we can except from this only phenomena of sound, and perhaps of electricity. As for the necessity of a certain mass, to manifest physical action, that is equally indispensable in chemistry. The best way of expressing the general fact which lies at the bottom of this distinction is, perhaps, that in chemistry, one at least of the bodies concerned must be in a state of extreme division; while this is so far from being a

Dealing with masses or molecules

necessary condition of physical action that it is rather an impediment to it. This is a proof of a real distinction between the two sciences, though it may not be a very marked one.

Changes of arrangement or composition of molecules. In the third place, the constitution of bodies,—the arrangement of their molecules,—may be changed, in exhibiting physical phenomena; but the composition of their molecules remains unchangeable: whereas, in Chemistry, not only is there always a change of state in one of the bodies concerned, but the mutual action of the bodies alters their nature; and it is this alteration which constitutes the phenomenon. Many physical agents can, no doubt, work changes of composition and decomposition, if their operation be very energetic and prolonged; and it is this which forms such connection as there is between the two sciences: but, at that point of activity, physical agencies pass the boundary, and become chemical.

Positive philosophy requires that we should draw off altogether from the study of *agents*, to which it may be imagined that phenomena are to be referred. Any number of persons may discover a supposed agent; as, for instance, the universal ether of modern philosophers, by which a variety of phenomena may be supposed to be explained; and we may not be able to disprove such an agency. But we have no more to do with modes of operation than with the nature of the bodies acted upon. We are concerned with phenomena alone, and what we have to ascertain is their laws. In departing from this rule, we leave behind us all the certainty and consistency of real science.

Keeping within our true limits, then, we see that if chemical phenomena should be reduced by analysis into the form of purely physical actions,—an achievement very possible to the present generation of scientific men,—our fundamental distinction between the two sciences will not be shaken. It will still be true that in a chemical fact something more is involved than in a simply physical one: namely, the characteristic alteration undergone by the molecular composition of the bodies, and therefore by the whole of their properties. Such a distinction is secure amid any scientific revolution that can ever happen.

Description of Physics. From these three considerations, taken together, we derive our description of Physics. This science consists in studying the laws which regulate the general properties of bodies, commonly regarded in the mass, and always placed in circumstances which admit of their molecules remaining unaltered, and generally in their state of aggregation. With a view to the great end of all science, we must add that the aim of physical theories is to foresee, as exactly as possible, all the phenomena that will be exhibited by a body placed in any set of given circumstances, excluding, of course, such as could alter its nature. This is not the less true because we can rarely attain the prescribed aim. The imperfection is in our knowledge alone. In estimating the true character of any science, the only way is, first, to suppose the science perfect, and then to study the fundamental difficulties presented by this ideal perfection.

Our description shows us how much more complexity we shall find in physical than in astronomical inquiries. In astronomy we study bodies, known to us only by sight, under two aspects only, their forms and motions. All considerations but these are excluded. But in Physics, on the contrary, the bodies we have to study are recognised by all our senses, and are regarded under an aggregate of general conditions, and therefore amid a complication of relations. It is clear, not only that this science is inferior to astronomy, but that it would be impracticable if the group of fundamental obstacles was not compensated for, up to a certain point, by the extension of our means of exploration. We meet here the law, before laid down, that in proportion as phenomena become complicated, they thereby become explorable under a proportionate variety of relations. *Instruments.*

Of the three procedures which constitute our art of observing, the last, Comparison, is scarcely more applicable here than with regard to astronomical phenomena. Its proper application is, in fact, to the phenomena of organized bodies, as we shall see hereafter. But the other two methods are entirely suitable to Physics. Observation was, in astronomy, restricted to the use of a single sense; but in Physics, all our senses find occupation. Yet would Observation effect little without the aid of Experiment, the regulated use of which is the great resource of physicists in all questions that involve any complexity. This procedure consists in observing beyond the range of natural circumstances;—in placing bodies in artificial conditions, expressly instituted to enable us to examine the action of the phenomena we wish to study under a particular point of view. We can see at once how eminently this art is adapted to physical researches; and how it must there find its triumphs: since there are hardly any bounds to our power of modifying bodies, for the purpose of studying their phenomena. In chemistry, experiment is commonly supposed to be more complete than in any other department: but I think it is of a higher order in physics, for the reason that in chemistry the circumstances are always artificially arranged, while in physics we have the choice of natural or artificial circumstances; and the philosophical character of experimentation consists in choosing the freest possible case that will show us what we want. We have a wider range, and a choice of simpler cases, in physics than in chemistry; and in physics, therefore, is experiment supreme. *Methods of inquiry.* *Observation.* *Experiment.*

The next great virtue of physics is its allowing the application of mathematical analysis, which enters into this science, and at present goes no further;—not yet, with real efficacy, into chemistry. It is less perfect in physics than in astronomy; but there is still enough of simplicity and fixedness in physical phenomena to allow of its extended use. Its employ- *Application of mathematical analysis.*

ment may be direct or indirect:—direct when we can seize the fundamental numerical law of phenomena, so as to make it the basis of a series of analytical deductions; as when Fourier founded his theory of the distribution of heat on the principle of thermological action between two bodies being proportionate to the difference of their temperatures: and indirect, when the phenomena have been referred to some geometrical or mechanical laws; when, however, it is not properly to physics that the analysis is applied, but to geometry or mechanics. Such are the cases of reflection or refraction in the geometrical relation; and in the mechanical, the investigation of weight, or of a part of acoustics. In either case extreme care is requisite in the first application, and the further development should be vigilantly regulated by the spirit of physical research. The domain of physics is no proper field for mathematical pastimes. The best security would be in giving a geometrical training to physicists, who need not then have recourse to mathematicians, whose tendency it is to despise experimental science. By this method will that union between the abstract and the concrete be effected which will perfect the uses of mathematical, while extending the positive value of physical science. Meantime, the uses of analysis in physics are clear enough. Without it we should have no precision, and no co-ordination: and what account could we give of our study of heat, weight, light, etc.? We should have merely series of unconnected facts, in which we could foresee nothing but by constant recourse to experiment; whereas, they now have a character of rationality which fits them for purposes of prevision. From the complexity of physical phenomena, however, the difficulty of the mathematical application is great. In some, a part of the essential conditions of the problem must be thrown out, to admit of its transformation into a mathematical question; and hence the necessity for reserve in the employment of analysis. The art of combining analysis and experiment, without subordinating the one to the other, is still almost unknown. It constitutes the last advance of the true method of physical study; and it will be developed when physicists, and not geometers, conduct the analytical process, and not till then.

Encyclopedical Rank of Physics. Having seen what is the object of Physics, and what the means of investigation, we have next to fix its position in the scientific hierarchy.

The phenomena of Physics are more complicated than those of Astronomy; and Astronomy is the scientific basis and model of Physics, which can not be effectually studied otherwise than through the study of the more simple and general science. In this, we individually follow the course of our race. It was by As-
Relation to Astronomy. tronomy that the positive spirit was introduced into natural philosophy, after it had been sufficiently developed by purely mathematical investigations. Our individual education is in analogy with this: for we have learned from astronomy what is the real meaning of the *explanation* of a phenomenon, with-

out any impracticable inquiry about its *cause*, first or final, or its mode of production. Physics should, more than the other natural sciences, follow closely upon astronomy, because, after astronomy, its phenomena are less complex than any.

Besides these reasons belonging to Method, there is the grand consideration that the theories of astronomy afford the only data for the study of terrestrial physics. Our position in the solar system, and the motions, form, size, and equilibrium of the mass of our world among the other planets, must be known before we can understand the phenomena going on at its surface. What could we make of weight, for instance, or of the tides, without the data afforded by astronomical science? These phenomena indeed make the transition from astronomy to physics almost insensible. In this way Physics is indirectly connected with Mathematics. There is also a direct connection, as some physical phenomena have a geometrical and mechanical character,—as much as those of astronomy, though under a great complication of the circumstances. The abstract laws of space and motion must prevail as much in the one science as in the other. If the relation is thus unquestionable in the doctrine, it is not less so in the spirit and method which we must bring to the study of physics. It must be ever remembered that the true positive spirit first came forth from the pure sources of mathematical science; and it is only the mind that has imbibed it there, and which has been face to face with the lucid truths of geometry and mechanics, that can bring into full action its natural positivity, and apply it in bringing the most complex studies into the reality of demonstration. No other discipline can fitly prepare the intellectual organ. We might further say that, as geometrical ideas are more clear and fundamental than mechanical ideas, the former are more necessary, in an educational sense, to physicists than the latter, though the use of mechanical ideas is the more immediate and extended in physical science. Thus we see how we must conclude that the education of physicists must be more complicated than that of astronomers. Both have need of the same mathematical basis, and physicists must also have studied astronomy, at least in a general way. And this, again, assigns the position of their science.

Its rank is equally clear, if we look in the opposite direction,—at the sciences which come after it.

It can hardly be by accident that, in all languages of thinking peoples, the word which originally indicated the study of the whole of nature should have become the name of the particular science we are now considering. Astronomy is, in fact, an emanation from mathematics. Every other natural science was once comprehended under the term Physics; and that to which it is now restricted must be supreme over the rest. Its relation to the rest is just this: that it investigates the general properties common to all bodies; that is, the fundamental constitution of matter; while the other sciences exhibit the modifications of those properties peculiar to each: and the study of those properties in the general must, of course, precede that of their particular cases. In regard to Physiology, for instance, it is clear that organized bodies are subject to the general laws of matter, those laws being modified in their manifestations by the characteristic circumstances of the state of Life. The same is the case with Chemistry. Without admitting the questionable hypothesis under which some eminent men of our time refer all chemical phenomena to purely physical action, it is yet evident that the concurrence of physical influences is indispensable to every chemical act. What could we make of any phenomenon of composition or decomposition, if we left out all data of weight, heat, electricity, etc.? And how could we estimate the chemical power of these various agents without first knowing the laws of the general influence proper to each? Chemistry is closely dependent on Physics; while Physics is wholly independent of Chemistry.

Relation to human progress. As for the direct operation of this science on the human intellect, it is less marked than that of the two natural sciences which occupy the extremities of the scale,—astronomy and physiology,—which immediately contemplate the two great objects of human interest,—the Universe and Man: but one striking fact with regard to physics is that it has been the great battle-ground between the old theological and metaphysical spirit and the positive philosophy. In Astronomy the positive philosophy took possession, and triumphed almost without opposition, except about the earth's motion; while in the domain of Physics the conflict has gone on for centuries; a circumstance attributable to the imperfection of physical, in comparison with astronomical science.

Human power of modifying phenomena. With this science begins the exhibition of human power in modifying phenomena. In Astronomy, human intervention was out of the question: in Physics it begins; and we shall see how it becomes more powerful as we descend the scale. This power counterbalances that of exact prevision which we have in astronomy, through its extreme simplicity. The one power or the other,—the power of foreseeing or of modifying,—is necessary to our outgrowth of theological philosophy. Our prevision disproves the notion that phenomena proceed from a supernatural will, which is the same thing as calling them variable: and our ability to modify them shows that the powers under which they proceed are subordinated to our own. The first is the higher order of proof; but both are complete in their way, and certain to command, sooner or later, universal assent. The proof which Franklin afforded of human control over the lightning destroyed the religious terror of thunder as effectually as the superstition about comets was destroyed by the prevision of their return; though the experiments by which Franklin established the identity of the lightning with the common electric discharge could be decisive only with physicists, while the generality of men could understand how the return of comets was foreseen. As the opposition between the theological and the positive philosophies becomes less simply evident, our power of intervention becomes more varied and extended; the amount of proof yielded in the two cases being equal in the eyes of men in general, though not strictly equivalent.

In regard to the speculative rank of Physics, it is clear that it does not admit of prevision to any extent at all comparable to that of astronomy, because it consists of numerous branches, scarcely at all connected with each other, and concurring only in a feeble and doubtful way in its chief phenomena. We can therefore see only a little way forward; often scarcely beyond the experiment in hand; but we shall see its speculative superiority to the sciences which come after it, when, in studying chemistry and physiology, we find another kind of incoherence existing among their phenomena, making prevision more imperfect still. The great distinction of Physics is that which has been referred to before,—that it instructs us in the art of Experiment. Philosophers must ascend to this source of experimentation, whatever their special objects may be, to learn what are the spirit and conditions of true experimentation, and what the necessary precautions.—Each of the sciences in the scale presents, besides the characters of the positive method which are common to them all, some indication appropriate to itself, which ought to be studied at its source, to be duly appreciated. Mathematical science exhibits the elementary conditions of postivity: astronomy determines the true study of Nature: physics teaches us the theory of experimentation: chemistry offers us the art of nomenclature: and physiology discloses the true theory of classification. *Prevision imperfect.* *Characteristics of each science.*

I have deferred till now what I have to offer on the important subject of the rational construction and scientific use of *hypotheses*, regarded as a powerful and indispensable auxiliary to our study of nature. It is in the region of astronomy that I must take my stand in discussing this subject, though it was not necessary to advert to it while we were surveying that region. Hypothesis is abundantly employed in astronomy; but there it may be said to prescribe the conditions of its own use,—so simple are the phenomena in question there. Thence do I think it necessary to derive, therefore, our conceptions of the character and rules of this valuable resource, in order to its employment in the other departments of natural philosophy. *Philosophy of hypotheses.*

There are only two general methods by which we can get at the law of any phenomenon,—the immediate analysis of the course of the phenomenon, or its relation to some more extended law already established; in other words, by induction or deduction. Neither of these methods would help us, even in regard to the simplest phenomena, if we did not begin by anticipating the results, by making a provisional supposition, altogether conjectural in the first instance, with regard to some of the very notions which are the object of the inquiry. Hence the necessary introduction of hypotheses into natural philosophy. The method of approximation employed by geometers first suggested the idea; and without it all discovery of natural laws would be impossible in cases of any degree of complexity; and in all, very slow. But the employment of this instrument must always be subjected to one condition, the neglect of which would impede the development of real knowledge. This condition is to imagine such hypotheses only as admit, by their nature, of a positive and inevitable verification at some future time,—the precision of this verification being proportioned to what we can learn of the corresponding phenomena. In other words, philosophical hypotheses must always have the character of simple anticipations of what we might know at once, by experiment and reasoning, if the circumstances of the problem had been more favorable than they are. Provided this rule be scrupulously observed, hypotheses may evidently be employed without danger, as often as they are needed, or rationally desired. It is only substituting an indirect for a direct investigation, when the latter is impossible or too difficult. But if the two are not employed on the same general subject, and if we try to reach by hypothesis what is inaccessible to observation and reasoning, the fundamental condition is violated, and hypothesis, wandering out of the field of science, merely leads us astray. Our study of nature is restricted to the analysis of phenomena in order to discover their *laws*; that is, their constant relations of succession or similitude; and can have nothing to do with their *nature*, or their *cause*, first or final, or the mode of their production. Every hypothesis which strays beyond the domain of the positive can merely occasion interminable discussions, by pretending to pronounce on questions which our understandings are incompetent to decide.—Every man of science admits this rule, in its simple statement; but it can not be practically understood,—so often as it is violated, and to such a degree as to alter the whole character of Physics. The use of conjecture is to fill up provisionally the intervals left here and here by reality; but practically we find the two materials entirely separated, and the real subordinated to the conjectural. It is necessary, therefore, to ascertain and explain the actual state of the question with regard to Physics.

Two classes of hypothesis. The hypotheses employed by physical inquirers in our day are of two classes; the first, a very small class, relate simply to the laws of phenomena: the other, and larger class, aim at determining the general agents to which different kinds of natural effects may be referred. Now, according to the rule just laid down, the first kind alone are admissible; the second have an antiscientific character, are chimerical, and can do nothing but hinder the progress of science.

First class indispensable. In Astronomy, the first class only is in use, because the science has a wholly positive character. A fact is obscure; or a law is unknown: we proceed to form an hypothesis, in agreement, as far as possible, with the whole of the data we are in possession of; and the science, thus left free to develop itself, always ends by disclosing new observable consequences, tending to confirm or invalidate, indisputably, the primitive supposition. We have before noticed frequent and happy examples of this method of discovering the primary truths of astronomy. But, since the establishment of the law of gravitation, geometers and astronomers have put away all their fancies of chimerical fluids causing planetary motions; or have, at least, indulged in them merely as a matter of personal taste, and not of scientific investigation. It would be well if, in a study so much more difficult as Physics, philosophers would imitate the astronomers. It would be well if they would confine their hypotheses to the yet unknown circum- stances of phenomena, or their yet hidden laws, and would entirely let alone their mode of production, which is altogether beyond the limit of our faculties. What scientific use can there be in fantastic notions about fluids and imaginary ethers, which are to account for phenomena of heat, light, electricity, and magnetism? Such a mixture of facts and dreams can only vitiate the essential ideas of physics, cause endless controversy, and involve the science itself in the disgust which the wise must feel at such proceedings. These fluids are supposed to be invisible, intangible, even imponderable, and to be inseparable from the substances which they actuate. Their very definition shows them to have no place in real science; for the question of their existence is not a subject for judgment: it can no more be denied than affirmed: our reason has no grasp of them at all. Those who, in our day, believe in caloric, in a luminous ether, or electric fluids, have no right to despise the elementary spirits of Paracelsus, or to refuse to admit angels and genii. We find them spurning Lamarck's notion of a resonant fluid; but the misfortune of this hypothesis was that it came too late,—long after the establishment of acoustics. If it had been put forth as early in the days of science as the hypotheses about heat, light, and electricity, this resonant fluid would no doubt have prospered as well as the rest. Without going into the history of more of these baseless inquiries, it is enough to point out that they are irreconcilable with each other; and when superficial minds witness the ease with which they destroy each other, they naturally conclude the whole science to be arbitrary, consisting more in futile discussion than in anything else. Each sect or philosopher can show how untenable is the hypothesis of another, but can not establish his own; and it would generally be easy to devise a third which might agree with both. It is true, physicists are now eager to declare that they do not attribute any intrinsic reality to these hypotheses; and that they countenance them merely as indispensable means for facilitating the conception and combination of phenomena. But we see here the working of an incomplete positivity, which feels the inanity of such systems, and yet dares not surrender them. But besides that it is scarcely possible to employ a fictitious instrument as a reality without at times falling into the delusion of its reality, what rational ground is there for proceeding in such a way, when we have before

Second class chimerical.

us the procedure and achievements of astronomical science, for a pattern and a promise? These hypotheses *explain* nothing. For instance, the expansion of bodies by heat is not *explained*,—that is, cleared up,—by the notion of an imaginary fluid interposed between the molecules, which tends constantly to enlarge their intervals; for we still have to learn how this supposed fluid came by its spontaneous elasticity, which is, if anything, more unintelligible than the primitive fact. And so on, through the whole range. These hypotheses clear away no difficulties, but only make new ones, while they divert our attention from the true object of our inquiries. As for the plea that habit has so taken hold of the minds of inquirers, that they would be adrift if deprived of all their moorings at once, and that their language must be superseded by a wholly new one, I think this kind of difficulty is very much exaggerated. We have seen, within half a century, how often men have contrived to pass from some physical systems to their opposites without being much hindered by obstacles of language. There would be scarcely more difficulty in casting aside futile hypotheses; and we might, as we see by existing examples, gradually substitute the real and permanent meaning of scientific terms for the fanciful and variable interpretation.

History of the second class. These fluids are nothing more than the odd entities materialized. Whichever way we look at it, what is heat apart from the warm body—light apart from the luminous body—electricity apart from the electric body? Are they not pure entities, like thought apart from the thinking body, and digestion apart from the digesting body? Here we have, instead of abstract beings, imaginary fluids deprived, by their very definition, of all material qualities, so that we can not even suppose in them the limit of the most rarefied gas. If the descent of these from the old entities be not recognised, what filiation of ideas can ever be admitted? The essential character of metaphysical conceptions is to attribute to properties an existence separate from the substance which manifests them. What does it matter whether we call these abstractions souls or fluids? The origin is always the same; and it is connected with that inquisition into the essence of things which always characterizes the infancy of the human mind, occasioning, first, the conception of gods, which grew into that of souls, which became in time imaginary fluids. In all positive science, our understandings, unable to pass abruptly from the metaphysical to the positive stage, have travelled through this transition state of development. Metaphysics itself is the transition stage from theology to positive science; but a secondary transition is also necessary, as we see by the fact; a transition from metaphysical to positive conceptions. The mathematicians and astronomers have attained the positive basis. The physicists, the chemists, the physiologists, and the social philosophers, are now in the last period of transition; the physical inquirers, ready to pass up to the level of the astronomers and geometers, and all the others held back for a while by the complexity of their respective subjects; as we shall see hereafter. This bastard positivism was the way out of the old metaphysical condition, in which men would, but for it, have been imprisoned to this day. Nascent science first humored the constitutional need, and then led us on by offering to our minds, in the place of the old scholastic entities, new entities, more tangible, which must by their nature introduce into our studies the contemplation of phenomena and their laws, restricting us to these more and more. This seems to have been the important temporary use of this system of hypotheses; to enable us to pass from the metaphysical to the positive stage.

Astronomy has not been exempted from this transition state, any more than the other sciences; but it was over so long ago that it is forgotten,—so few are those who are interested in the history of philosophy! If we look back to the action of the human mind in the seventeenth century, we shall see how geometers and astronomers were preoccupied with hypotheses of the kind we are considering. There is no better example of them than that famous conception, the Vortices of Descartes; for it presents clearly the three stages of existence common to them all; the creation of the hypothesis, its temporary use, and its rejection when its purpose is answered. These vortices, so ridiculed by men who believed in caloric, ether, and electric fluids, helped us to a sound philosophy by introducing the idea of mechanism, where even Kepler had imagined only the incomprehensible action of souls and genii. When the discussion had attained the firm ground of Celestial Mechanics, founded upon the Newtonian theory, the influence of the Cartesian hypothesis ceased to be progressive, and became retrograde. To the last, the Cartesian philosophers insisted, in arguments as plausible as those of our existing physicists, that it was impossible to philosophize without such a hypothesis. They were answered in the only effectual way, by philosophizing in another mode: and the vortices were heard of no more when geometers and astronomers apprehended the true object of scientific studies. The Cartesian hypothesis contributed to the education of the human mind by leading it to see that we have nothing to do with the primitive agents, or mode of production of phenomena, but only with their laws. If, in the other sciences, we have, as their professors assert, reached the stage of positivity, hypotheses like that of the vortices may be dismissed, as no longer needed to bring us out of the metaphysical state. As soon as they are needless, they become pernicious.

In Astronomy.

The transition has been obvious elsewhere than in astronomy. It has taken place in the most advanced departments of Physics; and especially with regard to Weight. There was scarcely a philosopher, even long after Galileo's time, who had not some system to offer about the causes of the fall of bodies. At that time, such hypotheses appeared the only method of studying weight; but who hears of them now? Acoustics was

In Physics.

emancipated about the same time. The labors of Fourier will evidently release thermology; and then there will remain only the study of light and of electricity;* and no reason can be assigned for their exclusion from the general rule. The question will be regarded as settled henceforward by all who believe that the historical development of the human mind is subject to natural laws, determinate and uniform; and such will admit, as the principle of the true theory of hypotheses, that every scientific hypothesis, to be a matter of judgment, must relate exclusively to the laws of phenomena, and never to their mode of production.

Here we find, as in every analogous case of difficulty, the use of the comparative historical method which I have just employed. We shall enlarge on this hereafter: meantime, I must offer the observation that the philosophy of the sciences can not be properly studied apart from their history; and, conversely, that the history apart from the philosophy would be idle and unintelligible.

Rule of arrangement in Physics. In reviewing the different departments of Physics, I shall follow the rule which determines the order of the sciences themselves: that is, I shall take them in the order of the generality of their phenomena, their simplicity, the relative perfection of our knowledge of them, and their mutual dependence. Under this rule, we shall find that the departments that offer themselves first border upon Astronomy, and those that come last upon Chemistry.

First will come Weight, in solids and fluids, regarded statically and dynamically. About this assignment, there are no two opinions; weight being absolutely universal. Its phenomena are simple, independent of others, and so exactly understood that science is here almost as positive as in astronomy, to which it is very nearly allied. Electric phenomena being the most opposite of all to those of weight, in all these particulars, will come last; and they are closely allied to chemistry. Between them will come thermology, acoustics, and optics. Fourier put away, in his study of heat, all fantastic notions about imaginary fluids, and brought his subject up to such a point of positivity as to place it next to the study of gravity. Acoustics might, perhaps, contest its place with thermology, but for the generality of the phenomena of the latter. In regard to positivity, there is little to choose between them: but there are gaps in our knowledge of acoustics which also indicate the lower place for it. Our order then is,—Barology,

Order. or the science of weight: thermology, or the science of heat: acoustics, optics, and electricity. But we must beware of attaching too much importance to this arrangement, which is really little better than arbitrary, though as good as our present knowledge admits. We shall now proceed to a philosophical review of them, exempt from details; having, in this chapter, analyzed the proper object of Physics; the modes of investigation

---

* In Electricity, the hypothesis of fluids is rapidly yielding before the rational idea of Polarity.—J. P. N.

appropriate to it; its position in regard to the other sciences; its influence upon the education of human reason; its degree of scientific perfection; its incomplete positivity at present; the means of remedying this by a sound institution of hypotheses; and finally the rational distribution of its different departments.

CHAPTER II.

BAROLOGY.

NOTWITHSTANDING the advanced state of our means for the study of Barology, we have no complete theory of weight, but only fragmentary portions of a theory, dispersed through treatises on rational mechanics or physics. It will be of great advantage to bring them together.

The division of the subject is into two principal sections, subdivided into three; the Statical and Dynamical consideration of weight, in its application to solids, liquids, and gases. Both philosophically and historically, this division is indicated. *Divisions.*

SECTION I.

STATICS.

TAKING the statics of gravity first, we must point out that we owe our elementary notions of positive barology to Archimedes. He first clearly established that the statical effort produced in a body by gravity—that is, *weight*—is entirely independent of the form of the surface, and depends only on the volume, as long as the nature and constitution of the body remain unchanged. This may appear to us very simple; but it is not the less the true germ of a leading proposition in natural philosophy, which was perfected only at the end of the last century; namely, that not only is the weight of a body independent of its form, and even of its dimensions, but of the mode of aggregation of its particles, and of all variations which can occur in their composition, even by the different vital operations; in a word, that this quality is absolutely unalterable, except by the circumstance of its distance from the centre of the earth. Archimedes could take none but geometrical circumstances into the account; but, in this elementary relation, his work was complete. He not only discovered that, in homogeneous masses, the weight is always in proportion to the volume; but he disclosed the best means of measuring, in solid bodies, by his famous hydrostatic principle, the specific co-efficient which enables us to estimate, according to this law, *Statics of Gravity. History.*

the weight and volume of the body, by means of each other. We owe to him, too, the idea of the centre of gravity, together with the first development of the corresponding geometrical theory. Under this view, all problems respecting the equilibrium of solids are included in the domain of rational mechanics; so that, except the important relation of weight to masses, which could be fully known only to the moderns, Archimedes ought to be regarded as the true founder of statical barology, in relation to solids. There is, however, another leading idea which was not clear in the time of Archimedes, though it became so soon afterward: that of the law of the direction of gravity, which men spontaneously considered to be constant, and which the school of Alexandria ascertained to vary from place to place, always being perpendicular to the surface of the terrestrial globe; a discovery which is evidently due to astronomy, by which alone the means are offered of manifesting and measuring, by comparison, the divergence of verticals.

Cases of liquids. The ancients had no accurate ideas about the equilibrium of liquids; for Archimedes contemplated only the equilibrium of solids sustained by liquids. His principle did not result, as with us, from an analysis of the pressure of the liquid against the vessel containing it, thus disclosing the total force exercised by the liquid in sustaining the weight. The theory of gravitating liquids is to be ascribed to the moderns.

The mathematical character of fluids is that their molecules are absolutely independent: and the geometrical character of liquids is that they are absolutely incompressible. Neither of these is strictly true. The mutual adherence of fluid molecules now forms an interesting section of physics; and, as for the compressibleness of liquids, it was indicated by several phenomena—especially the transmission of sound by water—and it is now proved by unquestionable experiments. The contraction, as yet producible, is very small; and we do not know what law the phenomenon follows, in its relation to variation of pressure. But this uncertainty does not affect the theory of the equilibrium of natural liquids, owing to the extreme smallness of the condensation. In the same way, imperfect fluidity is no hinderance, provided the mass has a certain extension. We may, therefore, put aside these exceptions, and proceed to consider the equilibrium of gravitating liquids, in the two cases in which they are studied: in a mass so limited that the verticals are parallel, which is the ordinary case; or in a great mass, as that of the sea, in which we have to allow for the variable direction of gravity.

First case. In the first case, there is clearly no difficulty about the surface; and the whole question is of the pressure against the enclosing vessel. Stevinus, following the principle of Archimedes, showed that the pressure upon a horizontal boundary, or floor, is always equal to the weight of the liquid column of the same base which should issue at the surface of equilibrium; and he afterward resolved into this the case of an inclined boundary, by decomposing it into horizontal elements, as we now do. From this it appeared that the pressure is always equal to the weight of a vertical column which should have the proposed boundary for its base, and for its height that of the surface of equilibrium above the centre of gravity of this boundary. According to that, the infinitesimal analysis enables us easily to calculate the pressure against any definite portion of any curved surface. The most interesting physical result is the estimate of the total pressure supported by the whole of the vessel, which is necessarily equivalent to the weight of the liquid it contains. The equilibrium of floating bodies is only a simple application of this rule of measurement of pressure. The immersed part of the solid is a boundary; and it is clear that the pressure of the liquid to sustain the body is equivalent to a vertical force equal to the weight of the displaced fluid, and applied to the centre of gravity of the immersed portion. This rule is precisely the principle of Archimedes. The main problem was geometrically treated by him. The only really difficult research in this matter relates to the conditions of stability of this equilibrium, and the analysis of the oscillations of the body floating round its stable position; and this is one of the most complex applications of the dynamics of solids. If the question was of the vertical oscillations of the centre of gravity, the study would be easy, because we can estimate at once the way in which the pressure increases as the body is further immersed, and diminishes as it rises, tending always to a return to the primitive state. But it is otherwise with oscillations from rotation, whether of rolling or pitching, the theory of which is a matter of much interest in naval art. Here, the mathematical difficulties of the problem can be met only by abstracting the resistance and agitation of the liquid; and the labors of geometers become merely mathematical exercises, of no practical use.

The question of the equilibrium of the vast liquid masses which compose the greater part of the earth's surface is clearly connected with the general theory of the form of the planets: but difficulties, unconnected with the figure of the planets, intervene, and can not be entirely surmounted. Rational hydrostatics shows us that equilibrium is possible when there is the same density at all points equally distant from the earth's centre; a condition which is impossible under our variety of temperatures in different positions. There is no rational result from any practicable study of currents, of varying temperatures, of the compressibleness of liquids, all of which, though following unknown laws, are necessary to the solution of the problem. We have no better resource, at present, than in empirical studies: and these, which belong more to the natural history of the globe than to physics, are very imperfect.

Second case.

The theory of tides will hereafter, when sciences and their arrangements are more perfected, take its place in the department of barology. The periodical disturbances of the equilibrium of the ocean are a proper subject for study in connection with terrestrial gravity; and it can make no difference that the cause of those perturbations is found in the planetary system.

<small>Case of gases.</small> In studying the third question of equilibrium, that of gases, we meet with a difficulty which does not pertain to that of solids and liquids;—we have to discover the gravity of the general medium in which we live. In the case of liquids, we have only to weigh an empty vessel, and then the same vessel filled; whereas, in the case of the atmosphere, a vacuum can not be created but by artificial means, which must themselves be founded on a knowledge of the weight of the medium to be weighed. The fact can be ascertained only by indirect means; and those are derived from the theories of pressure which we have been noticing. Stevinus was not thinking of the atmosphere in elaborating his theory; but, as it answered for heterogeneous liquids, it must answer for the atmosphere. From that date, the means for ascertaining the equilibrium of the atmosphere, in a positive manner, were provided. <small>History.</small> Galileo projected the work, in his last years; and it was well executed by his illustrious disciple, Torricelli. He proved the existence and measurement of atmospheric pressure by showing that this pressure sustained different liquids at heights inversely proportioned to their densities. Next, Pascal established the necessary diminution of this pressure at increasing heights in the atmosphere; and Guericke's invention of the airpump, an inevitable result of Torricelli's discovery, gave us direct demonstration in the power of making a vacuum, and consequently of estimating the specific gravity of the air which surrounds us, which had hitherto been only vaguely computed. The creation and improvement of instruments of observation is an invariable consequence of scientific discovery; and in this case, the fruits are the barometer and the air-pump.

One condition remained, before we could apply the laws of hydrostatics to atmospheric equilibrium. We had to learn the relation between the density of an elastic fluid and the pressure which it supports. In liquids (supposed incompressible) the two phenomena are mutually independent; whereas, in gases they are inevitably connected; and herein lies the essential difference between the mechanical theories of the two fluids. The discovery of this elementary relation was made about the same time by Mariotte in France and Boyle in England. These illustrious philosophers proved by their experiments that the different volumes successively occupied by the same gaseous masses are in an inverse ratio to the different pressures it receives. This law has since been verified by increasing the pressure to nearly that of thirty atmospheres; and it has been adopted as the basis of the whole Mechanics of gas and vapors. But we must beware of accepting it as the mathematical expression of the reality; for it is evidently the same thing as regarding elastic fluids as always equally compressible, however compressed they may already be; or, conversely, as always equally dilatable, however dilated they may already be:—suppositions which can not be indefinitely extended. But thus it is, more or less, with all the laws we have ascertained in our study of nature. They approximate, within narrow limits, we must suppose, to the mathematical reality: but they are not that reality itself, even in the grand instance of gravitation. These laws are sufficient for our use and guidance; and that is all the result that positive science pretends to.

Under the law of Mariotte and Boyle, the theory of atmospheric equilibrium falls into the department of Rational Mechanics. *Condition of the problem.* We see at once that the air can no more be in a state of real equilibrium than the ocean; and so much the further from it as heat expands air more than water. Yet we must conceive of the partial equilibrium of a very narrow atmospheric column, to form a just general idea of the mode of diminution in regard to the density and pressure of the different strata. Putting aside considerations of heat, and the small effects of gravitation in such a case, we see that density and pressure must diminish in a geometrical progression for altitudes increasing in arithmetical progression: but this abstract variation is retarded by the diminishing heat of the loftier atmospheric strata, which makes each stratum more dense than it would be from its position. Here, therefore, we are stopped by the intervention of a new element which we do not understand,—the law of the vertical variation of atmospheric temperatures—our ignorance of which can be supplied only by inexact and uncertain expedients. Great caution is necessary in using Bouguer's method of measuring altitudes by the barometer; a method very ingenious, but depending on such complex and uncertain conditions, and requiring sometimes so much delay, that it is even preferable, when circumstances permit, to enter upon a geometrical measurement, which has so greatly the advantage in certainty. Yet, considered by itself, the method of measurement by the barometer is valuable for its contributions to our knowledge of the surface reliefs of our globe.

SECTION II.

DYNAMICS.

WE have now to consider the laws of motion of gravitating bodies; and of Solids, in the first place.

The last elementary notion about gravity,—that of the necessary proportion between weight and mass,—which was still wanting to statical barology, was established by the admirable observation that all bodies in a vacuum fall through the same space in the same time. Proceeding from this, we must examine the discovery of the laws of motion produced by gravity. We shall find herein, not only the historical origin of physics, but *History.* the must perfect method of philosophizing of which the science admits. Aristotle observed the natural acceleration of bodies in their fall; but he could not discover the law of the case, for want of the elementary principles of rational dynamics. His hypothesis, that the velocity increases in proportion to the space traversed, was plausible, till Galileo found the true theory of varied motions. When Galileo had discovered the law that the velocity and the space traversed were necessarily proportioned, the one to the time, and the other to its square, he showed how it could be verified in two ways; by the immediate observation of the fall, or by retarding the descent at will by the aid of a sufficiently inclined plane,—allowance being made for the friction. An ingenious instrument, which affords a convenient verification, was afterward offered by Atwood: it retards the descent, while leaving it vertical, by compelling a small mass to move a very large one in that direction.

By this one law of Galileo, all the problems relating to the motions of falling bodies resolve themselves into questions of rational dynamics. They, indeed, compelled its formation, in the seventeenth century; as, in the eighteenth, questions of celestial mechanics thoroughly developed it. In all that relates to the motion of translation of a body in space, this study is due to Galileo, who established the theory of the curvilinear motion of projectiles—allowance being made for the resistance of the air. All attempts, however, to ascertain the effect of this resistance have hitherto been in vain; and therefore the study of the real motion of projectiles is still extremely imperfect.

As for the motions that gravity occasions in bodies that are not free in space,—the only important case is that of a body confined to a given curve. It constitutes the problem of the pendulum, which we have already considered, as the immortal achievement of Huyghens. Its practical interest, as the basis of the most perfect chronometry, presented itself first: but it has besides furnished two general consequences, very essential to the progress of barology. First, it enabled Newton to verify the proportion of weight to mass with much more exactness than would have been possible by the experiment of the fall of bodies in a vacuum; and, in the second place, the pendulum has enabled us, as we saw before, to observe the differences of the universal gravity at different distances from the earth's centre. It is by the use of this method that we are continually adding to our knowledge of the measure of gravity at various points on the globe, and, therefore, of the figure of the earth.

In these different sections of dynamical barology, solid bodies are regarded as points,—considerations of dimensions being discarded: but a new order of difficulties arises when we have to consider the particles of which the body is really formed. With regard to cases of restricted motion, as the pendulum, the thing to be done is to find out under what laws the different points of the body modify the unequal time of their respective oscillations, so that the whole may oscillate as one point, real or ideal. This law, discovered by Huyghens, and afterward obtained by James Bernouilli in a more scientific manner, easily transforms the compound into the simple pendulum, when the moment of inertia of the body is known. The study of the pendulum is involved in all the questions of the dynamics of solids. To give it the last degree of precision, it is necessary to consider the resistance of the air, though that resistance is small in comparison with the case of projectiles. This is done, with ease and certainty, by comparing theoretical oscillations with real ones, exposed to the resistance of the air; when, of course, the difference between the two gives the amount of that resistance.

We have seen enough of the difficulties of hydrodynamics to understand that the part of dynamical barology which relates to fluids must still be very imperfect. In the cases of the gases, and especially of the atmosphere, next to nothing has been attempted, from the sense of the impracticable nature of the inquiry. The only analysis which has been proceeded with, in regard to liquids, is that of their flow by very small orifices in the bottom or sides of a vessel: that is, the purely linear motion, mathematically presented by Daniel Bernouilli, in his celebrated hypothesis of the parallelism of laminæ. Its principal result has been to demonstrate the rule, empirically proposed by Torricelli, as to the estimate of the velocity of the liquid at the orifice, as equal to that of a weight falling from the entire height of the liquid in the vessel. Now, this rule has been reconciled with observation, even on the supposition of an invariable level, only by an ingenious fiction, suggested by the singular phenomenon of the *contraction* of the liquid filament. The case of a variable level is scarcely entered upon yet; or any which involves the form and size of the orifice. As for more complex cases,—their theory is yet entirely in its infancy.*

Fluids.

Case of liquids.

From this cursory review of Barology we may carry away some idea of its spirit, and of the progress of which it admits. Imperfect as our survey has been, we may perceive that this first province of Physics is not only the purest, but the richest. We may observe in it a character of rationality, and a degree of co-ordination which we shall not meet with in other parts of the science. It is because we look for a consistency and precision almost like those that we find in astronomy that we consider barology so imperfect as we do. It has long attained its position of positivity: there is no one of its subdivisions which is not at least sketched out: all the general means of investigation, Observation, Experiment, and Comparison, have been successively applied to it; and thus its future progress depends only on a more complete harmony between these three methods, and on a more uniform and close combination between the mathematical spirit and the physical.

Existing state of Barology.

---

* The subjects here spoken of by M. Comte have recently received remarkable elucidation through experiments entered upon chiefly by the instrumentality of the "British Association for the Advancement of Science." There can not be much hope, in the present state of our knowledge, of ascending otherwise than empirically toward any general law or rules which will either comprehend phenomena, or be of use in practice.—J. P. N.

CHAPTER III.

THERMOLOGY.

NEXT to the phenomena of gravity, those of heat are, unquestionably, the most universal in the province of Physics. Throughout the economy of terrestrial nature, dead or living, the function of heat is as important as that of gravity, of which it is the chief antagonist. The consideration of gravity presides over the geometrical and mechanical study of bodies: while that of heat prevails in its turn, when we investigate deeper modifications, relating to either the state of aggregation or the composition of molecules: and finally, vitality is subordinated to it. The intelligent application of heat constitutes the chief action of man upon nature. Thus, after barology, thermology is, of all the parts of physics, the one which most deserves our study.

<small>Its nature.</small>

The earliest scientific observations in thermology are almost as old as the discoveries of Stevinus and Galileo on gravity, the invention of the thermometer having taken place at the beginning of the seventeenth century; but, owing to the complication of its phenomena, it has always been far distanced by barology. At the end of the seventeenth century, the indications of the thermometer could not be compared, for want of two fixed points, the necessity for which was, at that time, pointed out by Newton. The greatest difference between the two studies, however, was in their spirit. While philosophers were already inquiring, with regard to weight, what were its phenomena and their laws, those who were studying heat were looking for the nature of fire, and reducing the facts of the case to something merely episodical. It was only at the end of the last century that the great discovery of Black imparted anything of a scientific character to thermology, while barology was almost as much advanced as it is now. Our philosophers still entertain some of the old chimeras; but now very loosely, and, as they say, to facilitate the study of the phenomena. The labors of Fourier, however, must soon establish a thoroughly scientific method; and this result can not but be aided by the fact that the two great modern hypotheses about the nature of heat are in direct collision. It is certain that, of all the provinces of physics, thermology is the nearest to a complete emancipation from the anti-scientific spirit.

<small>Its history.</small>

Of all the branches of Physics which admit Mathematical Analysis, this is the one which exhibits the most special application of it. Barology enters into the province of rational mechanics; and so, in a less degree, does acoustics. The

<small>Relation to Mathematics.</small>

analytical theory of heat now offers a scientific character as satisfactory as that of gravity and sound; and it may be treated as a dependency of abstract Mechanics, without any resort to chimerical hypotheses. Our present business is, however, with the purely physical study of Heat.

SECTION I.

MUTUAL THERMOLOGICAL INFLUENCE.

PHYSICAL thermology consists of two parts, distinct, but nearly connected. The first relates to the mutual influence of two bodies in altering their respective temperatures. The second consists in the study of those alterations: that is, of the modifications or entire change which the physical constitution of bodies may undergo in consequence of their variations of temperature, stopping at the point at which chemistry intervenes,—that is, where the molecular composition becomes affected. The first of these consists in the theory of warming and cooling. *Two parts.*

No thermological effect is produced between bodies of precisely the same temperature. The action begins when the temperatures become unequal. The warmer body then raises the temperature of the cooler; and the cooler depresses that of the warmer, till, sooner or later, they reach a common temperature. Though the final state may usually be at an unequal distance from the two extremes, action is not, properly speaking, the less truly equivalent to reaction in a contrary direction. This case again divides itself into two. The bodies may act at a distance, greater or smaller; or they may be in contact. The first case constitutes what is called the *radiation* of heat. *First: Mutual Influence.*

The direct communication of heat between two isolated bodies was long denied by philosophers, who regarded the air, or some other medium, as indispensable to the effect. But there is now no doubt about it; as thermological action takes place in a vacuum: and the small density and conducting power of the air could not account for the effects observed in the majority of ordinary cases. This action, like that of gravity, extends, no doubt, to all distances, in conformity to the fundamental approximation between these two great phenomena, pointed out by Fourier: for we can conceive of the planets of our system as exerting an appreciable mutual influence in this respect: and it seems as if the temperature of the whole solar system were attributable to the thermometrical equilibrium to which all the parts of the universe are for ever tending.

The first law relating to such an action consists in its rectilinear propagation. Though the term *radiation* has been connected with untenable hypotheses, we may retain it, provided we carefully restrict it to the meaning that it is in a right line that two points act thermologically on each other. It thus implies that in placing bodies to prevent this mutual action between two others, the absorbing body must be placed in a right line.— *Radiation of Heat.*

This radiating heat can be reflected like light, and in conformity with the same rule: and it undergoes the same refractions as light, with some modifications.—Another question about this action relates to the influence of the direction of radiation considered in regard to the surface of either the warming or the warmed body. The experiments of Leslie, confirmed by the mathematical results of the case, have established that, in either case, the intensity of the action is greater as the rays approach the perpendicular, and that it varies in proportion to the sine of the angle that they form with each surface.—The last consideration, and the most important, is the difference of temperatures between the two bodies. When this difference is not very great, the intensity of the action is in precise proportion to it: but this relation appears to cease when the temperatures become very unequal.

When the radiation is not direct, but transmitted through an intermediate body, the conditions just noticed become complicated with new circumstances relating to the action of the interposed body. We owe to Saussure a fine series of experiments—hardly varied enough, however—upon the effect of a set of transparent coverings in changing remarkably the natural mode of accumulation or dispersion of heat, whether radiant or obscure. More recently, M. Melloni has pointed out an essential distinction between the transmission of heat and that of light, in proving that the most translucent bodies are not always those through which heat passes most easily, as was previously supposed.

Propagation by Contact. It is well for physicists to separate the radiation of heat from its propagation by contact, for analytical purposes; but it is evident that the separation can not be found in nature. They are always found in connection, however unequally. Besides that the atmosphere can hardly be absent, and is always establishing an equilibrium of heat between bodies that are apart, it is clear that it is only the state of the surface that can be determined by radiation. The interior parts, which have as much to do with the final condition as the surface, can grow warmer or cooler only by contiguous and gradual propagation. Thus, no real case can be analyzed by the study of its radiation alone. And again, the action by contact of two bodies can take place only in those small portions which are in contact while the bodies are acting upon each other by the radiation of all the rest of their surfaces. Thus, though the two modes of action are really distinct, the analysis of either is rendered extremely difficult by their perpetual combination.

Of the three conditions noticed above, relating to the action exerted at a distance, the only one which applies equally to the propagation of heat by contact is the difference of temperatures. The temperature of the parts in question can have so little inequality, that the law which makes the action increase in proportion to the difference may be regarded almost as an exact expression of the fact. As for the law relating to direction, it probably subsists here too, though we can not be perfectly assured of it. But that law which relates to the distance must be totally changed; for, on the one hand, the action of the almost contiguous particles can not be nearly so great as the variations which we are able to estimate would lead us to suppose; and, on the other hand, when we compare the various small intervals between them, the diminution is certainly much more rapid than in the case of distant bodies.

Whatever may be the mode of the warming and cooling of respective bodies, the final state which is established under the laws just noticed is numerically determined by three essential coefficients, proper to every natural body, as its specific gravity is in barology.

Under the old term *conductibility*, two properties were confounded, which Fourier separated, giving them the names of *penetrability* and *permeability;* the first signifying that by which heat is admitted at the surface, or dispersed from it, and the other that by which the changes at the surface are propagated through the interior. Permeability depends altogether on the nature of the body and its state of aggregation. The differences of bodies in this respect have always been open to observation; for instance, the rapid propagation of heat in metallic bodies in comparison with coal, which may be burning at one point and scarcely warm a few inches off, while the heat rapidly pervades the whole body of metal. It varies with the physical constitution of bodies, so diminishing in fluids that even Rumford went so far as to deny permeability in them altogether, ascribing the propagation of heat in them to interior agitation. This was a mistake; but permeability is very weak in liquids, and weaker still in gases. As to penetrability, while partly depending on the state of aggregation of bodies, it depends much more on the state of their surfaces,—on color, polish, and the regularity in which radiation in various directions can take place, and divers other modifications: and it changes in the same surface as it is exposed to the action of different media.

<small>Conductibility.</small>
<small>Permeability.</small>
<small>Penetrability.</small>

Strictly speaking, the different degrees of these two kinds of conductibility can not affect the final thermological state of the two bodies, but only the time at which it is reached. Yet, as real questions often become mere questions of time, it is clear that if the inequalities are very marked, they must affect the intensity of the phenomena under study; for instance, where the permeability is so feeble that the requisite interior heat can not be obtained in time, but by applying such heat to the surface as will break or burn it; in which case, the phenomenon can not take place; or, not within any practicable time. In general, the more perfect both kinds of conductibility, the better the bodies will obey the laws of thermological action, at a distance or in contact. It would therefore be very important to measure the value of these two coefficients for all bodies under study: but unhappily such estimates are at present extremely imperfect. The business was vague enough, of course, when the two kinds of conductibility were confounded; but Fourier has taught us how to estimate permeability directly, and of course, penetrability indirectly, by subtracting the permeability from the total of conductibility. But the application of his methods is as yet hardly initiated.

Specific heat.

One consideration, that of *specific heat*, remains to be noticed, as concurring to regulate the results of thermological action. Whether under conditions of equal weight, or of equal volume, the different substances consume distinct quantities of heat to raise their temperature equally. This property, of which little was known till the latter half of the last century, depends, like permeability, only in a less degree, on the physical constitution of bodies, while it is independent of the state of their surfaces. It must considerably affect the equalized temperature common to two bodies, which can not be equally different from the primitive temperature of each, if they differ from each other in the point of their specific heat. Physicists have achieved a good deal in the estimate of specific heat. The best method is that of experiment with the calorimeter, invented by Lavoisier and Laplace, for its measurement. The quantity of heat consumed by any body at a determinate elevation of temperature, is estimated by the quantity of ice melted by the heat it gives out in its passage from the highest to the lowest temperature. The apparatus is so contrived as to isolate the experiment from all thermological action of the vessel and of the medium; and thus the results obtained are as precise as can be desired.

There are the three coefficients which serve to fix the final temperatures which result from the thermological equilibrium of bodies. Till we know more of the laws of their variations, it is natural to suppose them essentially uniform and constant: but it would not be rational to conceive of conductibility as identical in all directions, in all bodies, however their structure may vary in different directions; and specific heat probably undergoes changes at extreme temperatures, and especially in the neighborhood of those which determine a new state of aggregation, as some experiments already seem to indicate. However, these modifications are still so uncertain and obscure, that physicists can not be blamed for not keeping them, at this day, perpetually in view.

SECTION II.

CONSTITUENT CHANGES BY HEAT.

Second part. Constituent changes by heat.

THE second part of thermology is that which relates to the alterations caused by heat in the physical constitution of bodies. These alterations are of two kinds; changes of volume, and the production of a new state of aggregation; and this is the part of thermology of which we are least ignorant.

These phenomena are independent of those of warming and cooling, though they are found together. When we heat any substance, the elevation of temperature is determined solely by the portion of heat consumed, the rest of which (often the greater part), insensible to the thermometer, is absorbed to modify the physical constitution. This is what we mean when we say that a portion of heat has become latent; a term which we may retain, *Latent heat.* though it was originally used in connection with a theory about the nature of heat. This is the fundamental law discovered by Black, by observation of the indisputable cases in which a physical modification takes place without any change of temperature in the modified body. When the two effects co-exist, it is much more difficult to analyse and apportion them.

Considering first the laws of change of volume, it is a general truth that every homogeneous body dilates *Change of volume.* with heat and contracts with cold; and the fact holds good with heterogeneous bodies, such as organized tissues especially, in regard to their constituent parts. There are very few exceptions to this rule, and those few extend over a very small portion of the thermometrical scale. The principal anomaly however being the case of water, it has great importance in natural history, though not much in abstract physics, except from the use that philosophers have made of it to procure an invariable unity of density, always at command. These anomalies, too rare and restricted to invalidate any general law, are sufficient to discredit all *à-priori* explanations of expansions and contractions, according to which every increase of temperature should cause an expansion, and every diminution a contraction, contrary to the facts.

Solids dilate less than liquids, under the same elevation of temperature, and liquids than gases; and not only when the same substance passes through the three states, but also when different substances are employed. The expansion of solids proceeds, as far as we know, with perfect uniformity. We know more of the case of liquids, which is rendered extremely important from its connection with the true theory of the thermometer, without which all thermological inquiries would be left in a very dubious state. Experiments, devised by Dulong and Petit, have shown that for above three hundred centigrade degrees the expansion of the mercury follows an exactly uniform course,—equal increase of volume being produced by heat able to melt equal weights of ice at zero. This is the only case fully established; but we have reason to believe that the rule extends to that of all liquids. The most marked case of such regularity is that of gases. Not only does the expansion take place by equal gradations, as usually in liquids and solids, but it affects all gases alike. Gases differ from each other, like liquids and solids, in their density, their specific heat, and their permeability; yet they all dilate uniformly and equally, their volume increasing three eighths, from the temperature of melting ice to that of boiling water. Vapors are like gases in this particular, as in so many others. These are the simple general laws of the expansion of elastic fluids, discovered at once by Gay-Lussac at Paris, and Dalton at Manchester.

Change in state of aggregation. Next, we have to notice the changes produced by heat in the state of aggregation of bodies.

Solidity and fluidity used to be regarded as absolute qualities of bodies; whereas, we now know them to be relative, and are even certain that all solid bodies might be rendered fluid if we could apply heat enough, avoiding chemical alteration. In the converse way, we used to suppose that gases must preserve their elasticity, through all degrees of cooling and of compression; whereas Bussy and Faraday have shown us that most of them easily become liquid, when they are seized in their nascent state; and there is every reason to believe that by a due combination of cold and pressure, they may be always liquefied, even in their developed state. Under this view, therefore, different substances are distinguished only by the different parts of the indefinite thermometrical scale to which their successive states, solid, liquid, and gaseous, correspond. But this simple inequality is an all-important characteristic, which is not yet thoroughly connected with any other fundamental quality of each substance. Density is the relation which is the least obscure and variable—gases being in general less dense than liquids, and liquids than solids. But there are striking exceptions in the second case, and might be in the first, if we knew more of gases in regard to compression, and in varied circumstances of other kinds. As for the three states of the same substance, there is always, except in some cases of scarce anomaly, rarefaction in the fusion of solids and in the evaporation of liquids. All these changes have been brought by the illustrious Black under one fundamental law, which is, both from its importance and its universality, one of the finest discoveries in natural philosophy. It is this: that in the passage from the solid to the liquid state, and from that **Law of engagement and disengagement of heat.** to the gaseous, every substance always absorbs a more or less distinguishable quantity of heat, without raising its temperature; while the inverse process occasions a disengagement of heat, precisely correspondent to the absorption. These disengagements and absorptions of heat are evidently, after chemical phenomena, the principal sources of heat and cold. In an experiment of Leslie's, an evaporation, rendered extremely rapid by artificial means, has produced the lowest temperatures known. Eminent natural philosophers have even believed that the heat which is so abundantly disengaged in most great chemical combinations could proceed only from the different changes of state which commonly result from them. But this opinion, though true in regard to a great number of cases, has too many exceptions to deal with to become a general principle.

Vapors. We have now done with physical thermology. But the laws of the formation and tension of vapors now form an appendix to it; and also of course hygrometry. The theory is, in fact, the necessary complement of the doctrine of changes of state; and this is its proper place.

Before Saussure's time, evaporation was regarded as a chemical fact occasioned by the dissolving action of air upon liquids. He showed that the action of the air was adverse to evaporation, except in the case of the renovation of the atmosphere. The test was found in the formation of vapor in a restricted space. Saussure found that, in such a case, with a given time, temperature, and space, the quantity and elasticity of the vapor were always the same, whether the space was a vacuum or filled with gas. The mass and tension of the vapor increased steadily with the temperature; whereas it appears that no degree of cold suffices to stop the process entirely, since ice itself produces a vapor appreciable by very delicate means of observation. We do not know by what law the increase of temperature accelerates the evaporation, at least while the liquid remains below boiling point; but the variations in elasticity of the vapor produced have been successfully studied.

One term common to all liquids is the boiling point. At that point, the rising tension of the vapor formed has become equal to the atmospheric pressure; a fact which can be ascertained by direct experiment. Proceeding from this point, Dr. Dalton discovered the important law that the vapors of different liquids have tensions always equal between themselves to temperatures equi-distant from the corresponding boiling points, whatever may be otherwise the direction of the difference. Thus, the boiling of water taking place at one hundred degrees, and that of alcohol at eighty degrees, the two vapors, having at that point the same tension, equal to the atmospheric pressure, will then have equal elasticities, superior or inferior to the preceding, when their two temperatures are made to vary in the same number of degrees. The many new liquids discovered by chemists since this law was found have all tended to confirm it. It is very desirable that some harmony should be discovered between the boiling temperatures of different liquids and their other properties; but this remains to be done, and these temperatures appear to us entirely incoherent, though there is every reason to believe that they are not so. *Temperatures of ebullition.*

It is evident that this law of Dalton's simplifies prodigiously the inquiry into the variation of the tension of vapors, according to their temperatures; since the analysis of these variations in one instance will serve for all the rest. The experiments undertaken for this purpose by Dr. Dalton and his successors have not fully established the rule of proportion between the tension and the temperature: but an empirical law proposed by Dulong has thus far answered to the observed phenomena. All *à-priori* determinations of the law have utterly failed.

The study of hygrometrical equilibrium between moist bodies seems a natural adjunct to the theory of evaporation. Saussure and Deluc have given us a valuable instrument for this inquiry; but we know scarcely anything of the laws which regulate the equilibrium of moisture. Prevision, which is the exact measure of science of every kind, is almost non-existent in the case of hygrometry. The small part that it plays in the *Hygrometical equilibrium.*

inorganic departments of nature is, no doubt, the reason of the little attention that physicists have devoted to it; but we shall hereafter find how important is its share in vital phenomena. According to M. de Blainville, hygrometrical action constitutes the first degree and elementary mode of the nutrition of living bodies, as capillarity is the germ of the most simple organic motions. In this view, the neglect is much to be regretted. It is one instance among a multitude of the mischiefs arising from the restricted training of natural philosophers. In this case, two important studies, which can be accomplished only by physical inquirers, are neglected, merely because their chief destination concerns another department of science.

After this survey, we can form some idea of the characteristics of this fine section of Physics. We see the rational connection of the different questions comprised in it; the degree of perfection which each of them has attained; and the gaps which remain to be filled up. A vast advance was made when, by the genius of Fourier, the most simple and fundamental phenomena of heat were attached to an admirable mathematical theory.

SECTION III.

CONNECTION WITH ANALYSIS.

Thermology connected with Analysis. This theory relates to the first class of cases—those in which an equalization of heat takes place between bodies at a distance or in contact; and not at all to those in which the physical constitution is altered by heat. It is only by an indirect investigation that we can learn how heat, once introduced into a body from the surface, extends through its mass, assigning to each point, at any fixed moment, a determinate temperature; or the converse—how this interior heat is dissipated, by a gradual dispersion through the surface. As direct observation could not help us here, we must remain in ignorance, if Fourier had not brought mathematical analysis to the aid of observation, so as to discover the laws by which these processes take place. The perfection with which this has been done opens so wide a field of exploration and application, unites so strictly the abstract and the concrete, and is so pure an example of the positive aim and method, that future generations will probably assign to this achievement of Fourier's the next place, as a mathematical creation, to the theory of gravitation. Many contemporaries have hastened into the new field thus opened; but most of them have used it only for analytical exercises which add nothing to our permanent knowledge; and perhaps the labors of M. Duhamel are hitherto the only ones which afford really any extension of Fourier's theory, by perfecting the analytical representation of thermological phenomena.

According to the plan of this work, we ought not to quit the limits of natural philosophy to notice any concrete considerations of natural history—the secondary sciences being only derivatives from the primary. It is departing from our rule, therefore, to bring forward the important theory of terrestrial temperatures; yet this most important and difficult application of mathematical thermology forms so interesting a part of Fourier's doctrine, that I can not refrain from offering some notice of it.

SECTION IV.

TERRESTRIAL TEMPERATURES.

THE temperature of each point of our globe is owing (putting aside local or accidental influences) to the action of three general and permanent causes variously combined: first, the solar heat, affecting different parts unequally, and subjected to periodicul variations; next, the interior heat proper to the earth since its formation as a planet; and thirdly the general thermometrical state of the space occupied by the solar system. The second is the only one of the three which acts upon all the points of the globe. The influence of the two others is confined to the surface. The order in which they have become known to us is that in which I have placed them. Before Fourier's time, the whole subject had been so vaguely and carelessly regarded, that all the phenomena were ascribed to solar heat alone. It is true, the notion of a central heat was very ancient; but this hypothesis, believed in and rejected without sufficient reason, had no scientific consistency—the question having never been raised of the effect of this original heat on the thermological variations at the surface. The theory of Fourier afforded him mathematical evidence that at the surface the temperatures would be widely different from what they are, both in degree and mutual proportion, if the globe were not pervaded by a heat of its own, independent of the action of the sun; a heat which tends to dispersion from the surface, by radiation toward the planets, though the atmosphere must considerably retard this dispersion. This original heat contributes very little, in a direct way, to the temperatures at the surface; but without it, the solar influence would be almost wholly lost, in the total mass of the globe; and it therefore prevents the periodical variations of temperature from following other laws than those which result from the solar influence. Immediately below the surface, the central heat becomes preponderant, and soonest in the parts nearest the equator; and it becomes the sole regulator of temperatures, and in a rigorously uniform manner, in proportion to the depth. As to the third cause, Fourier was the first to conceive of it. He was wont to give, in a simple and striking form, the results of his inquiries in the saying that if the earth left a thermometer behind it in any part of its orbit, the instrument (supposing it protected from solar influence) could not fall indefinitely: the column would stop at some point or other, which would indicate the temperature of the space in which we revolve. This is one way of saying that the state of the temperatures on the surface of the globe would be inexplicable, even considering the interior heat, if the surrounding space had not a determinate temperature differing but little from that which we should find at the poles, if we could precisely estimate it. It is remarkable that, of the two new thermological causes discovered by Fourier, one may be directly observed at the equator, and the other at the poles; while, for all the intermediate points, our observation must be guided and interpreted by mathematical analysis.

New as this difficult inquiry is, our progress in it depends only on the perfecting of the observations which Fourier's theory has marked out for us. When the data of the problem thus becomes better known, this theory will enable us to lay hold of some certain evidences of the ancient thermological state of our globe, as well as of its future modifications. We have already learned one fact of high importance; that the periodical state of the earth's surface has become essentially fixed, and can not undergo any but imperceptible changes by the continuous cooling of the interior mass through future ages. This rapid sketch will suffice to show what a sudden scientific consistency has been given, by the labors of one man of genius, to this fundamental portion of natural history, which, before Fourier's time, was made up of vague and arbitrary opinions, mingled with incomplete and incoherent observations, out of which no exact general view could possibly arise.

CHAPTER IV.

ACOUSTICS.

Its nature. This science had to pass, like all the rest, through the theological and metaphysical stages; but it assumed its positive character about the same time with Barology, and as completely, though our knowledge of it is, as yet, very scanty, in comparison with what we have learned of gravity. The exact information which was obtained in the middle of the seventeenth century about the elementary mechanical properties of the atmosphere, opened up a clear conception of the production and transmission of sonorous vibrations. The analysis of the phenomena of sound shows us that the doctrine of vibrations offers the exact expression of an incontestable reality. Besides its philosophical interest, and the direct importance of the phenomena of Acoustics, this department of Physics appeals to special attention in two principal relations, arising from its use in perfecting our fundamental ideas regarding inorganic bodies, and Man himself.

By studying sonorous vibrations, we obtain some insight into the interior mechanical constitution of natural bodies, manifested by the modifications undergone by the vibratory motions of their molecules. Acoustics affords the best, if not the only means for this inquiry; and the small present amount of our acquisitions seems to me no reason why we should not obtain abundant results when the study of acoustics is more advanced. It has already revealed to us some delicate properties of natural bodies which could not have been perceived in any other way. For instance, the capacity to contract *habits*—a faculty which seemed to belong exclusively to living beings (I mean the power of contracting fixed dispositions, according to a prolonged series of uniform impressions)—is clearly shown to exist, in a greater or smaller degree, in inorganic apparatus. By vibratory motions, also, two mechanical structures, placed apart, act remarkably upon each other, as in the case of two clocks placed upon the same pedestal. *Relation to the study of inorganic bodies.*

On the other hand, acoustics forms a basis to physiology for the analysis of the two elementary functions which are most important to the establishment of social relations—hearing and the utterance of sound. Putting aside, in this place, all the nervous phenomena of the case, it is clear that the inquiry rests on a knowledge of the general laws of acoustics, which regulate the mode of vibration of all auditory apparatus. It is remarkably so with regard to the production of the voice,—a phenomenon of the same character with that of the action of any other sonorous instrument, except for its extreme complication, through the organic variations which affect the vocal system. Yet, it is not to physicists that the study of these two great phenomena belongs. The anatomists and physiologists ought not to surrender it to them, but to derive from physics all the ideas necessary for conducting the research themselves: for physicists are not prepared with the anatomical data of the problem, nor yet to supply a sound physiological interpretation of the results obtained. Science has indeed suffered from the prejudices which have grown out of the introduction into physics of superficial theories of hearing and phonation, from physical inquirers having intruded upon the province of the physiologists. *Relation to Physiology.*

After Barology, there is no science which admits of the application of mathematical doctrines and methods so well as Acoustics. In the most general view, the phenomena of sound evidently belong to the theory of very minute oscillations of any system of molecules round a situation of stable equilibrium; for, in order to the sound being produced, there must be an abrupt perturbation in the molecular equilibrium; and this transient derangement must be followed by a quick return to the primitive state. Once produced, in the body directly shaken, the vibrations may be transmitted at considerable intervals, by means of an elastic medium, by exciting a gradual succession of expansions and con- *Relation to Mathematics.*

tractions which are in evident analogy with the waves formed on the surface of a liquid, and have given occasion to the term sonorous *undulations*. In the air, in particular, so elastic as it is, the vibration must propagate itself, not only in the direction of the primitive concussion, but in all directions, in the same degree. The transmitted vibrations, we must observe, are always necessarily isochronal with the primitive vibrations, though their amplitude may be widely different.

It is clear from the outset that the science of acoustics becomes, almost from its origin, subject to the laws of rational mechanics. Since the time of Newton, who was the first to attempt to determine the rate of propagation of sound in the air, acoustics has always been more or less mixed up with the labor of geometers to develop abstract Mechanics. It was from simple considerations of acoustics that Daniel Bernouilli derived the general principle relating to the necessary and separate coexistence, or independency, of small and various oscillations occasioned at the same time in any system, by distinct concussions. The phenomena of sound afford the best realization of that law, without which it would be impossible to explain the commonest phenomenon of acoustics— the simultaneous existence of numerous and distinct sounds, such as we are every moment hearing.

Though the connection of acoustics with rational Mechanics is almost as direct and complete as that of Barology, this mathematical character is far less manageable in the one case than the other. The most important questions in barology are immediately connected with the clearest and most primitive mechanical theories; whereas the mathematical study of sonorous vibrations depends on that difficult and delicate dynamical theory, the theory of the perturbations of equilibrium, and the differential equations which it furnishes relative to the highest and most imperfect part of the integral calculus. Vibratory motion of one dimension is the only one, even in regard to solids, whose mathematical theory is complete. Of such motion of three dimensions we are, as yet, wholly ignorant.

To form any idea of the difficulties of the case, we must remember that vibratory motions must occasion certain physical modifications of another nature in the molecular constitution of bodies; and that these changes, though affecting the vibratory result, are too minute and transient to be appreciable. The only attempt that has been made to analyze such a complication is in the case of the thermological effects which result from the vibratory motion. Laplace used this case to explain the difference between the velocity of sound in the air as determined by experiment and that prescribed by the dynamic formula, which indicated a variation of of about one sixth. This difference is accounted for by the heat disengaged by the compression of the atmospheric strata, which must make their elasticity vary in a greater proportion than their density, thereby accelerating the propagation of the vibratory motion. It is true, a great gap is left here; since, as it is impossible to measure this disengagement of heat, we must assign to it conjecturally the value which compensates for the difference of the two velocities. But we learn from this procedure of Laplace the necessity of combining thermological considerations with the dynamical theory of vibratory motions. The modification is less marked in the case of liquids, and still less in that of solids; but we are too far behind with our comparative experiments to be able to judge whether the modification is or is not too inconsiderable for notice.

Notwithstanding the eminent difficulties of the mathematical theory of sonorous vibrations, we owe to it such progress as has yet been made in acoustics. The formation of the differential equations proper to the phenomena is, independent of their integration, a very important acquisition, on account of the approximations which mathematical analysis allows between questions, otherwise heterogeneous, which lead to similar equations. This fundamental property, whose value we have so often to recognise, applies remarkably in the present case; and especially since the creation of mathematical thermology, whose principal equations are strongly analogous to those of vibratory motion.—This means of investigation is all the more valuable on account of the difficulties in the way of direct inquiry into the phenomena of sound. We may decide upon the necessity of the atmospheric medium for the transmission of sonorous vibrations; and we may conceive of the possibility of determining by experiment the duration of the propagation, in the air, and then through other media; but the general laws of the vibrations of sonorous bodies escape immediate observation. We should know almost nothing of the whole case if the mathematical theory did not come in to connect the different phenomena of sound, enabling us to substitute for direct observation an equivalent examination of more favorable cases subjected to the same law. For instance, when the analysis of the problem of vibrating cords has shown us that, other things being equal, the number of oscillations is in inverse proportion to the length of the cord, we see that the most rapid vibrations of a very short cord may be counted, since the law enables us to direct our attention to very slow vibrations. The same substitution is at our command in many cases in which it is less direct. Still, it is to be regretted that the process of experimentation has not been further improved.

Acoustics consists of three parts. We might perhaps say four, including the *timbre* (ring or tone) arising from the particular mode of vibration of each resonant body. This quality is so real that we constantly speak of it, both in daily life and in natural history: but it would be wandering out of the department of general physics to inquire what constitutes the ring or tone peculiar to different bodies, such as stones, wood, metals, organized tissues, etc., whose properties lie within the scope of concrete physics. But, if we regard this quality as capable of modification by changes of circumstances, then we bring it into the domain of acoustics, and recognise its proper position, though we know nothing else about it. That part of the science presents a mere void.

The three parts referred to are, first, the mode of propagation of sounds: next, their degree of intensity; and thirdly, their musical tone. Of these departments, the second is that of which our knowledge is most imperfect.

SECTION I.

PROPAGATION OF SOUNDS.

Propagation of sounds. As to the first, the propagation of sounds, the simplest, most interesting, and best known question is the measurement of the duration, especially when the atmosphere is the medium. Newton enunciated it very simply, apart from all modifying causes;—that the velocity of sound is that acquired by a gravitating body falling from a height equal to half the weight of the atmosphere,—supposing the atmosphere homogeneous. In an analogous way, we may calculate the velocity of sound in the different gases, according to their respective densities and elasticities. By this law the speed of sound in the air must be regarded as independent of atmospheric vicissitudes, since, by Mariotte's rules, the density and elasticity of the air always vary in proportion; and their mutual relation alone influences the velocity in question. Of Laplace's rectification of Newton's formula, we took notice just now.—One important result of this law is the necessary identity of the velocity of different sounds, notwithstanding their varying degrees of intensity or of acuteness. If any inequality existed, we should be able to establish it, from the irregularity which must take place in musical intervals at a certain distance.

Effect of atmospheric agitation. All mathematical calculations about the velocity of sound suppose the atmosphere to be motionless, except in regard to the vibrations under notice, and it is one of the interesting points of the case to ascertain what effect is produced by agitations of the air. The result of experiments for this purpose is that, within the limits of the common winds, there is no perceptible effect on the velocity of sound when the direction of the atmospheric current is perpendicular to that in which the sound is propagated; and that when the two directions coincide, the velocity is slightly accelerated if the directions agree, and retarded if they are opposed: but the amount and, of course, the law of this slight perturbation are unknown.—It is only in regard to the air that the velocity of sound has been effectually studied.

SECTION II.

INTENSITY OF SOUNDS.

Intensity of sounds. We can not pretend to be any wiser about the intensity of sounds,—which is the second part of acoustics. Nor only have the phenomena never been analyzed or estimated, but the labors of the student have added nothing essential to the results of popular experience about the influences which regulate the intensity of sound; such as the extent of vibrating surfaces, the distance of the resonant body, and so on. These subjects have therefore no right to figure in our programmes of physical science; and to expatiate upon them is to misconceive the character of science, which can never be anything else than a special carrying out of universal reason and experience, and which therefore has for its starting-point the aggregate of the ideas spontaneously acquired by the generality of men in regard to the subjects in question. If we did but attend to this truth, we should simplify our scientific expositions not a little, by stripping them of a multitude of superfluous details which only obscure the additions that science is able to make to the fundamental mass of human knowledge.

With regard to the intensity of sound, the only scientific inquiry,—a very easy one,—which has been accomplished, relates to the effect of the density of the atmospheric medium on the force of sounds. Here acoustics confirms and explains the common observation on the attenuation of sound in proportion to the rarity of the air, without informing us whether the weakening of the sound is in exact proportion to the rarefaction of the medium, as it is natural to suppose. In my opinion, we know nothing yet of a matter usually understood to be settled,—the mode of decrease of sound, in proportion to the distance of the sounding body; as to which science has added nothing to ordinary experience. It is commonly supposed that the decrease is in an inverse ratio to the square of the distance. This would be a very important law if we could establish it: but it is at present only a conjecture; and I prefer admitting our ignorance to attempting to conceal a scientific void, by arbitrarily extending to this case the mathematical formula which belongs to gravitation. A natural prejudice may dispose us to find it again here; but we have no proof of its presence.

It would be strange if we had any notion of the law of the case, when we have not yet any fixed ideas as to the way in which intensity of sound may be estimated; nor even as to the exact meaning of the term. We have no instrument which can fulfil, with regard to the theory of sound, the same office as the pendulum and the barometer with regard to gravity, or the thermometer and electrometer with regard to heat and electricity. We do not even discern any clear principle by which to conceive of a *sonometer*. While the science is in this state, it is much too soon to hazard any numerical law of the variations in intensity of sound.

SECTION III.

THEORY OF TONES.

THE third department of acoustics,—the theory of tones,—is by far the most interesting and satisfactory to us in its existing state.

Theory of Tones.

The laws which determine the musical nature of different sounds, that is, their precise degree of acuteness or gravity, marked by the number of vibrations executed in a given time, are accurately known only in the elementary case of a series of linear, even rectilinear, vibrations produced either in a metallic rod, fixed at one end and free at the other, or in a column of air filling a very narrow cylindrical pipe. It is by a combination of experiment and of mathematical theory that this case is understood. It is the most important for the analysis of the commonest inorganic instruments, but not for the study of the mechanism of hearing and utterance. With regard to stretched chords, the established mathematical theory is that the number of vibrations in a given time is in the direct ratio of the square root of the tension of the chord, and in the inverse ratio of the product of its length by its thickness. In straight and homogeneous metallic rods this number is in proportion to the relation of their thickness to the square of their length. This essential difference between the laws of these two kinds of vibrations is owing to the flexibility of the one sounding body and the rigidity of the other. Observation pointed it out first, and especially with regard to the effect of thickness. These laws relate to ordinary vibrations, which take place transversely; but there are vibrations in a longitudinal direction much more acute, which are not affected by thickness, and in which the difference between strings and rods disappears, the vibrations varying reciprocally to the length; a result which might be anticipated from the inextensibility of the string being equivalent to the rigidity of the rod. A third order of vibrations arises from the twisting of metallic rods, when the direction becomes more or less oblique. It ought to be observed however that recent experiments have shown that these three kinds are not radically distinct, as they can be mutually transformed by varying the direction in which the sounds are propagated. As for the sounds yielded by a column of air, the number of vibrations is in inverse proportion to the length of each column, if the mechanical state of the air is undisturbed; otherwise, it varies as the square root of the relation between the elasticity of the air and its density. Hence it is that changes of temperature which alter this relation in the same direction have here an action absolutely inverse to that which they produce on strings or rods: and thus it is explained by acoustics why it is impossible, as musicians have always found it, to maintain through a changing temperature the harmony at first established between stringed and wind instruments.

Thus far the resonant line has been supposed to vibrate through its whole length. But if, as usually happens, the slightest obstacle to the vibrations occurs at any point, the sound undergoes a radical modification, the law of which could not have been mathematically discovered, but has been clearly apprehended by the great acoustic experimentalist, Sauveur. He has established that the sound produced coincides with that which would be yielded by a similar but shorter chord, equal in length to that of the greatest common measure between the two parts of the whole string. The same discovery explains another fundamental law, which we owe to the same philosopher—that of the series of harmonic sounds which always accompanies the principal sound of every resonant string, their acuteness increasing with the natural series of whole numbers; the truth of which is easily tested by a delicate ear or by experiment. The phenomenon is, if not explained, exactly represented by referring it to the preceding case; though we can not conceive how the spontaneous division of the string takes place, nor how so many vibratory motions, so nearly simultaneous, agree as they do.

These are the laws of simple sounds. Of the important theory of the composition of sounds we have yet very imperfect notions. It is supposed to be indicated by the experiment of the musician Tartini, with regard to resulting sounds. He showed that the precisely simultaneous production of any two sounds, sufficiently marked and intense, occasions a single sound, graver than the other two, according to an invariable and simple rule. Interesting as this fact is, it relates to physiology, and not to acoustics. It is a phenomenon of the nerves; a sort of normal hallucination of the sense of hearing, analogous to optical illusions. *Composition of sounds.*

The vibrations of resonant surfaces have exhibited some curious phenomena to observation, though the mathematical theory of the case is still in its infancy: and M. Savart's observations on the vibratory motions of stretched membranes must cast much light on the auditory mechanism, in regard to the effects of degrees of tension, the hygrometrical state, etc.

The study of the most general and most complicated case, that of a mass which vibrates in three dimensions, is scarcely begun, except with some hollow and regular solids. Yet this analysis is above all important, as without it it is clearly impossible to complete the explanation of any real instrument; even of those in which the principal sound is produced by simple lines, the vibrations of which must always be more or less modified by the masses which are connected with them. We may say that the state of acoustics is such that we can not explain the fundamental properties of any musical instruments whatever. Daniel Bernouilli worked at the theory of wind instruments; a subject which may appear very simple, but which really requires the highest perfection of the science, even putting aside those extraordinary effects, far transcending scientific analysis, which the art of a musician may obtain from any instrument whatever, and restricting ourselves to influences which may be clearly defined and durably characterized.

Imperfect as is our review of Acoustics, I hope we now understand something of its general character, the importance of its laws, as far as we know them, the connection of its parts, the development that they have obtained, and the intervals which are left void, to be filled up by future knowledge.

CHAPTER V.

OPTICS.

THE emancipation of natural philosophy from theological and metaphysical influence has thus far gone on by means of a succession of partial efforts, each isolated in intention, though all converging to a final end, amid the entire unconsciousness of those who were bringing that result to pass. Such an incoherence is a valuable evidence of the force of that instinct which universally characterizes modern intelligence; but it is an evil, in as far as it has retarded and embarrassed and even introduced hesitation into the course of our liberation. No one having hitherto conceived of the positive philosophy as a whole, and the conditions of positivity not having been analyzed, much less prescribed, with the modifications appropriate to different orders of researches, it has followed that the founders of natural philosophy have remained under theological and metaphysical influences in all departments but the one in which they were working, even while their own labors were preparing the overthrow of those influences. It is certain that no thinker has approached Descartes in the clearness and completeness with which he apprehended the true character of modern philosophy; no one exercised so intentionally an action so direct, extensive, and effectual, on this transformation, though the action might be transitory; and no one was so independent of the spirit of his contemporaries; yet Descartes, who overthrew the whole ancient philosophy about inorganic phenomena, and the physical phenomena of the organic, was led away by the tendency of his age in a contrary direction, when he strove to put new life into the old theological and metaphysical conceptions of the moral nature of man. If it was so with Descartes, who is one of the chief types of the progress of the general development of humanity, we can not be surprised that men of a more special genius, who have been occupied rather with the development of science than of the human mind, should have followed a metaphysical direction in some matters, while in others not very remote they have manifested the true positive spirit.

Hypothesis on the nature of Light.

These observations are particularly applicable to the philosophical history of Optics—the department of Physics in which an imperfect positivism maintains the strongest consistence—chiefly through the mathematical labors which are connected with it. The founders of this science are those who have done most toward laying the foundations of the Positive Philosophy—Descartes,

Huyghens, and Newton; yet each one of them was led away by the old spirit of the absolute to create a chimerical hypothesis on the nature of light. That Newton should have done this is the most remarkable, considering how his doctrine of gravitation had raised the conception of modern philosophy above the point at which Descartes had left it, by establishing the radical inanity of all research into the nature and mode of production of phenomena, and by showing that the great end of scientific effort is the reduction of a system of particular facts to one singular and general fact. Newton himself, whose favorite saying was, "O! Physics, beware of Metaphysics!" allowed himself to be seduced by old habits of philosophizing to personify light as a substance distinct from and independent of the luminous body: a conception as metaphysical as it would have been to imagine gravity to have an existence separate from that of the gravitating body.

After what has been said about the philosophical theory of hypotheses, there can be no occasion to expose the fictitious character of the respective doctrines of philosophers on the nature of light. Each one has exposed the untenableness of those of others; and each explorer has confined himself to the evidence which favored his own conception. Euler brought fatal objections against the doctrine of emission; yet, at the present day, our instructors conceal the fact that the advocates of the emission doctrine have offered equally fatal objections to that of undulations. To take the most simple instance—Has the fact of propagation in all directions, characteristic of the vibratory motion, ever been reconciled with the common phenomenon of night; that is, of darkness produced by the interposition of an opaque body? Does not the fundamental objection of the Newtonians about this matter hold its ground against the system of Descartes and Huyghens, untouched at this hour as it was above a century ago, after all the subterfuges that have been in use ever since? The case is made clearer by the fact that there are phenomena which the two theories will suit equally well. If the laws of reflection and refraction issue with equal ease from the hypotheses of emission and undulation, it is pretty clear that our business is with the laws, and not with the hypotheses. The mathematical labors expended on the opposite theories will not have been thrown away; they will show, in a very short time, that the analytical apparatus is no certain instrument of truth, as it has served the purpose of both hypotheses equally well; as it would, quite as easily, of many others, if the progress of positivity was not excluding, more and more, this vicious method of philosophizing. It is true, the most enlightened advocates of both systems are ready to give up the reality of emission and of undulation, and hold to them only as a matter of logical convenience—as a rallying-point of ideas. But if we can pass from the one hypothesis to the other without affecting the science at all, it is clear that such an artifice is needless. We must admit, as we before said, that the combination of scientific ideas would be extremely difficult to minds trained under the prevalent habits of thought, if they were suddenly deprived of such a mode of connection as they here contend for; but it is not the less true that the next generation of scientific thinkers would combine their ideas more easily, and much more perfectly, if they were trained to regard directly the relations of phenomena, without being troubled by artifices like these, which only obscure scientific realities.

The history of Optics, regarded as a whole, seems to show that these hypotheses have not sensibly aided the progress of the theory of light, since all our important acquisitions have been entirely independent of them. This is true not only of the laws of reflection and refraction, which were discovered before these hypotheses were created, but with regard to all the other leading truths of Optics. The hypothesis of emission no more suggested to Newton the notion of the unequal refrangibility of the different colors, than that of undulation disclosed to Huyghens the law of double refraction proper to certain substances. Great discoveries like these observe a connection of facts, and then create an hypothesis to account for the connection; and then those who come after them conclude that the chimerical conceptions must be inseparable from the immortal discoveries. There is a use, as I have before asserted, in these imaginary conceptions, which, in regard to their one function, are indispensable. They serve, transiently, to develop the scientific spirit by carrying us over from the metaphysical to the positive system. They can do this and nothing more, and they accomplished their task some time ago. Their action can henceforth be only injurious, and especially in the case of Optics, as any one may see who will inquire into the state of this science—particularly since the almost universal adoption of the undulatory in the place of the emissive system.

Excessive tendency to systemize.
One more error must be noticed before we leave the subject of the unscientific pursuit of Optics. Some enlightened students imagine that the science acquires a satisfactory rationality by being attached to the fundamental laws of universal mechanics. The emission doctrine, if it means anything, must suppose luminous phenomena to be in analogy with those of ordinary motion; and if the doctrine of undulation means anything, it means that the phenomena of light and sound are alike in their vibratory agitation; and thus the one party likens optics to barology and the other to acoustics. But not only is nothing gained by the supposition, but if either was the case, there would be no room for imagination or for argument. The connection would be at once apparent to all eyes on the simple view of the phenomena. Such a reference of phenomena to those general laws has never been a matter of question or of conjecture. The only difficulty has been to know those laws well enough to admit of the application. No one doubted the mechanical nature of the principal effects of gravity and sound long before the progress of rational dynamics admitted of their exact analysis. The application pow-

erfully tended, as we have seen, to the perfecting of barology and acoustics; but this was precisely because there was nothing forced or hypothetical about it. It is otherwise with Optics. Notwithstanding all arbitrary suppositions, the phenomena of light will always constitute a category *sui generis*, necessarily irreducible to any other: a light will be for ever heterogeneous to a motion or a sound.

Again, physiological considerations discredit this confusion of ideas, by the characteristics which distinguish the sense of sight from those of hearing, and of touch or pressure. If we could abolish such distinctions as these by gratuitous hypotheses, there is no saying where we should stop in our wanderings. A chemical philosopher might make a type of the senses of taste and smell, and proceed to explain colors and tones by likening them to flavors and scents. It does not require a wilder imagination to do this, than to issue as a supposition, now become classical, that sounds and colors are radically alike. It is much better to leave such a pursuit of scientific unity, and to admit that the categories of heterogeneous phenomena are more numerous than a vicious systemizing tendency would suppose. Natural philosophy would, no doubt, be more perfect if it were otherwise; but co-ordination is of no use unless it rests on real and fundamental assimilation. Physicists must then abstain from fancifully connecting the phenomena of light and those of motion. All that Optics can admit of mathematical treatment is with relation, not to mechanics, but to geometry, which is eminently applicable to it, from the evidently geometrical character of the principal laws of light. The only case in which we can conceive of a direct application of analysis is in certain optical researches in which observation would immediately furnish some numerical relations; and in no case must the positive study of light give place to a dynamical analysis. These are the two directions in which geometers may aid the progress of Optical science, which they have only too effectually impeded by prolonging the influence of anti-scientific hypotheses through inappropriate and ill-conceived analyses.

The genius of Fourier released us from the necessity of applying the doctrine of hypotheses, as previously laid down, to the case of thermology; and neither barology nor acoustics required it. As to electrology, there are abundance of chimerical conceptions preponderant in that department; but their absurdities are so obvious, that almost all their advocates acknowledge them. It is in Optics that the plausibility and consistence of such chimeras give them the most importance; and I have therefore chosen that department as the ground on which they should be judged.

We will now pass from these useless hypotheses to the real knowledge that we are in possession of about the theory of light. The whole of Optics is naturally divided into four departments, as light, whether homogeneous or colored, is direct, reflected, refracted, or diffracted. These elementary ef- *Divisions of Optics.*

fects usually co-exist in ordinary phenomena; but they are distinct, and must therefore be separately considered. These four parts comprehend all optical phenomena which are rigorously universal; but we must add, as an indispensable complement, two other sections, relating to double refraction and polarization. These orders of phenomena are proper to certain bodies; but, beside that, they are a remarkable modification of fundamental phenomena, they appear in more and more bodies, as the study proceeds, and their conditions refer more to general circumstances of structure than to incidents of substance. For these reasons they ought to be exactly analyzed. As for the rest, it is not our business to classify the application of these six departments either to natural history, as in the beautiful Newtonian theory of the rainbow, or to the arts, as in the analysis of optical instruments. These applications serve as the best measure of the degree of perfection of the science; but they do not enter into the field of optical philosophy, with which alone we are concerned.

Irrelevant matters.

For the same reasons which have led us to condemn theories of hearing and utterance, in connection with Physics, we must now refuse to include among optical phenomena the theory of vision, which certainly belongs to physiology. When physicists undertake the study of it, they bring only one of the special qualifications necessary, being otherwise on a level with the multitude; and, however important their one qualification may be, it can not fulfil all the conditions. It is in consequence of so many conditions being unfulfilled, that the explanations hitherto offered have been so incomplete, and therefore illusory. There is scarcely a single law of vision which can be regarded as established on a sound basis, even where the simplest and commonest phenomena are in question. The elementary faculty of seeing distinctly at unequal distances remains without any satisfactory explanation, though physicists have attempted to refer it to almost every part of the ocular apparatus in succession. This humbling ignorance is no doubt owing to scientific men, both physiologists and physicists, having left the theory of sensations in the hands of the metaphysicians, who have got nothing out of it but some deceptive ideology; but before this time we should have approached to something like positive solutions, but for the bad organization of scientific labor among us. If, from the time of these questions beginning to assume a positive character, anatomists and physiologists had occupied themselves with a theory of vision grounded on the materials furnished by Optical science, instead of looking to physicists for solutions which they could not furnish, our condition in regard to this important subject would be somewhat less deplorable than it is.

Theory of vision.

Another study which must be excluded from Optics, and from all natural philosophy, is the theory of the color of bodies. I need not explain that I am not referring to the admirable Newtonian experiments on the decomposition of light, which have supplied a fundamental idea, common to all the departments of Optics.

Specific Color of Bodies.

I refer to the attempts made to ascertain, now through the theory of emission, and now through that of undulation, the inexplicable primitive phenomenon of the elementary color proper to every substance. The so-called explanations, about the supposed faculty of reflecting or transmitting such and such a kind of rays, or of exciting such and such an order of ethereal vibrations, in virtue of certain supposed derangements of the molecules, are more difficult to conceive than the fact itself, and are, in truth, as absurd as the explanations that Molière puts into the mouth of his metaphysical doctors. It is lamentable that we should have such comments to make in these days. Nobody now tries to explain the specific gravity proper to any substance or structure: and why should we attempt it with regard to specific color, which is quite as primitive an attribute? In physiology, the consideration of colors is of high importance, in connection with the theory of vision; and in natural history, it may prove a useful means of classification: but, in optics, the object of the true theory of colors is merely to perfect the analysis of light, so as to estimate the influence of structure or other circumstance upon transmitted or reflected color, without entering into the causes of specific coloring. The field of inquiry is vast enough, without any such illusory research as this.

SECTION I.

STUDY OF DIRECT LIGHT.

THE first department is that of Optics, properly so called, or the study of direct light. This and catoptics are the only part of the science cultivated by the ancients; but this branch is as old as the knowledge of the law of the rectilinear propagation of light in every homogeneous medium. This primary law makes purely geometrical questions of problems relating to the theory of shadows; questions difficult to manage in many cases, but not in the most important,—those of very distant luminous bodies, or bodies of extremely small dimensions. The theory depends, both for the shadow and the penumbra, on the determination of an extensible surface, circumscribed at once by the luminous and the illuminated body.—Whatever its real antiquity may be, this first part of Optics is still very imperfect, regarded from the second point of view; that is, with regard to the laws of the intensity of light, or what is called *photometry*. Important as it is to have a clear knowledge, our notions are as yet either vague or precarious as to how the intensity of light is modified by such circumstances as its direction, whether emergent or incident; its distance; its absorption by the medium; and, finally, its color.

Optics proper.

Imperfections.

We are met by a grand difficulty at the outset. We have no photometrical instruments that can be depended on for enabling us to verify our conjectures on the different modes of gradation of light. All our photometers rest on a sort of vicious circle, being devised in accordance with the laws which they are

Photometry.

destined to verify, and generally according to the most doubtful of all, in virtue of its metaphysical origin,—that which relates to distance. We have called light an emanation; have calculated its intensity by the square of its distance: and then, without confirming this conjecture by any experiment whatever, we have proceeded to found the whole of photometry upon it. And when this conjecture was replaced by that of undulations, we accepted the same photometry, neglecting the consideration that it must require revision from its very basis. It is clear what our present photometry must be, after such treatment as this. The law relating to direction, in the ratio of the sine of the angle of emergence or of incidence, is no better demonstrated than that of distance, though it comes from a less suspicious source. It has nothing about it at present like Fourier's labors on radiating heat; and yet it seems as if it would admit of an analogous mathematical elaboration. The only part of photometry which has, as yet, any scientific consistency is the mathematical theory of gradual absorption of light by any medium. Bouguer and Lambert have given us some interesting knowledge about this: but even here we are on unstable ground, for want of precise and unquestionable experiments. Again, the photometrical influence of color has been the subject of some exact observations; but we are not yet in possession of general and precise conclusions, unless it be the fixing of the maximum of brightness in the middle of the solar spectrum. Thus, to sum up, in this first, oldest, and simplest department of optics, philosophers have scarcely outstripped popular observation,—leaving out what belongs to geometry, and the measurement of the velocity of the propagation of light, which is furnished by astronomy.

SECTION II.

CATOPTRICS.

It is otherwise with regard to catoptrics, and yet more, dioptrics, if we discard questions about the first causes of reflection and refraction. Scientific studies have largely extended and perfected universal ideas about those two orders of general phenomena; and the varied effects belonging to them are now referred with great precision to a very small number of uniform laws, of remarkable simplicity.

Great law of reflection. The fundamental law of catoptrics, well known by the ancients, and abundantly confirmed by experiment, is, that whatever may be the form and nature of the reflecting body, and the color and intensity of the light, the angle of reflection is always equal to the angle of incidence, and in the same normal plane. Under this law, the analysis of the effects produced by all kinds of mirrors is reduced to simple geometrical problems, which might, it is true, involve some long and difficult calculations, according to the forms of some bodies, if it were not usually sufficient to examine the simple forms of the plane, the sphere, and, at most, the circular cylinder. If we pretended to absolute precision in the analysis of images, we might encounter considerable geometrical difficulties: but this is not necessary. This analysis depends, in general, mathematically speaking, on the theory of *caustic curves*, created by Tschirnhausen. But even in the application of this theory, some conjectures are hazarded; and the want of direct and exact experiments, and the uncertainty which attends almost all the parts of the theory of vision, prevents our depending too securely on the reality of the remote results of any general principle that we can yet employ.

Every luminous reflection upon any body whatever is accompanied by an absorption of more or less, but always of a great part of the incident light; and this gives rise to a second interesting question in catoptrics. But our knowledge about it amounts to very little, from our backwardness in photometry; so that we have not yet laid hold of any law. We do not know whether the loss is the same in all cases of incidence: nor whether it is connected with the degree of brightness: nor what is the influence of color upon it: nor whether its variations in different reflecting bodies are in harmony with other specific, and especially optical characters. These questions are not only untouched—they have never been proposed. All that we know is simply that the absorption of light appears to be always greater (but to what degree we are ignorant) by reflection than by transmission. From this has resulted, in recent times, the use of lenticular beacons, introduced by Fresnel. *Laws of absorption not found.*

A more advanced kind of inquiry belongs to the study of transparent substances; but here, again, the laws are ill understood. In these bodies, reflection accompanies refraction, and we have the opportunity of inquiring by what laws, general or special, the division between transmitted and reflected light takes place. We only know that the last is more abundant in proportion as the incidence is more oblique: and that reflection begins to become total from a certain inclination proper to each substance, and measured exactly with regard to several bodies. The inclination appears to be less in proportion as the substance is more refracting; but the supposed law of the case is connected with chance conjectures upon the nature of light, and requires to be substantiated by direct experiment.

SECTION III

DIOPTRICS.

OF all the departments of Optics, dioptrics is at present the richest in certain and exact knowledge, reduced to a few simple laws, embracing a large variety of phenomena. The fundamental law of refraction was wholly unknown to the ancients, and was discovered at the same time, under two distinct and equivalent forms, by Snellius and Descartes. It consists *Great law of refraction.*

of the constant proportion of the sines of the angles that the refracted ray and the incident ray, always contained in the same normal plane, form with the perpendicular to the refracting surface, in whatever direction the refraction may be. The fixed relation of these two sines, when the light passes from a vacuum into any medium whatever, constitutes the most important optical co-efficient of every natural body, and holds a real rank in the aggregate of its physical characteristics. The philosophers have labored at its determination with much care and success, by ingenious and exact processes: they have prepared very extensive tables, which may rival, as to precision, our tables of specific gravity—the uncertainty not exceeding a hundredth part of the numerical value of the refracting power. If the light passes from one medium to another, the relation of the refraction depends on the nature of both; but in every case, the inverse passage gives it always a precisely reciprocal value, as experiment has constantly shown. Again, while a body undergoes no chemical change, and becomes only more or less dense, the relation of refraction which belongs to it varies in proportion to the specific gravity; as may be easily shown, especially with regard to liquids, and yet more to gases, in which we can so extensively modify density by temperature and pressure. This is why philosophers have adopted, in preference to the proper relation of refraction, its quotient by the density, which they have named refracting power; in order to obtain more fixed and specific characters in the dioptric comparison of different substances. There is substantial ground for this distinction, though its origin was suspicious. But it must be observed that the refracting power varies when the substance does not undergo any chemical change, but passes, as we have seen in the case of water, through different states of aggregation. These variations in the refracting power have given occasion to conflicts between the advocates of the two hypothetical systems—each of which requires an invariability in the refracting power which we do not know to exist; and the difficulty of separating what is really established from what they require is one of the mischievous consequences of anti-scientific hypotheses, and one which may well render the actual character of the science itself doubtful to impartial minds.

Newton's discoveries on elementary colors. Newton's discoveries of the unequal refrangibility of the different elementary colors form an indispensable complement of the law of refraction. From the fact of the decomposition of light in a prism, it clearly follows that the relation of the sine of incidence, though constant for each color, varies in the different portions of the solar spectrum. The total increase which it undergoes from the red rays to the violet measures the *dispersion* proper to each substance, and must complete the determination of its refracting power in the common tables, where only the mean refraction can be inserted. This estimate constitutes, from its minuteness, one of the most delicate operations of optics, and does not admit of so much exactness as that of the refracting action properly so called, especially in bodies which bend the light but little, as the gases; but it is ascertained for a considerable number of substances, solid or liquid. In comparing the changes of the dispersive power as we pass from one body to another, we discover that the variations are not, as Newton supposed, in proportion to the refracting power; and indeed we find, in more than one case, that the light is least dispersed by substances which refract it most. The discovery of this discrepancy between two qualities which appear to be analogous, was made by Dollond, about the middle of the last century. It is an idea of high importance in Optics, as it indicates the possibility of achromatism by the compensation of the opposite action pertaining to two different substances which, without that, could not cease to disperse the light but by ceasing to bend it.

The laws of refraction show us that there can be none but purely geometrical difficulties in the analysis of the effects of homogeneous media upon the light which traverses them. The great complication which might arise from the form of the refracting body is diminished in ordinary cases by our satisfying ourselves with plane, spherical, or cylindrical surfaces; but we should yet find the inquiry embarrassing, and especially in regard to the dispersion, if we did not confine it to an approximate estimate of the few commonest circumstances.

SECTION IV.

DIFFRACTION.

THE modification called diffraction has now become one of the essential parts of Optics. It was entered upon by Grimaldi and Newton, advanced by the researches of Dr. Young, and completed by those of Fresnel. It consists of the deviation, always accompanied by a more or less marked dispersion, that light undergoes, in passing close by the edges of any body or opening. Its simplest way of manifesting itself is by the unequal and variously-colored fringes, some exterior and some interior, which surround the shadows produced in a darkened room. The famous general principle of *interferences*, discovered by Dr. Young, is the most important idea connected with this theory. It was not appreciated, remarkable as it is, till Fresnel made use of it to explain several interesting phenomena, difficult to analyze; and, among others, the celebrated phenomenon of the colored rings, which were by no means fully accounted for by Newton's admirable efforts. The law of interferences is this: that when two luminous cones emanate from the same point, and follow, for any reason, two distinct courses, but little inclined toward each other, the intensities proper to the two lights neutralize and augment each other alternately, increasing by equal and minute degrees, the value of which is determined, the difference in length between the entire paths traversed by the two cones. It is a pity that this important principle should have suffered, like the rest, from being implicated with chimerical conceptions on the nature of light.

We have done all that the nature of this work admits, in regard to Optics; and we must pass over the subjects of the double refraction proper to various crystals, the general law of which was discovered by Huyghens. We must also omit the phenomena of *polarization*, disclosed by Malus. In what I have brought forward, I hope that, while I have pointed out the gaps in this science, of which we are too little conscious at present, I have also placed in a clear light the great and numerous results obtained during the last two centuries, notwithstanding the disastrous preponderance of vain hypotheses about the nature of light over the spirit of rational experimentation.

CHAPTER VI.

ELECTROLOGY.

History. This last branch of Physics, relating as it does to the most complex and least manifest phenomena, could not be developed till after the rest. The electrical machine indeed is as old as the air-pump; but it was not till a century later that the study assumed a scientific character, through the distinction of the two electricities, Muschenbroek's experiments with the Leyden jar, and then through Franklin's great meteorological discovery, which was the first manifestation of the influence of electricity in the general system of nature. Up to that time, the isolated observations of philosophers had only suggested the character of generality inherent in this part of Physics, as in all others, by continually adding to the number of substances susceptible of electrical phenomena; and it was not till the end of the last century that this department of Physics presented anything like the rational character which belongs to the others. It is owing to the labors of Coulomb that it takes its place, and still an inferior place, with the rest.

Condition. No other science offers so great a variety of curious and important phenomena; but facts do not constitute science, though they are its foundation and material. Science consists in the systemizing of facts under established general laws; and, regarded in this way, Electrology is the least advanced of all the branches of Physics, imperfect as they all are. In the absence of ascertained laws, arbitrary hypothesis has run riot.

Arbitrary hypotheses. The simple confidence with which students have explained all phenomena by endowing imaginary fluids with new properties for every fresh occurrence, reminds us of the old metaphysical explanations—the ancient entities being merely replaced by supposed fluids. But the delusion is less mischievous here than in Optics, where the arbitrary conjectures are closely connected with real laws, and share their imposing character. In electrology the hypotheses, standing alone, exhibit their barrenness; and everybody can see that they have borne no share in the great discoveries of the last half-century, though the discoveries, once made, have been afterward attached to the hypotheses. Most people regard them now as a sort of mnemonic apparatus, useful for connecting facts in the memory, though originally designed for a very different purpose. They are a bad apparatus for even this object, which would be much better answered by a system of scientific formulas especially adapted to that use. And, though less mischievous than in Optics, hypotheses of this order do harm in electrology, as everywhere else, by concealing from most minds the real needs of the science. It should be remembered, moreover, that anti-scientific action like this extends its influence over the succeeding and more complex sciences, which, on account of their greater difficulty, require the severest method, the type of which will naturally be looked for in the antecedent sciences. It is a serious injury to transmit to them a radically vicious model. While physicists are using these hypotheses as having avowedly no intrinsic reality, their very use leads students of the successive sciences, and especially physiologists, to consider them the very sublimity of physics, and to proceed to take them for the bases of their own labors. We see how the notion of magnetic and electric fluids tends to confirm that of a nervous fluid, and to encourage wild dreams about the nature of what is called animal magnetism, in which even eminent physicists have shared. Such consequences show how a study which is naturally favorable to the positive development of human intelligence may, by vicious methods of philosophizing, become fatal to our understandings.

From the complex nature of the phenomena, there can be but little application of mathematics in electrology. *Relation to Mathematics.* It has as yet borne only a small share in the progress of the science: but it is as well to point out the two ways,—the one illusory, the other real,—in which the application of mathematics has been attempted.

Those who have occupied themselves with imaginary fluids as the causes of electrical and magnetic phenomena, *Unsound application.* have transferred the general laws of rational mechanics to the mutual action of their molecules; thus making the body under notice a mere *substratum*, necessary for the manifestation of the phenomenon, but unconcerned in its production; with which office the fluid is charged. It is clear that mathematical labors, so baseless, can serve no other purpose than that of analytical exercise, without adding a particle to our knowledge. In the other case,— *Sound appplication.* of a sound application,—the mathematical process has been based on some general and elementary laws, established by experiment, according to which the study of phenomena proper to the bodies themselves has been pursued,—all chimerical hypotheses being discarded. This is the character of the able researches of M. Ampère and his successors, on the mathematical investigation of electro-magnetic phenomena, in which the laws of abstract dynamics have been efficaciously applied to certain cases of mutual action between electric conductors or magnets.

In examining the principal parts of electrology, we must exclude all that belongs to the chemical or physiological influence of electricity, and all connection of electricity with concrete physics; and especially with meteorology.

Divisions. Thus limited to the physical and abstract, electrology at present comprehends three orders of researches. The first relates to the production, manifestation, and measurement of electrical phenomena: the second, to the comparison of the electric state proper to the different parts of the same mass, or to different contiguous bodies: the third, to the laws of the motions which result from electrization: we may add, as a fourth head, the application of the results under the other three to the special study of magnetic phenomena, which can never henceforth be separated from them.

SECTION I.

ELECTRIC PRODUCTION.

THE sum of our observations leads us to regard the electric condition of bodies as being, more or less evidently, an invariable consequence of almost all the modifications they can undergo: but the *Causes of electrization.* chief causes of electrization offer themselves, in the order of their power and scientific importance, thus: chemical compositions and decompositions: variations of temperature: friction: pressure: and, finally, simple contact. This distribution differs widely from that first indicated by inquiry,—friction being long supposed the only, and then the most powerful means of producing the electric condition. The comparison of means is very far from being exhausted; but we may be assured that the order specified above will never be radically changed.

Chemical action. There is no doubt that chemical actions are the most general sources of electricity, as well as the most abundant; as they are with regard to Heat. In the most powerful electrical apparatus, and especially in the Voltaic pile, the chemical action which at first passed unnoticed, is now recognised, thanks to the labors of Wollaston and others, as the principal source of electrization, which becomes indeed almost insensible when care is taken to exclude chemical action.—After this, the next most powerful *Thermological action.* cause is thermological action, though, till recently, it was recognised only in the single case of heated tourmalin. We now know that marked differences of temperature between consecutive bars of different kinds, whether homogeneous or otherwise in the particular case, suffice to induce a marked electrical condition, the more intense as the elements are more numerous,

CAUSES OF ELECTRICITY.

—the thermometrical conditions remaining the same.—These two causes are so powerful, and so difficult to exclude, that the estimate of the others becomes a very delicate matter. It is difficult to determine how much influence to ascribe to any cause after these two, while yet they are almost unavoidably present. Thus, even about friction, which used to be regarded as so powerful a cause, it is now doubtful whether the friction itself has any influence, and whether the electrization is not due to the thermometrical, and even the chemical effects which always accompany friction, but which always used to be overlooked in this instance. *Friction.*

The case is nearly the same with Pressure, the electric influence of which however is, if less marked, more unquestionable, from our being able to isolate it more. But the remark is above all applicable to the production of the electric state by the simple contact of heterogeneous bodies. It was by this contact that Volta brought out the power of his wonderful instrument, while it is well known now that chemical action bears a chief part in it, and that contact contributes to it in only a secondary manner, if even it be not altogether doubtful. *Pressure.* *Contact.*

Besides these leading causes of electrization, there are many less important,—as changes in the mode of aggregation, the fusion of solids, and the evaporation of liquids. Even simple motion suffices, under special conditions, to induce an electric state, as M. Arago has shown in the experiment of the influence of the rotation of a metallic disk upon a magnetized needle, near but not contiguous. Our philosophers however must beware of passing into the other extreme from that with which they justly reproach their predecessors. It is, no doubt, prejudicial to electrology to neglect all sources of electrization but the most conspicuous: but it may not be less so to carry analysis too far, and see causes of electrization in all sorts of minute phenomena.* *Other causes.*

A special instrument, or class of instruments, naturally corresponds to each of the general modes of electrization, in order to realize the most favorable conditions for the production and support of the electric state. However important these may be, it is clear that we can not here enter upon the consideration of them. But we must not pass over the instruments invented for the manifestation and measurement of the electric condition,—the electroscope and the electrometer. The most eminent philosophers have always attached the highest importance to the perfecting of these instruments, in the invention of which real genius has often been exhibited. Their perfection is of more consequence than that of electric producers; because very weak electric powers often answer best in delicate experiments, from their simplicity; while the utmost ingenuity is required in instituting means of manifesting and measuring the minutest electric effects.— *Instruments.*

* In this paragraph, M. Comte alludes to the now most fertile, but when he wrote the comparatively unknown subject of the development of Electricity by Induction.—J. P. N.

Though the electric condition can not be measured without being first manifested, and the manifestation leads to some sort of estimate, there is a real distinction between electroscopes and electrometers. Among simple electroscopes, the most remarkable for use in very delicate researches, is that kind called *condensers*, which render feeble electrical effects sensible through their gradual accumulation: and all these instruments are so arranged as to show, by the method of experimentation itself, the positive or negative character of the electricity under notice.—Coulomb's electrical balance is certainly the most perfect of electrometers. It was by its means that he discovered, and that we every day demonstrate, the fundamental law of the variation of electric action, repulsive or attractive, inversely to the square of the distance; a law which could not be unquestionably obtained by any other means. As we have advanced in the science of electro-magnetism, a new class of electrometers has been introduced, for purposes of measurement, for which Coulomb's balance would not answer. These are the class of *multipliers*. Valuable and delicate as they are, they have not yet been applied, with so much certainty as the balance, to exact measurements, from the difficulty of proportioning the graduation to the intensity of the observed phenomenon.

SECTION II.

ELECTRICAL STATICS.

THE second part of electrology includes what is improperly called *electrical statics:* a term imputable to illusory hypotheses about the nature of electricity: yet it is not a wholly absurd title, as it relates, in fact, to the distribution of electricity in a mass, or in a system of bodies, the electric state of which is regarded as invariable. We may therefore continue to use this abridged term, if we carefully keep clear of all mechanical notions of the equilibrium of any supposed electric fluid, and attach to it a sense analogous to that of Fourier, when he spoke of an equilibrium of heat, and of economists when they speak of an equilibrium of population.

Great law of distribution. Considering first the case of an isolated body, Coulomb has established a fundamental law which is (metaphorically expressed) the constant tendency of electricity to the surface, or, in rational language, that after an inappreciable instant of time electrization is always limited to the surface, however it may have been in the first place produced. As for the distribution of the electric state among the different parts of the surface, it depends on the form of bodies, being uniform for the sphere alone, unequal for all other forms, but always subject to regular laws. The analysis of these may be supposed to present insurmountable difficulties; nevertheless, Coulomb has established a general fact of great importance, by comparing the electric states proper to the extremities of an ellipsoid gradually elongated: he has perceived that their electrization increases rapidly as the figure is elongated, diminishing in the rest of the body; whence he deduced an explanation of that remarkable power of points, disclosed by Franklin.*

The laws of electric equilibrium between several contiguous bodies afford a yet more difficult and extensive inquiry. *Electric equilibrium.* Coulomb studied them only in the limited and insufficient single case of spherical masses. However we learn from his labors that the nature of substances exercises no influence over the electric distribution established among them, the mode depending merely on their form and their magnitude; only, the electric state assumed by each surface is more or less persistent, and manifests itself with more or less rapidity, according to the degree of conductibility in the body. Coulomb analyzed completely the mutual action of two equal spheres; discovering that the electric condition is always null and the point of contact, scarcely sensible at 20 degrees from that point, fast increasing from 60 to 90 degrees, and then more slowly increasing up to 180 degrees, which is its maximum. If the globes are unequal, the smallest is the most strongly affected: and it makes no difference whether they are electrized together, or the one before the other. The question becomes more complex when more than two bodies are concerned. Coulomb examined only a series of globes ranged in a straight line; but if they had been so placed as that each should touch three or four others, the mode of electric distribution would inevitably have undergone great changes. The subject must be regarded as merely initiated by this great philosopher; and no one has added anything to it since his time. It offers to electricians a subject of almost inexhaustible research.†

SECTION III.

ELECTRICAL DYNAMICS.

The third part of electrology is very properly called Electrical Dynamics, because it relates to the motions *Ampère's experiments.* which result from electrization. Recent as is its origin, it is superior to the others in its scientific condition, through the labors of M. Ampère; always supposing conjectures about the nature of electric phenomena to be discarded. M. Ampère has referred the analysis of the effects observed in this branch of electrology to one great and general phenomenon, the laws of which he has fully ascertained; the direct and mutual action of two threads, charged with electricity by Voltaic piles, habitually reduced to their greatest simplification; that is, almost always composed of a single element.

M. Ampère so arranged his experiment as to guard the conducting threads from the perturbing influence of the earth's electricity, and this done, he could easily seize the elementary laws of the phenomenon under his notice. He found that when the two con

---

* Much has since been added to this class of investigations.—J. P. N.
† These specific facts are now comprehended within general laws.—J. P N.

ductors are sufficiently mobile, they tend to place themselves in directions parallel to each other; and that they then attract or repel each other, according to the conformity or contrariety of the two electric currents. In looking for the laws of the case, it is necessary, for the sake of generality and simplicity, to keep in view only infinitely small portions of the different conductors. These laws, mathematically considered, relate either to the influence of the direction, or to that of the distance.

As to the direction, there are the two cases to be considered of the conducting elements being in the same plane, or in different planes. In the first case, the intensity of the action depends only on the angle formed by each of the two elements with the line which joins their middle points: it is null at the same time with this angle, and increases with it, attaining its maximum when it becomes right. All phenomena, direct or indirect, appear to be exactly represented if this intensity is made to vary in proportion to the sine of the inclination, according to the formula adopted by all the successors of M. Ampère. In the other case,—of the conductors not being in the same plane,—the action depends moreover on the mutual inclination of the planes indicated by each of them, and by the common line of their middle points; and the result of this second relation is wholly different. The perpendicularity of the two planes determines the absence of all action: there is attraction while the angle is acute, and it increases as the angle diminishes, its *maximum* taking place at the moment of coincidence; when the angle is obtuse, the action becomes repellent, and increases as each plane approaches toward the prolongation of the other, a situation which produces the *maximum* of repulsion. The supposition which arises in this case is that the action is in proportion to the cosine of the angle of the two planes; but we have not yet attained such certainty as in the former case.

As for the influence of distance, M. Ampère supposed that, in analogy with Coulomb's law of common electric attraction and repulsion, the action of two conducting elements is always reciprocal to the square of the distances of their middle points. But analogy is not sufficient to conclude upon; and direct observation is out of the question when the parts taken are infinitely small, and the result sought must be affected by the form and magnitude of the conductors. However, it may be mathematically demonstrated that, in the hypothesis adopted by M. Ampère, the action of a rectilinear conductor, of an indefinite length, upon a magnetized needle, must vary exactly in the inverse ratio of their shortest distance. This consequence has been precisely verified by experiment; and it places beyond a doubt the reality of the proposed law.

Under this law, electric action would seem to be, mathematically, in analogy with that of gravitation. But this case affords a lesson against incaution in transferring to the study of these singular movements the ordinary procedure of abstract dynamics. Gravitation is independent of mutual direction, which is the determining influence in electrical dynamics: and thus the parallel fails. We see, further, how many more difficulties are in the way of the analysis of the electric forces than in that of molecular gravitation. If this last is, from its complexity, unmanageable except in the simplest cases, it is no wonder that electrical dynamics has not been mathematically studied further than in one dimension, and never at all in surface. Even this much would be hardly effected but for a last fundamental idea, established by M. Ampère; that in an infinitely small extent, and as long as the distance is not sensibly changed, the electric action is identical for two conducting elements issuing at the same extremities, whatever may be otherwise their difference of form. Such a property must introduce valuable analytical simplifications, tending to establish a remarkable analogy between electric, and ordinary dynamic decompositions.

These are the grounds on which the study of the various action of electrized threads proceeds. Among the many dispositions of these conductors, the most interesting case is that of the spiral form; and especially when the turns are very close together. M. Ampère has shown the high importance of this form, in order to imitate, as exactly as possible, the phenomena characteristic of magnetized bodies.*

* * * * * * * *

We have now reviewed the philosophy of Physics, noticing in turn the aspects presented by the study of the properties common to all substances and all structures. *Conclusion of Physics.* These are not so much branches of a single study as distinct sciences. Part of our business has been to carry on a philosophical operation, hardly necessary in astronomy, but becoming more and more so as we descend to the more complex sciences;—that of disengaging real science from the influence of the old metaphysical philosophy, under which it still suffers deplorably, and which manifests itself in Physics through illusory and arbitrary conceptions about the primitive agents of phenomena. I have been able only to indicate the mischief, and where it resides; and I must leave the work of purification to rational philosophers, whose attention will, we must hope, be more and more drawn to this vital question. It is with the same view that I have endeavored to assign the true application of mathematical theories to the principal branches of physics, pointing out by the way the danger of the excessive systemization which is too often sought by carrying the use of this powerful instrument further than the complex nature of the corresponding phenomena would fairly allow. While giving my chief attention throughout to the method, I have pointed out, in brief, the principal natural laws relating to each department of science, discovered by human effort during the two centuries which have

* M. Comte concludes the section on Electricity by a slight reference to the discoveries of Oersted, Arago, and others, regarding its virtual identity with all we term the magnetic forces. But as the whole of this most interesting and important part of Physics has taken a new form since the date of his work, it has not, for reasons assigned in the Preface, been thought necessary to reproduce his remarks in this place.—J. P. N.

elapsed since the birth of Physics, properly so called: and I have shown what gaps are disclosed in the course of such a survey.

Our next study will be of the last science which belongs to the class of general knowledge, or that of inorganic nature. Chemistry relates to the molecular and specific reactions which different substances exert upon each other. It is a more complex, and consequently more imperfect science than those which we have reviewed: but its general character may be perfected, through the means afforded by its subordination to the anterior sciences.

BOOK IV.
CHEMISTRY.

CHAPTER I.

WE have now to review the last of the sciences which relate to the inorganic world. Chemistry has for its object the modifications that all substances may undergo in their composition in virtue of their molecular reactions. Without this new order of phenomena, the most important operations of terrestrial nature would be incomprehensible to us; and there is no other class of phenomena so intimate and so complex. Inert bodies can never appear so nearly like vital ones as when they produce in each other those rapid and profound perturbations which characterize chemical effects. We shall see hereafter that the spirit of all theological and metaphysical philosophy consists in conceiving of all phenomena as analogous to the only one which is known by immediate consciousness—Life: and we can easily understand that the primitive method of philosophizing must have exerted a more powerful and obstinate dominion over chemical phenomena than any other, in the inorganic world.—We must consider, too, that direct and spontaneous observation must have been applied in the first place only to very complicated phenomena, such as vegetable combustions, fermentations, etc., the analysis of which now requires all the resources of our science: and that the most important chemical phenomena are produced only in artificial circumstances, which were long in being devised, and very difficult at first to institute. Easy as it is now for even the most ordinary inquirers to use known substances for the disclosure of new relations, we can hardly imagine the difficulty there must have been, in the infancy of chemistry, in creating suitable subjects for observation: and we can not suppose that the ancient investigators of nature could have had energy and perseverance to discover the principal phenomena of the science if they had not been constantly stimulated by the unbounded hopes arising from their chimerical notions of the constitution of matter.

Its nature.

Great imperfection. The complex and doubtful nature of the phenomena, in the first place, and next, the difficulty of getting at them, are quite enough to account for the tardy and incomplete positivity of chemical conceptions, in comparison with all others in the inorganic region of nature. If, as we have seen, Physics is defective in several respects, much more must that science be so which, being at once more difficult and more recent, seeks the laws of composition and decomposition. Whichever way we look at it, whether speculatively, as to the value of its explanations, or actively, as to the provisions which they admit of, this science is evidently the least advanced of all the branches of inorganic philosophy. Indeed, it is hardly possible to call chemistry a science at all while it scarcely ever leads to that precise prevision which is the criterion of perfection in speculative knowledge. We can rarely tell what will be the result of the smallest and fewest modifications introduced among the best explored chemical operations; and while that is the case, however important and numerous may be the facts collected, we are in possession of only erudition, and not science. To suppose otherwise is to mistake a quarry for an edifice.

Capacities. It is not to be hoped that chemistry can ever attain a state of rationality so satisfactory as that of the sciences which relate to phenomena of a more simple character; and especially that of the eternal type of natural philosophy—Astronomy. But so much of its inferiority seems to be due to a vicious philosophy, and to the defective education of philosophers, that I can not but hope that a judicious philosophical analysis may contribute to a speedy perfecting of so important a science. This is the conviction that I desire to awaken by the rapid sketch which I propose to offer of chemical philosophy, regarded in all its essential aspects. Little as can be done within the bounds of this section, it is possible that some one eminent inquirer may be impressed by the necessity of submitting to a new and more rational elaboration the fundamental conceptions which constitute the science.

Object of Chemistry. First,—what is the general object of Chemistry? Vast and complex as is its subject, the definition of Chemistry is easier than that of Physics. We are already prepared for it, indeed, by having contrasted that of Physics with it. It is easy to characterize the phenomena of chemistry, in a direct and marked manner; for all indicate an alteration, greater or smaller, in the constitution of bodies: that is, a composition or decomposition, and generally both, taking into the account the whole of the substances which participate in the action. Thus, at all epochs of scientific development, since chemistry first became an object of speculative study, chemical researches have steadily manifested a remarkable originality, which has prevented their being confounded with other parts of natural philosophy; even while Physics itself was mixed up, as its title shows, with physiology; which was the case up to a very recent time.—It is by this general character of its phenomena that Chemistry is distinguished from Physics which precedes it, and Physiology which follows it. The three sciences may be considered as having for their object the molecular activity of matter, in all the different modes of which it is susceptible. Each corresponds to one of three successive degrees of activity, which are essentially and naturally distinguished from each other. The chemical action obviously presents something more than the physical action, and something less than the vital. The physical activity modifies the arrangement of particles in bodies; and these modifications are usually slight and transient, and never alter the substance. The chemical activity, on the contrary, besides these alterations in the structure and the state of aggregation, occasions a profound and durable change in the very composition of the particles: the bodies which occurred in the phenomenon are no longer recognizable,—so much has the aggregate of their properties been disturbed.—Again, physiological phenomena show us the molecular activity in a much higher degree of energy; for, as soon as the chemical combination is effected, the bodies become, once more, completely inert; while the vital state is characterized, over and above all physical and chemical effects, by a double continuous motion of composition and decomposition, adapted to maintain, within certain limits of variation and of time, the organization of the body by incessantly renewing its substance. This is the gradation, which no sound philosophy can ever confound, of the three modes of molecular activity.

Two more characteristics of this science must be pointed out: one relating to its nature, and the other to its general conditions.

Chemistry would not be classed among the inorganic sciences unless its phenomena were general; that is, unless every substance were susceptible of chemical action, more or less. And it is because chemistry is thus radically different from physiology that it ranks as the last of the inorganic sciences, —physiological phenomena being, by their nature, peculiar to certain substances, organized in certain modes. Nevertheless, it is incontestable that chemical phenomena present, in every case, something specific, or, to use Bergmann's energetic expression, *elective*. Not only does each material element produce chemical effects which are altogether peculiar to it, but it is the same with their innumerable combinations of different orders, among the most analogous of which certain fundamental differences are observable, even so as to be adopted as their characteristics. While, therefore, physical differences among different bodies are those of degree only, chemical properties are specific. Physical properties afford the common foundation of material existence; and it is by chemical properties that individuality is manifested. *Specific character of its action.*

The other characteristic relates to the mode of chemical action. The immediate contact of antagonistic particles is absolutely necessary to chemical action; and therefore one at least of the substances concerned must be fluid or gaseous. When this condition does not already exist, it must be artificially *Condition of action.*

procured by liquefying the substance. It is the earliest axiom in the science, that combination can not take place, except under this condition; and there is not an instance upon record of chemical action between two solids, unless at a temperature which obscures the true state of aggregation of substances; and the action is never so powerful as when both substances are liquid. These facts establish the eminently molecular character of chemical effects, and especially in comparison with physical effects. The distinction from physiological effects is, though less marked, as real, the latter requiring, as we shall see hereafter, the junction of solids with fluids.

Definition. The definition of Chemistry, then, is that it relates to the laws of the phenomena of composition and decomposition, which result from the molecular and specific mutual action of different substances, natural or artificial.

It will be long, we must fear, before a more precise definition than this can be given. Meantime, however incomplete, the most rational that can as yet be offered is of importance as far as it goes. In this view, and connecting, as usual, the consideration of science with that of prevision, the aim proposed should be this: the characteristic properties of substances, simple or compound, being given, and those properties being placed in a chemical relation in well-defined circumstances, to determine in what their action will consist, and what will be the chief properties of the new products. This problem is, at all events, determinate; and nothing contained in it could be omitted without its ceasing to be so; and the formula, therefore, contains nothing superfluous. On the other hand, if we could obtain such solutions as are indicated, the application of chemistry to the three great objects, vital phenomena, the natural history of the globe, and industrial operations, would be rationally organized, instead of being, as now, the almost accidental result of the spontaneous development of science. Each question would at once be referred to our formula, the data of which would be supplied by the circumstances peculiar to the application. Far distant as we are from being able thus to conduct our inquiries, this is the end to be kept in view: and chemists all agree that the most advanced portions of their science are those few and simple questions in which this aim has been more or less completely attained.

By a continued application of this method, all the data must finally be reducible to the knowledge of the essential properties of simple substances, which would lead to that of the different immediate principles; and, consequently, to the most complex and remote combinations. As for the study of the elements, that must, of course, be a matter of direct, experimental elaboration, divided into as many parts as there are undecomposed substances. Whether or not it may be possible to discover, by rational methods, relations between the chemical properties of each element and its aggregate physical properties, we must lay down as indispensable a direct exploration of the chemical characters of each element. This general basis once obtained from experiment, all other chemical problems must be susceptible of a rational solution, under a small number of invariable laws.

The classes of combinations naturally divide themselves into two, according, first, to the simplicity, or the greater or less degree in which the immediate principles are compounded, and, secondly, the number of elements combined. Chemical action is observed to become more difficult the more substances are compounded: the greater part of compound atoms belong to the first two orders; and beyond the third their composition seems almost impossible: and, in the same way, in regard to the number of elements, combinations lose their stability in proportion as the elements are multiplied:—there are usually only two; and scarcely anybody involves more than four. Thus, the number of chemical classes must always be very small in regard to the distinction under notice: and each of them must have a corresponding law of combination, according to which the result might be certainly anticipated through a knowledge of the data. This would be the scientific perfection of chemistry. Our prodigious remoteness from such a state is ascribable to the feebleness of our faculties, and, in an accessory way, to their vicious direction. We must remember that the great aim has begun to be fulfilled in one secondary department of chemical research,—the study of proportions, as we shall see hereafter. What has been done in that one category makes us ask why an analogous perfection should not be attained in other departments. We may sum up this account of the requisites, with the fully rational definition of Chemistry, that it has for its object,—the properties of all simple bodies being given, to find those of all the compound bodies which may be formed from them. Every science falls short of its definition: but a real definition is the first evidence that a science has attained some consistency: it then measures its own advancement from one epoch to another; and it always keeps inquirers in a right direction, and supports them in a philosophic progress. *Combination.*

Rational definition.

Looking now at our means of investigation, we shall find that in chemistry the law holds good that the complication of phenomena coincides with the extension of our means of inquiry. *Means of investigation.*

Here *Observation* begins to find its full development. Up to this time it has been more or less partial. In astronomy it is confined to the sense of sight: in physics we use hearing and touch also; and chemistry employs, besides these, taste and smell. How much is thus gained we may know by imagining what would become of chemistry, if we were without taste and smell, which are often the only means by which we can recognise effects produced. The important thing to observe under this head is that there is nothing accidental, nor even empirical, in such a correspondence; for, as we shall hereafter see, the sound physiological theory of sensation shows that the apparatus of taste and *Observation.*

smell, unlike that of the other senses, operates in a chemical manner, and thus shows these two senses to be specially adapted for the perception of phenomena of composition and decomposition.

Experiment. As for *Experiment*, it is enough to say that the greater number of chemical phenomena, and especially the most instructive, are of artificial production. Still, we must remember that the essential character of experimentation consists in the institution, or the choice of the circumstances of the phenomena, in order to a more evident and decisive investigation. This process is more difficult in chemistry than in physics, because it is more difficult to institute two parallel cases, undisturbed by the intrusion of irrelevant influences; and yet this is the fundamental condition of experimentation. On this account, I dissent from the ordinary supposition that the experimental method is more appropriate to chemical than to physical researches. Though this is my view, and though the greater advancement of physics gives it the advantage over chemistry in the use of experiment, I can have no doubt of the powerful influence of experimentation in chemistry, independently of its having supplied new subjects of observation. From the early days of the science, the immortal series of Priestley's experiments, and yet more, those of Lavoisier, have offered admirable models, almost comparable to the most perfect researches in physics, and quite enough to prove that there is nothing in the nature of chemical phenomena to prevent the extended and luminous employment of the experimental method.

Comparison. The third means, *Comparison*, which we have before seen to be inapplicable in Astronomy, and of especial use in Physiology, begins to have a real use in Chemistry. The essential condition of this valuable method is that there shall be an extended series of cases, analogous but distinct, in which a phenomenon shall be modified more and more, whether by successive simplifications or gradations. It is evident that this can take place fully only with regard to vital phenomena; accordingly, it is only by physiological analysis that a clear idea of its value can be obtained. But chemical phenomena approach those of physiology nearly enough, not only to demand this method, but to indicate that without it the science can never find the road to perfection. The existence of natural families in chemistry is now admitted by the best inquirers: but the classification remains to be made. The need of the classification must lead to the use of the comparative method, both being based on the common consideration of the uniformity of certain preponderant phenomena in a long series of different bodies. There is even such a connection between the two orders of ideas that the construction of a natural chemical classification is impossible without a large application of the comparative art, as the physiologists understand it; and conversely, comparative chemistry cannot be regularly cultivated without the guidance of some sketch of a natural classification. Chemistry is at present only a nascent science; general methods are as yet scarcely recognised in connection with it; and only a very few researches afford an example of the comparative method; but I am persuaded, not only of the fundamental suitability of that method in chemistry, but of its application, before very long, to the perfecting of the science. Such an anticipation, somewhat preceding the spontaneous development of any science, may be a contribution to its actual progress.

All the means employed are subject, especially, but not solely in chemistry,—to a verification by the precise collation of the two procedures of *analysis and synthesis;*— or, (as these terms have been corrupted by metaphysical uses) composition and decomposition.—Every substance which has been decomposed must evidently be capable of recomposition, whether the process be otherwise practicable or not. If the inverse operation reproduces precisely the primitive substance, the chemical demonstration is complete. Unfortunately, the vast extension of chemical resources in this century has had a much stronger bearing on analytical powers than synthetical means; so that there is at present little proportion and harmony between the two methods.— Such harmony is indispensable to the establishment of certainty in some cases, as we see when we duly distinguish two widely differing kinds of chemical analysis: the preliminary, consisting of the simple separation of the immediate principles; and the final, leading to the determination of the *elements*, properly so called. Though both are essential to chemical research, the first is of the most important and extensive use. The elementary analysis might be spared a synthetical verification,—because the composition of the reacting substances may be compared with the results obtained, thus indicating the composition of the proposed substance, the different elements of which will in this way have been in some sort separated. The impossibility of recombining the elements, to reproduce the primitive body, ought not to throw any doubt on the solution, unless there is some special reason for suspecting the simplicity of any one of the elements. Synthesis can, in this case, only add a valuable confirmation, to what was before not doubtful. But the case is very different when we have to determine only the immediate principles. As the elements concerned can produce combinations of different orders, we can never be sure that one or more of the supposed immediate principles obtained does not result from the reactions caused by the analysis itself. It is only synthesis which, by reconstructing the proposed substance with the materials concerned, can decide the question conclusively, though in some cases of feeble agency in the reactives, and strong analogical induction, there is no room left for reasonable doubt. In immediate analyses of great complexity, when the agreement of various analytical means strongly corroborates the conclusions obtained, we can not rely on real chemical demonstration without the synthetical confirmation. This maxim of chemical philosophy is abundantly exemplified in the analysis of mineral waters, and yet more of organic substances.—It is noticeable that synthesis is easiest where it is most necessary, and would be most difficult in the case of elementary analysis, where it can, as we have seen, be dispensed with. This is owing to the combinations becoming less tenacious as the order of composition of the constituent particles is higher; and if the decomposition is easy, so is the recomposition. The cases of immediate analysis require only feeble antagonisms, offering no great obstacles to the synthetical operations indispensable for their demonstration.

<small>Rank of the science.</small> We have next to consider the encyclopedical position of Chemistry, to justify the rank assigned to it in our scale.

It is from no vain and arbitrary consideration that Chemistry is placed between Physics and Physiology in our scale. By the important series of electro-chemical phenomena Chemistry becomes, as it were, a prolongation of Physics: and at its other extremity, it lays the foundations of physiology by its research into organic combinations. These relations are so real that it has sometimes happened that chemists, untrained in the philosophy of science, have been uncertain whether a particular subject lay within their department, or ought to be referred either to physics or to physiology.

The phenomena of Chemistry are more complex than those of Physics, and are certainly dependent on them. Their degree of generality is inferior,—chemical effects requiring a much more extended concurrence of varied conditions. Physical properties belong not only to all substances, but, with simple modifications, to all the states of aggregation, and even of combination, of each of them: whereas, it is only in a more or less determined and restricted condition that each body manifests its chemical properties. In a word, nature often shows us physical effects apart from the chemical, while there can be no chemical effects apart from certain physical phenomena. Thus Chemistry can not be rationally studied without a previous knowledge of physics. Besides, the most powerful chemical agents are derived from physics, which presents, in its different orders of phenomena, the first distinctive characters of different substances. It is impossible in our day to conceive of scientific chemistry without giving it the whole of physics for its basis: and thus is its first relation in the scale established. And, as physics is dependent on astronomy and mathematics, so must its own dependent be. But it must be owned that, with regard to doctrine, the connection of Chemistry with the first two sciences is neither extensive nor very important.

Every attempt to refer chemical questions to mathematical doctrines must be considered, now and always, profoundly irrational, as being contrary to the nature of the phenomena. In the case of <small>Relation to Mathematics.</small> physics, the mischief would be, as we have seen, merely from the misuse of an instrument which, properly directed, may be of admirable efficacy: but if the employment of mathematical analysis should ever become so preponderant in chemistry (an aberration which is happily almost impossible) it would occasion vast and rapid retrogradation, by substituting vague conceptions for positive ideas, and an easy algebraic verbiage for a laborious investigation of facts. The direct subordination of chemistry to astronomy is also slight, but more marked. It is almost insensible in regard to abstract chemistry, *To Astronomy.* which alone is cultivated in our day. But, when the time shall come for the development of concrete chemistry,—that is, the methodical application of chemical knowledge to the natural history of the globe,—astronomical considerations will no doubt enter in where now there seems no point of contact between the two sciences. Geology, immature as it is, hints to us such a future necessity, some vague instinct of which was probably in the minds of philosophers in the theological age, when they were fancifully and yet obstinately bent on uniting astrology and alchemy. It is, in fact, impossible to conceive of the great intestinal operations of the globe as radically independent of its planetary conditions.—Inconsiderable as are the relations of chemistry with mathematics and astronomy, in regard to doctrine, it is far otherwise with regard to method. It is easy to see how the perfection of chemistry might be secured and hastened by the training of the minds of chemists in the mathematical spirit and astronomical philosophy. Besides that mathematical study is the necessary foundation of all positive science, it has a special use in chemistry in disciplining the mind to a wise severity in the conduct of analysis: and daily observation shows the evil effects of its absence. Yet, it can never be said that chemists have so much need of a mathematical education as physicists, because they do not need it as an instrument in daily use, but as an intellectual preparation for the rational study of nature. As to astronomy, we have seen that it constitutes the most perfect type of the study of nature; and this at once establishes its relation of superiority to chemistry. The more complex the phenomena, the more important is the influence of such a model; and it is only by having always before their eyes such an exemplification of the true spirit of natural philosophy, that chemists can rightly estimate the inanity of the metaphysical explanations which vitiate their doctrine, and can acquire an adequate sense of the true character, conditions, and destiny of chemical science. Under this point of view, astronomy is more useful to chemists than even physics, in proportion to the superiority of its method.

So much for the sciences which precede chemistry. *To Physiology.* As for those that follow, physiology depends upon chemistry both as a point of departure and as a principal means of investigation. If we separate the phenomena of life, properly so called, from those of animality, it is clear that the first, in the double intestinal movement which characterizes them, are essentially chemical. The processes which result from organization have peculiar characteristics; but apart from such modifications, they are necessarily subjected to the general laws of chemical effects. Even in studying living bodies under a simply statical point of view, chemistry is of indispensable use in enabling us to distinguish with precision the different anatomical elements of any organism.—We shall see hereafter that the new science of Social Physics is subordinated to chemical science. In the first place, it depends on it by its immediate and manifest connection which physiology: but, besides that, as social phenomena are the most complex and particular of all, their laws must be subject to those of all the preceding orders, each of which manifests, in social science, its own peculiar influence. In regard to Chemistry especially, it is evident that among the conditions of man's social existence several chemical harmonies between man and external circumstances are involved. Even if individual existence could be sustained, society could not, if these harmonies were destroyed, or even only somewhat disturbed,—as by changes in the atmospheric medium, or in the waters or the soil.

To Sociology.

The position of Chemistry among the sciences being thus determined, the next inquiry is about the degree of scientific perfection that its nature admits, in comparison with others. As for the method, if physics suffers from the intrusion of hypotheses, we may say that chemistry has been their absolute prey, through its more difficult and tardy development. The doctrine of *affinities* appears to me more ontological than that of fluids and imaginary ethers. If the electric fluid and the luminous ether are, as I called them before, materialized entities, *affinities* are at bottom pure entities, as vague and indeterminate as those of the scholastic philosophy of the Middle Age. The pretended solutions that they yield are of the usual character of metaphysical explanations,—a mere reproduction, in abstract terms, of the statement of the phenomenon. The advance of chemical knowledge which must at last discredit for ever such vain philosophy has as yet only modified it, so far as to disclose its radical futility. While affinities were regarded as absolute and invariable, there was at least something imposing in them; but since facts have compelled the belief of their being variable according to a multitude of circumstances, their use has only tended to prove, more and more, their utter inanity. Thus, for instance, it is known that at a certain temperature, iron decomposes water, or protoxide of hydrogen: and yet, it has been since discovered that, under the influence of a higher temperature, hydrogen in its turn decomposes oxide of iron. What signifies, in this case, any order of affinity that we may ascribe to iron and hydrogen with oxygen? If we make the order vary with the temperature, we have a merely verbal, and therefore pretended explanation. Chemistry affords us now many such cases, apparently contradictory, independently of the long series of decisive considerations that have made us reject absolute affinities,—the only ones, after all, that have any scientific consistency whatever. The old habit is, however, so strong that

Degree of possible perfection.

Intrusion of hypotheses.

even Berthollet, in the very work in which he overthrows the old doctrine of invariable or elective affinities, proposes vague affinities under many modifications. The strange doctrine of predisposing affinity is to be found in the work, among others, of the most rational of recent chemists, the illustrious Berzelius. When for instance, water is decomposed by iron through the action of sulphuric acid, so as to disengage the hydrogen, this remarkable phenemenon is commonly attributed to the affinity of the sulphuric acid for the oxyde of iron which *tends* to become formed. Now, can anything be imagined more metaphysical, or more radically incomprehensible, than the sympathetic action of one substance upon another which does not yet exist, and the formation of the last by virtue of this mysterious affection? The strange fluids of physicists are rational and satisfactory in comparison with such notions. These considerations justify the desire that chemists should have a sufficient training in mathematical, astronomical, and then in physical philosophy, which have already put an end to such chimerical researches within their own domain, and would discard them speedily from the more complex part of natural philosophy. It is only by having witnessed the purification in the anterior sciences that chemists could realize it in their own: and there could not be complete positivity in chemistry if metaphysics lingered in astronomy or physics. This, again, justifies the place assigned to chemistry among the sciences. The individual must follow the general course of his race in his passage to the positive state. He must find that true science consists, everywhere, in exact relations, established among observed facts, allowing the deduction of the most extensive series of secondary phenomena from the smallest possible number of original phenomena, putting aside all vain inquiry into causes and essences. And this is the spirit which has to be made preponderant in chemistry,—dissolving for ever the metaphysical doctrine of affinities.

The inferiority of chemistry to physics, in regard to method and doctrine, explains its relative imperfection with regard to actual science. We have only to compare with the formula which told us what chemistry ought to be what it actually is, to see that it is at an immense distance—much further than physics—from its true scientific aim. Chemical facts are at this day essentially incoherent, or, at best, feebly co-ordinated by a small number of partial and insufficient relations, instead of those certain, extended, and uniform laws of which physics is so justly proud. As for prevision, if it is imperfect in physics in comparison with astronomy, it can hardly be said to exist in chemistry at all; the issue of each chemical event being usually known only by specially consulting the immediate experiment, when, as it were, the event is already accomplished. *Actual imperfection.*

Imperfect as chemistry is, in regard to method and doctrine, it is yet superior to physiology, and still more to social science, not only because, from the comparative simplicity *Comparative imperfection.*

of its phenomena, the facts and investigations are clearer and more decisive, but because it has a few, though very few, real theories, capable of affording complete previsions; a thing as yet impracticable, except in a general manner, with living bodies. We shall have occasion to notice the theory of proportions, the equivalent of which is not, in any sense, to be looked for in physiology. We must remember, while estimating the comparative imperfection of the sciences, that the importance to us of their perfection is in proportion to their simplicity; our available means being always found to correspond with our reasonable wants. I hope, too, that this severe estimate of the actual state of each science will stimulate rather than discourage the student; for it is more gratifying to our human activity to conceive of the sciences as susceptible of vast, varied, and indefinite progress, than to suppose them perfect, and therefore stationary, except in their secondary developments.

This leads us to consider the function of Chemistry in the education of the human mind.

Relation to human progress. It may be said to train us in the great art of experimentation; not as being our exclusive teacher, for, as we have seen, physics is superior to it in this; and it is more the art of observing than of experimenting that Chemistry is chiefly distinguished for. But there is an important part of the positive method which chemistry seems destined to carry to the highest perfection. I do not mean the theory of classifications, of which *Art of nomenclature.* chemists know too little at present; but the art of rational nomenclatures, which is quite unconnected with classifications. Since the reform in chemical language, attempts have been incessantly made, to this hour, to form a systematic nomenclature in anatomy, in pathology, and especially in zoology; but these endeavors have not had, and never can have, any success to compare with that of the reformers of chemical language; for the nature of the phenomena does not admit of it. It is not by accident that the chemical nomenclature is alone in its perfection. The more complex phenomena are, and the more varied and less restricted the comparisons of objects, the more difficult it becomes to subject them to a system of denominations, at once rational and abridged, so as to facilitate the habitual combination of ideas. If the organs and tissues of the living body differed only from one point of view; if maladies were sufficiently defined by their seat; if, in zoology, genera, or at least families, could be established by a homogeneous consideration, the corresponding sciences might at once admit of systematic nomenclatures as rational and as efficacious as that of Chemistry. But the diversity of aspects, rarely reducible to one head, renders such an arrangement extremely difficult and not very advantageous.

The case of chemistry is the only one in which, by its nature, the phenomena are simple, uniform, and determinate enough to allow of a rational nomenclature at once clear, rapid, and complete, so as to contribute to the general progress of the science. The idea of composition, the great end of the science, is always preponderant. Thus, the systematic name of each body, expressing its composition, indicates first a correct general view, and then, the sum of its chemical history; and, by the nature of the science, the more it advances toward perfection, the more must this double property of the nomenclature be developed. In another view, dualism being the commonest constitution in chemistry, and the most essential, and that to which all other modes of composition are more and more referred by science, we see that the conditions of the problem are as favorable as possible to a rapid and expressive nomenclature. Thus, there has always been some system of nomenclature, more or less rough, though none to be compared to that so happily founded by Guyton-Morveau. Though the art can manifest its excellence only in proportion to the advance of chemistry, it is in such harmony with the nature of the science that, in its present imperfect state, it upholds it, by provisionally supplying, as it were, the almost absolute deficiency of true rationality. Thus chemistry may be regarded as specially adapted to develop one of the few fundamental means, the aggregate of which constitutes the general power of the human mind. The formation of a similar aid in the more complex sciences offers a real and strong interest; and I have only desired to show that we must resort to chemistry for the true principles and general spirit of the art of nomenclature, according to the rules so often set forth in this work, that each great logical artifice should be directly studied in the department of natural philosophy where it is found in the greatest perfection, that it may be afterward applied in aid of the sciences to which it less specially belongs.

The high philosophical properties of Chemistry are more striking in regard to doctrine than to method. However imperfect our chemical science is, its development has operated largely in the emancipation of the human mind. Its opposition to all theological philosophy is marked by the two general facts in which it has a share with all the rest of positive philosophy—first, the prevision of phenomena, and next, our voluntary modification of them. We have already seen that the more the complexity of phenomena baffles our prevision, the greater becomes our power of modifying them, through the variety of resources afforded by the complexity itself; so that the anti-theological influence of science is infallible, in the one way or the other. In chemistry, our modifying power is so strong that the greater part of chemical phenomena owe their existence to human intervention, by which alone circumstances could be suitably arranged for their production; and if the phenomena of physiology and social science admit of modification in a yet greater degree, chemistry will always, in this particular, hold the first rank, since the highest order of modifications is that which we here find—those which are most important for the amelioration of the condition of man. In the system of the action of man upon nature, chemistry must ever be regarded as the chief source of power, though all the fundamental sciences participate in it more or less.

In this way, chemistry effectually discredits the notion of the rule of a providential will among its phenomena. But there is another way in which it acts no less strongly; by abolishing the idea of destruction and creation in nature. Before anything was known of gaseous materials and products, many striking appearances must inevitably have inspired the idea of the real annihilation or production of matter in the general system of nature. These ideas could not yield to the true conception of decomposition and composition till we had decomposed air and water, and then analyzed vegetable and animal substances, and then finished with the analysis of alkalies and earths, thus exhibiting the fundamental principle of the indefinite perpetuity of matter. In vital phenomena, the chemical examination of not only the substances of living bodies, but their functions—imperfect as it yet is—must cast a strong light upon the economy of vital nature by showing that no organic matter radically heterogeneous to inorganic matter can exist, and that vital transformations are subject, like all others, to the universal laws of chemical phenomena. Chemical analysis seems to have fulfilled its function in this direction: henceforth it must be by the more difficult, but more luminous method of synthesis that this great philosophical revolution must be completed: and attempts enough have been successfully made to prove the possibility of it.

Divisions of the science. The divisions of the science have not been clearly and permanently settled, partly because of its very recent origin, and partly on account of its nature. In the first place, students have been more occupied in multiplying observations than in classifying them; and in the next, the homogeneous character of chemical phenomena causes essential differences to be less profound, and therefore less marked, than in any other of the fundamental sciences. In astronomy, there can be no question of a division into geometrical and mechanical phenomena. Physics is less a unique science than a group of almost isolated sciences; and they indicate their own arrangement. We shall see hereafter that nearly the same thing happens, though from a different cause, in physiology. But in chemistry, the conditions are less favorable, the distinctions being scarcely more marked than those which exist in a single department of physics—as thermology, and yet more, electrology. The imperfection and small importance of its present divisions are easily explained: and there are strong symptoms of an approaching discussion of this great subject; for the majority of eminent chemists are more or less dissatisfied with the provisional division which they have been hitherto obliged to accept as guidance in their labors.

No organic chemistry. The general division of *organic* and *inorganic* chemistry can not be sustained, on account of its evident irrationality. What is at present called organic chemistry has an essentially bastard character, half chemical, half physiological, and not, in fact, either the one or the other, as we shall have occasion to see. The division can not even be sustained under another form, as equivalent to the general distinction between cases of dualism and of other composition. For if inorganic combinations are usually binary, there are some which are composed of three elements and even of four; while, conversely, we very often meet with a true dualism in bodies which are called organic. For a genuine division we must look to general ideas relating to composition and decomposition; and in this form, attending to the rule of following the gradual complication of phenomena: first, the growing plurality of constituent principles (mediate or immediate), according as the combinations are binary, ternary, etc.; and secondly, the higher or lower degree of composition of the immediate principles, each of which may (as in the case of a continual dualism) be decomposable into two others, for a greater or smaller number of consecutive times. Though each of these two points of view is of high importance, the preponderance of the one or the other must be agreed upon before the rational division of chemistry can be organized. Though this is not the place to discuss this new question of high chemical philosophy, it may be well to state that I regard it as solved; and that the consideration of the degree of composition is, in my eyes, evidently superior to that of the number of elements, inasmuch as it affects more profoundly the aim and spirit of chemical science, as they have been characterized in this chapter. As for the rest, whatever the decision may be, we may remark that the two classifications differ from each other much less than we might at first be tempted to suppose; for they necessarily concur, whether in the preliminary or in the final case, and diverge only in the intermediate parts.

Principles of composition and decomposition.

We have now reviewed the nature and spirit of chemical science; the means of investigation proper to it; its true encyclopedical position; the kind and degree of perfection of which it is susceptible; its philosophical properties in regard to method and to doctrine; and, finally, the mode of division which would be suitable to it. We must complete the survey of the science by a special and direct notice of the few essential doctrines which have been disclosed by the spontaneous development of chemical philosophy. It must be remembered that the object of this work is not to present a treatise on each science, or to enlarge upon it in proportion to its proper importance, or the multiplicity of its facts; but to ascertain its relative importance, as one head of positive philosophy. No one will expect that chemical philosophy, in its present state, can be examined here as fully or satisfactorily as, for instance, astronomical philosophy, the perfection of which admits of a methodical analysis, clear and complete, though summary, such as befits that immutable type of natural philosophy.

CHAPTER II.

INORGANIC CHEMISTRY.

Mode of beginning the study. WHATEVER may be the principles of division and classification preferred in the general system of chemical studies, it is agreed by almost all chemists that the preliminary and fundamental study should be the successive and continuous history of all the simple bodies. The plan of M. Chevreul is an exception to this, his method being to proceed at once from the study of each element to all the combinations, binary, ternary, etc., that it can form with those already examined; confining himself, however, to compounds of the first order. This plan has the advantage that simple bodies are more completely known from the beginning than by the usual method, which scatters through the different parts of the science the most important chemical properties of each of them. But, on the other hand, the history of any element remains incomplete; a factitious inequality is established among chemical researches into different elementary substances; and the didactic inconvenience which M. Chevreul proposed to escape seems to be unavoidable, under any method. On no plan can any chemical history be completed by a first study. The provisional information obtained by a first study must be followed by a revision which allows us to take into consideration the whole series of phenomena relative to each substance. The question is merely a didactic one, only of secondary importance in this work, though of great practical interest. On any scheme, it remains certain that the preliminary study of elementary substances is, by the nature of the science, the necessary foundation of chemical knowledge.

On account of the considerable and always increasing number of substances regarded as simple, some modern philosophers, possessed Plurality of elements. with the notion of the simplicity and economy of nature, have concluded *à priori* that most substances must be the various compounds of a much smaller number of others. But, while endeavoring to conceive of nature under the simplest aspect possible, we must do so under the teaching of her own phenomena, not substituting for that instruction any thoughtless desires of our own. We have no right to presume beforehand that the number of simple substances must be either very small or very large. Chemical research alone should settle this; and all that we are entitled to say is that our minds are disposed to prefer the smaller number, even, if it were possible, so far as there being but two. But not the less are we bound to suppose all substances which have never in any way been decomposed to be simple, though we should not pronounce them to be for ever undecomposable. All chemists now admit this rule as the first axiom of sound chemical philosophy.

Aristotle first saw this rule, though he did not conceive of its rational grounds. His doctrine of the four elements, popularly cried down in our time, should be judged of as the first attempt of the true philosophical spirit to conceive of the composition of natural bodies, amid the then existing deficiency of all suitable means of research. To appreciate it we must compare it with anterior notions. Now, up to that time, all the schools, however they might differ about other things, agreed that there was only one elementary substance; and their dispute was about the choice of the principle. Aristotle, with his rational character of mind, put an end to all those barren controversies by establishing the plurality of elements. This immense progress must be considered the true origin of chemical science, which would be radically impossible on the supposition of a single element, excluding all idea of composition and decomposition. Whatever appearances may be, there is no doubt that it must be much more difficult for the human mind to pass from the absolute idea of unity of principle to the relative idea of plurality, than to rise gradually, by means of research, from the four elements of Aristotle to the fifty-six simple bodies of our chemistry of this day. Our *Naturists*, who are all for simplicity and economy without caring much for reality, have no right to appeal to the authority of Aristotle, who had so much reverence for reality as to infringe the notion of simplicity which he found prevailing. They should go back further than Aristotle,—to Empedocles or Heraclitus,—and attain the utmost simplicity at once, by admitting only a single principle.

Other philosophers, among whom was Cuvier, have objected to the simplicity of most of the elements now admitted by chemists, that some of them seem to be extremely abundant in nature, while others are scantily and partially distributed: whereas, it seems natural to presume that the different elements must be almost equally diffused throughout the globe, and that therefore chemical analysis will, sooner or later, prove the rare ones to be compound substances, requiring peculiar and rare influences for their formation. It would be enough to say that the presumption, though plausible, is nothing more than a presumption: but it may be added that we know nothing of our planet beyond the upper strata; and we can form no prejudgment of the composition of the whole. It would be too much to say that there should be an equality of elements on the surface, even the probability being the other way; for the heaviest elements are the rarest at the surface, and the commonest are those which go to the composition of living bodies; and the probability is strong that the preponderance is reversed in the interior of the globe, to make up the mean density, which is not to be found among the solids, liquids, and gases, which are required for the existence of life. Thus the objection seems to be converted by a chemical analysis into a sort of confirmation.

Since the time—recent, it is true—of the decomposition of the elements of Aristotle, there has not been a single instance of a substance having passed from the class of simple to that of compound bodies, while the inverse case has been frequent. Yet, no chemist disputes the possibility of a reduction of the elements by a more thorough analysis; for chemical simplicity, as it is to us, is a purely negative quality, not admitting of those irreversible demonstrations proper to positive compositions and recompositions. The great general example of substances, called organic, the chemical theory of which is so complex, notwithstanding the small number of their elements, might lead us to suppose that such a reduction would not be, after all, so very great an advantage: but in this case, the difficulty seems to me to be referrible to the deficiency of duality. Notwithstanding this example, we can not but think that chemistry would become more rational and more systematic, if the elements were fewer, from the closer and more general relation which must then subsist among the different classes of phenomena. But the apparent perfection could be only barren and illusory if we were to assume it by conjecture anticipating the real progress of chemical analysis.

Classification of elements. This profusion of elements has naturally led to endeavors to classify them. The high importance of the question has become manifest through the deep persuasion that the rational classification of simple bodies must determine that of compound substances, and therefore that of the whole chemical system. The first principle to be laid down is that the hierarchy of elementary substances is not to be determined only by their proper essential characters, but by the less direct consideration of the principal phenomena of the compounds which they form. Without this requisition, the classification would have little use or interest; for it would be of small consequence in what conventional order we studied fifty-six bodies all independent of each other: whereas, with its proper condition, this question is as important as any that chemical philosophy can present.

The old division of the elements into the comburent and combustible (those which burn in the active and in the neuter sense), and the subdivision of these into metallics and non-metallics, are evidently too artificial to be maintained, except provisionally. For many years, endeavors have been made made to supersede it; but no irreversible classification has been yet obtained. M. Ampère seems to have been the first who pointed out the necessity; and he proposed a system in 1816; but it was not one which induced the chemists to abandon their ancient distribution, the binary structure of which made it easy of application, whatever might otherwise be its defects. A few years after, Berzelius offered, in a simple and almost incidental form and manner, a far superior system of classification. He first understood the necessity of rising finally to a unique series, constituting, by a uniform and preponderant character, a true hierarchy; whereas, M. Ampère saw only the importance of natural groups, which might be arbitrarily co-ordinated. Both conditions are imposed by the general theory of classifications; but that which Berzelius had chiefly in view is unquestionably superior to the other; and especially in the present case, when the small number of objects to be classified renders the formation of groups a matter of secondary importance, provided the series be naturally ordained.

M. Berzelius's conception is grounded on the consideration of electro-chemical phenomena. Its simple and lucid principle is that the new elements are to be so disposed as that each shall be electro-negative to those which precede it, and electro-positive to those which follow it. The series thus derived appears, thus far, to be in conformity with the whole of the known properties of both the elements themselves and their principal compounds. It is too soon, however, to speak decisively of this: and, on the other hand, the chemical preponderance of electric characters is by no means so logically established as to compel us to seek the bases of a natural classification in that order of phenomena. It must, it seems to me, be clearly proved at the outset that the point of departure is a real one,—that is, that a constant order of electrization exists among the different elements, which is maintained under all conditions of exterior circumstances, of aggregation and decomposition: but, not only has this never been adequately undertaken, but there is some reason to apprehend that its result would be opposite to the proposed principle. Whatever may be the issue of future labors, Berzelius has secured the eternal honor of having first exhibited the true nature of the problem, and the aggregate of its principal conditions, and perhaps the order of ideas in which its solution is to be sought. Wherever this solution is obtained, chemistry will have made a great stride toward a truly rational state: for, under a hierarchy of the elements, the systematic nomenclature of compound substances will almost suffice to give a first indication of the general issue proper to each chemical event: or, at least, to restrict the uncertainty within narrow limits. Yet, through this very connection of such a research with the whole of chemical studies, I do not think it can be efficaciously pursued while we separate it, as has hitherto been done, from the general question about the establishment of a complete system of chemical classification for all bodies, simple and compound. Now, this great question seems to me at present premature. The preliminary conditions, both of method and of doctrine, are, as we have seen, far from being completed. As such a general system of classification must constitute both the summing up and the fundamental view of the whole of chemical philosophy, I shall further expand my idea about it in this place.

Classification of Berzelius.

Premature effort.

As for the method, it requires perfecting in two ways, for which chemists must resort to physiology.

Requisite preparation as to method.

They must understand the fundamental theory of natural classifications, which can be obtained nowhere else: and they must, for the same reason, study in the same school the general spirit of the comparative method, of which chemists have very little idea, and without which they can never proceed properly in search of a rational classification. These two improvements must be derived from biological philosophy; the one to lay down the problem of chemical classification, and the other to undertake its solution. It will be by perceiving these harmonies and mutual applications among the sciences commonly treated as isolated and independent, that philosophers in all departments will at length become aware of the reality and utility of the fundamental conception of this work; the cultivation of the different branches of natural philosophy under the impulsion and direction of a general system of positive philosophy, as a common basis and uniform connection of all scientific labors. We have little idea what we lose by the narrow and irrational spirit in which the different sciences are cultivated, and especially with regard to method. When the great scientific relations of the future shall be regularly organized, men will scarcely be able to imagine, otherwise than historically, that the study of nature could ever have been conceived and directed in any other way.

As to doctrine. As to doctrine, we have seen that the desired classification can not take place till we have settled the preponderance of the one or the other consideration—the order of composition of the immediate principles, or their degree of plurality. Now, such a problem has not yet been rationally proposed. If we suppose it resolved, adopting the rule which I think almost incontestable, as I explained before, of treating the first point of view as necessarily superior to the second, we must still attend to two special conditions, before we can proceed to the rational construction of the system of chemical substances.

First condition. By the first of these conditions, we must dismiss the irrational distinction of substances into organic and inorganic. We shall see hereafter that organic chemistry must soon dissolve, parting with some of its questions to chemistry proper, and others to physiology. When any combination is susceptible of a chemical examination, it must be subjected to a fixed order of homogeneous considerations, whatever may have been its origin and mode of concrete existence, with which chemistry has nothing to do, unless as a source of information. As long as any classification must be adapted to the strange conception of a sort of double chemistry, established upon a false division of substances, it must be precarious and artificial in its details, because it is vitiated in principle. The evil is felt, as is shown more and more by the tendency to refer organic combinations to the general laws of inorganic combinations; but it would not be enough, as might be supposed, that a distinguished chemist should take the initiative, in a large and direct manner, to accomplish this important reform.

Such a work demands a special and difficult operation, requiring a delicate combination of the chemical and physiological point of view, in order to make a true division of what should remain with chemistry, and what should return to physiology.

The second condition is closely connected with the first. It requires that all combinations should, if possible, be submitted to the law of dualism, erected into a constant and necessary principle of chemical philosophy. Great as would be such an improvement in the way of simplification of chemical conceptions, it must, however, be admitted that it is not so indispensable to classification as the preceding. Without the first condition, rational classification would be impossible; whereas, it might take place, with imperfection and difficulty, without the second. As for the prospects of the case, the tendency to improvement is as real and marked in the one case as the other; as any one may observe for himself. *Second condition.*

It is of the more importance to set the consideration of the order of composition of immediate principles above that of their degree of plurality, as before proposed, because the first is, by its nature, clear and incontestable, while the other is always more or less obscure and dubious. The one is, in fact, the simple appreciation of an analytical or synthetical fact; the second has always a certain hypothetical character, since we then pronounce upon the mode of agglomeration of elementary particles; which is a thing radically inaccessible to us. Thus, for example, a chemist may establish with certainty that such or such a salt is a compound of the second order, and that certain acids and alkalies are, on the contrary, of the first order; for analysis and synthesis can demonstrate that each of the last bodies is composed of two elementary substances, and that, on the contrary, the immediate principles of the salt are decomposable into two elements. But, in another view, when the analysis of any substance has established the existence in it of three or four elements, as in the case of vegetable or animal matters, we can not without resort to hypothesis, pronounce that this combination is really ternary or quaternary, instead of being simply binary; for we can never assert that we could not, by a preliminary analysis less violent than this final one, resolve the proposed substance into two immediate principles of the first order, each of which should be further susceptible of a new binary decomposition. *Method of analysis. Chemical dualism.*

If an unskilled chemist should at this day apply unduly strong means to the analysis of saltpetre, the results might authorize him, following our present erroneous procedure, to conceive of this substance as a ternary combination of oxygen, azote, and potassium: and yet we know that such a conclusion would be false, as the substance may be easily reconstructed by a direct combination between nitric acid and potash, which might have been separated by a less disturbing analysis, without occasioning their decomposition. How do we know that it may not be so with every combination habitually classed as ternary or quaternary? Immediate analysis being as yet so imperfect in comparison with elementary analysis, especially with regard to these substances, would it be rational to proclaim, for the time to come, its necessary and eternal impotence with regard to them? Such judgments seem to be founded on a confusion between these two kinds of analysis, so really different in themselves, and so characterized in their operations by delicacy in the one case and energy in the other. One important consideration, relating to the synthetical point of view, is evidence of this confusion between the two analyses; and that is, the extreme difficulty, if not impossibility, of verifying by synthesis the analytical results proper to these substances. We have seen that immediate synthesis is usually very easy, while elementary synthesis is scarcely practicable. Thus, reciprocally, it seems to me rational to suppose that when the recomposition can not be effected, the analysis has not been immediate—there being no other objection to such a conclusion. For example, we exhibit the impossibility of reproducing by synthesis vegetable and animal substances; and this has been even set up as a sort of empirical principle. But is not this impossibility owing to our persisting in an elementary synthesis when we ought to proceed by an immediate synthesis, the materials of which might in many cases be discovered beforehand? This remark is true with regard to a multitude of combinations the dualism of which is, however, very certain, with the sole difference that the immediate principles are better known. If we tried to recompose saltpetre by directly combining oxygen, azote, and potassium, we should succeed no better than in reproducing organic substances by throwing together their three or four elements; the obstacles which we admit in the last case apply equally to the first. The most striking achievement is that of M. Wœhler, in producing the animal substance urea. He could not have done this if he had tried, according to the common prejudice, to combine directly oxygen, hydrogen, carbon, and azote, which concur in the elementary constitution of this substance, instead of uniting only its two immediate principles, till then unknown in this quality. Is there any reason to suppose that it is otherwise in any other case? It appears, then, that chemists will be safe in attributing an entire generality to the fundamental principle of dualism of all combinations, under the one easy condition of regarding as still very imperfect the analysis of substances exceeding the binary composition; and especially the substances called organic, the true immediate principles of which would thus remain to be discovered. These principles can be conceived of only by imagining a considerable number of new binary combinations, of the first and second orders, between oxygen, hydrogen, carbon, and azote; and the realization of this may seem, in the present state of our knowledge, almost impossible. But we have no right to conclude it to be so, while our analytical procedures are what they are; and there is no scientific objection to our supposing that there may be many more direct and binary combinations among the elements of ternary or quaternary substances than chemistry has yet established.

It must be observed, however, that universal and indefinite dualism can not be maintained unless chemists will scientifically determine the sense of the word *substance*; that is, restrict it to mean real *combination*: for it would be easy to cite, and especially in physiological chemistry, very marked cases of the defect of dualism. But we can not regard as a true chemical substance an accidental assemblage of heterogeneous substances, whose agglomeration is evidently mechanical, such as sap, blood, a biliary calculus, etc., unless we confound the notion of dissolution, and even of mixture, with that of combination. If we extend in this way the use of the term *substance*, so valuable in chemistry, we might as well treat, as so many chemical substances, the waters of different seas, different mineral waters, soils, etc.: and even more, artificial mixtures of a variety of salts dissolved together in water or alcohol. We shall see hereafter that all difficulties in this subject may be disposed of by our learning that they proceed from our not having clearly and rigorously separated the chemical from the physiological point of view. We may be assured that the most elementary notions of chemical philosophy can not be rationally established, in their due clearness, generality, and stability, without being founded on a full comparison with biology; a comparison which can be organized only under a complete system of positive philosophy.

Meantime, there is a marked tendency in the present movement of chemical ideas, toward a complete dualism. The increasing assimilation attempted between organic and inorganic substances is an indirect advance in that road: but much more striking, in this view, are experiments like those of Mr. Wœhler, which refer the most refractory compounds to dualism, either by analysis or synthesis. A binary formula is adopted, too, to represent the proportion of elements proper to the most complete substances: and, though this is not a true dualism, it helps to prepare minds for the establishment of a real and general one. The sum of what has been said on this important subject of chemical dualism is this:—the real mode of agglomeration of elementary particles is, and ever must be, unknown to us, and therefore no proper object of our study:—our positive researches being thus circumscribed, we may rationally conceive of the *immediate* composition of any substance as binary; but so as to represent all the phenomena that chemistry can offer to us, in any future state of perfection. Thus, I do not propose universal dualism as a law of nature; for this we could never establish: but I declare it to be a fundamental artifice of true chemical philosophy, destined to simplify our elementary conceptions, by using our optional intellectual liberty in accordance with the true end and aim of positive chemistry.

These are the conditions necessary to the institution of a system of natural classification, answering in chemistry to the universal hierarchy of living bodies in biology, if the complication of phenomena would admit of our obtaining such a system. Up to this time, perhaps no one has formed an adequate idea of the nature and spirit of such an operation: but in my view, chemical classification, thus conceived of, is the science itself, condensed into the most substantial summary. All I claim to have done is to have introduced into chemical science the special kind of philosophical spirit which is naturally developed by biological science, as it has been conceived of by all its great masters, from Aristotle downward.

It is because I have high expectations of what Chemistry will become, that I attach so much importance to the preceding discussion. The science is now weak and desultory, notwithstanding its rich collection of facts: but, extended and complex as it is, there is no fundamental science, except astronomy, whose phenomena are so homogeneous, and therefore so fit for a true systemization, in the positive spirit. Now, this future constitution of chemical science must, it seems to me, consist in a complete system of natural classification, which can not be obtained till all combinations, whatever their origin, are subjected to a fixed order of homogeneous consideration, and, on the other hand, constantly referred to a fundamental dualism.

We cannot form any certain expectation of the future condition of Chemistry from its present state: but, before proceeding to examine the two doctrines which at this day approach nearest to positive rationality,—that of definite proportions and the electro-chemical theory,—I will indicate two points of doctrine which seem by their nature, to indicate with precision the true dogmatic formation toward which, the science as a whole, must tend.

Law of double saline decompositions. First, there is a great law of double saline decompositions, discovered by Berthollet, and completed by M. Dulong's investigations on the reciprocal action of soluble and insoluble salts. The case of double solubility, considered by Berthollet, is this: two soluble salts, of any kind, mutually decompose each other whenever their reaction may produce an insoluble salt, or one less soluble than either of the two. This theorem stands first among general propositions in chemistry, and is the only one which can as yet give an exact idea of what, in chemistry, constitutes a true *law*. It has all the characters of a law: it relates to the proper subject of chemical science; it establishes a relation between two classes of phenomena before independent; and, above all, it admits of prevision of phenomena according to their positive relations. In establishing this law, Berthollet escaped some metaphysical snares, rejecting hypotheses of affinities; but he fell into one when he attempted to explain the law which he had just discovered. No law can be explained otherwise than by showing that it enters into another more general than itself: but this law of Berthollet's is alone of its kind; and it therefore admits of no explanation. It may hereafter be attached to a fundamental theory of the reciprocal action of all compounds of the same order; and such a relation will truly explain it: but at present it is simply a general fact which, inexplicable itself, serves to explain each of the particular facts which it comprehends.

The influence of air and water in the production of chemical phenomena is another of the most perfect doctrines of chemistry as it stands. *Chemical theory of air and water.* The importance of the action of air and water in the terrestrial economy has induced some German philosophers irrationally to set up the system of these two fluids into a sort of a third reign, between the inorganic and the organic: but abstract chemistry has nothing to do with natural history, and regards the study of air and water from a different point of view, while aware of its fundamental importance.

All chemical phenomena takes place in the presence of air; and they almost invariably require the intervention of water; it is clear therefore that before we study any chemical reaction, we must be able to analyze the participation of these two fluids. Thus the chemical theory of air and water is a sort of necessary introduction to the system of chemistry, properly so called, as belonging more to method than to doctrine, and as immediately following the study of simple bodies. It is an historical fact that the double analysis of air and water marked the first great advance in modern chemistry.

The influence of the air, not less important than that of water in chemical phenomena, was less difficult to characterize: *Air.* for the air is simply a mixture, and its chemical action is merely that of the gases which compose it, each of which acts as if it were isolated, allowing for the diminution of intensity from its diffusion, and for the very few cases in which the accomplishment of the proposed phenomenon determines the combination of the gases in an accessory way. Chemistry has only to analyze it, leaving all other study of it to the department of natural history. This analysis was effected in the early days of modern chemistry, except that there is still some uncertainty about the proportion of carbonic acid gas, and perhaps of some other more considerable principles, as, for instance, hydrogen, the existence of which begins to be generally suspected. Though no appreciable change in the composition of the atmosphere has taken place within half a century, it is impossible to conceive that some alteration must not happen, in some direction, in course of time, among the many perturbing influences which act upon the mixture. Their antagonism, and that of vegetable and animal action, partly neutralizes them: but the equilibrium can not be precise and continuous. Geological considerations and botanical fossils lead us to suppose that at some remote periods the composition of the air must have been sensibly different: and chemists themselves have actually established some slight periodical variations, dependent on the proportion of carbonic acid at different seasons. Our analytical resources are, however, very imperfect with regard to the accessory principles of the atmosphere; for chemists can ascertain nothing of the distinctions which are proved to exist in the best-marked localities, by their influences on living beings. The study of these variations, all-important in its way,—even as possibly indicating the limits of human life in a remote future,—belongs to natural history; and that is probably the reason why chemists trouble themselves so little about it: and if there is neglect, it should be charged upon the naturalists. It is true that a preparation is required for their order of study, like all others,—a provision of knowledge, rising from physiology to astronomy itself: but the research is not especially a matter of chemical duty.

Water. The study of water requires much more extended and complex researches than that of the air; and it is indispensable to the general system of chemical science: for water being a real combination, and perhaps the most perfect known to us, may exercise chemical effects proper to itself, independently of those attributable to its elements, and apart from its importance as a solvent,—to say nothing of it as a simple mixture. Thus there are three aspects under which water must be considered by chemists, all distinct and all essential; and the appreciation of them has been slow and difficult, if even we may say that this fundamental examination is yet complete.

The analysis of water, represented by a quantity of hydrogen double in volume that of oxygen, and unquestionably confirmed by synthesis, is the finest of the early discoveries of modern chemistry, not only from the light it casts upon the whole of chemical phenomena and the general economy of nature, but also from its conquest of prodigious difficulties. In regard to the first view, chemical science leaves nothing to desire. Yet, a notion has arisen, in recent times, of the existence of a new and more highly oxygenated combination between the two elements of water, which may raise some interesting questions, not about the irreversible composition of water, but about the kind of chemical influence which is taken for granted in its decomposition and recomposition in a multitude of phenomena; and especially, about the true mode of union of oxygen and hydrogen in all substances, and above all in liquids, which can not be obtained without water. Some doubts have lately been proposed about this, which seem to me to deserve mature examination.

The dissolving action of water has been the subject of a long series of laborious researches, much less difficult, and not far from complete. Yet more attention ought to be paid than is paid to the fine experiment of Vauquelin, in which it is shown that water, saturated with one salt, remains capable of receiving another, and even acquires by that the singular property of dissolving a new quantity of the first. This experiment, which has been in a manner despised, seems to me of the first order in its way, and a fit basis of a series of interesting researches about the apparently capricious laws of solubility, the study of which is yet essentially empirical.

Chemists were long in conceiving that water, besides being a solvent, might act in a really chemical manner, otherwise than by its elements. It seemed as if a combination so eminently neuter must be inoffensive, and inoperative, except by its decomposition. It was Proust who thought that this neutrality itself afforded a presumption of certain chemical affections, independently of its composition. This was the rational consideration which led him to create the important study of the *hydrates*, regarded as a sort of new salts, in which water plays the part, with regard to the alkalies, of a kind of hydric acid.

The examination of these combinations, and of all others that water can form with any substances without being decomposed, constitutes the third and last part of the fundamental study of water, regarded as an indispensable preliminary to the general system of chemical studies.

CHAPTER III.

DOCTRINE OF DEFINITE PROPORTIONS.

There are two general doctrines in chemistry, as it now exists, which present a systematic appearance, and invest the science with such rationality as it has attained. The first of these is the important doctrine of Definite Proportions.

Even if this doctrine were complete, it could exert only a secondary influence on the solution of the great problem of the science,—the study of the laws of the phenomena of composition and decomposition. The essential question is, what separations and new combinations must take place under determinate circumstances; and the theory of definite proportions affords no assistance to this kind of prevision. It proceeds, indeed, on the supposition that the question is already solved; and that it is to be taken as the point of departure for the estimate of each of the new products,—of their quantity and the proportion of their elements. Thus, the theory of the definite proportions present the singular scientific character of rendering rational, in its numerical details, a solution which usually remains empirical in its most important aspect.

Scope of the doctrine.

It was natural that the founders of modern chemistry should have attended to the laws of composition and decomposition, in preference to a study which they regarded as subordinate; and it was natural also that, as the advance of science disclosed to them the fast difficulties of the main problem, they should attend more and more to the secondary study, which promised an easier and more speedy success. But the most important office of this subordinate theory,—that of supplying the defect of immediate experiment,—can be but very imperfectly fulfilled, while it is regarded apart from the principal theory; and thus, the doctrine of definite proportions will never require its full scientific value till it is connected with an unquestionable basis of chemical laws, of which it will be the indispensable numerical complement.

Meanwhile, however, it affords a real, though secondary assistance to chemists, in rendering their analyses more easy and more precise. Moreover, it restricts the number of cases of combination logically possible, by exhibiting the very small number of distinct proportions; and by thus diminishing the uncertainty in cases of chemical action, it is, in fact, a natural preliminary to the establishment of those chemical laws to which it will be, under another view, a necessary supplement.

As to doctrine. In regard to doctrine, this theory offers a perfect type of the precise kind of rationality which must hereafter belong to Chemistry as a whole. In regard to method, the inquirers who have devoted themselves to establish the theory have advanced chemical science while appearing to diverge from it; simplifying the vast problem which their successors will solve, and preparing for the disclosure of the great laws of composition and decomposition, which would be undiscoverable amidst the infinity of products, if substances could combine, within certain limits, in all imaginable proportions. Such are the claims of this theory, as to doctrine and to method.

Its history. It assumed its existence and present form during the first quarter of this century: and it arose from a phenomenon discovered by Richter, and a speculative discussion established by Berthollet.—During the latter half of the last century, several chemists had observed that, in the mutual decomposition of two neutral salts, the two new salts thus formed are always equally neuter. Bergmann, among others, had steadily and specially dwelt upon this. Yet the fact was neglected or underrated till Richter, at the end of the century, generalized the observation, saw what it imported, and derived from it the fundamental law which bears his name. The law is this: that the ponderable quantities of the different alkalies requisite to neutralize a given weight of any acid are always proportionate to those required for the neutralization of the same weight of every other acid. This is, in fact, evidently the immediate consequence of the maintenance of neutrality after the double decomposition. Such a transformation would appear almost spontaneous if it related to a simpler and more developed science than Chemistry; but amidst its complications and the imperfection of our intellectual habits, the closest deductions are difficult if they have any character of generality, and therefore of abstraction; and this achievement of Richter's is, in consequence, eminently meritorious, on other grounds than its high utility.—His law, with the complements it has since received, is the original basis of the general doctrine of definite proportions. It exhibited, in the case of a considerable number of compounds, the great end of this doctrine;

viz. the assignment to every substance of a certain chemical coefficient, invariable and specific, indicating the proportions in which it can combine with each of those that have been similarly characterized. When it had been determined, by a double series of trials, what was the numerical composition of all the salts that may be formed by any one acid with the different alkalies, and any one alkali with the different acids, Richter's law enabled us to deduce immediately the proportions relating to all the compounds that can result from the binary combination of these two orders of substances. Richter himself brought his discovery up to this result, and prepared (but on a basis of experiment too narrow and imperfect) the first table of what were afterward called *chemical equivalents*.

These neutral salts constituted a particular case, which could hardly have led on to a general theory of definite proportions. *Berthollet's extension.* The idea of perfect neutralization must probably, at all times, have suggested to chemists that of a single proportion, on either side of which the neutrality must be destroyed; and thus the neutral salts were a natural first stage of the general theory; but they could not in themselves involve such a theory. It was Berthollet who extended the consideration of proportions to the whole of chemical phenomena. Some years after Richter's discovery he established as a fundamental principle, in his " Chemical Statics," the necessary existence of definite proportions for certain compounds of all orders; and he assigned the essential conditions of this characteristic property, which he attributed to all causes which can release the product of chemical reaction, as it forms, from the ulterior influence of the primitive agents. He thus added to Richter's restricted case the idea of a great number of cases subjected to the same principle, and able to lead on to its entire generalization. It is assigning much too little honor to Berthollet to recognise only the influence of his controversy with Proust, eminent as was the service rendered by Proust in that conflict, in establishing directly the general principle of determinate and invariable proportions.

Such was the double origin, experimental and speculative, of numerical chemistry. The next development had also a double character, arising from the harmony between the conception of Dr. Dalton and the experimental researches *Dalton's further extension.* of Berzelius, Gay-Lussac, and Wollaston. The inquiry was in a nascent state when Dalton's philosophic mind discerned its possible generality. He proposed the great Atomic theory, under which the doctrine of definite proportions was *Atomic theory.* developed to the whole extent that it has reached, and which serves as the basis of its daily application. The general principle of the theory is this: all elementary bodies are conceived of as formed of individual atoms, the different species of which unite, generally by twos, in a small number of groups, constituting compound atoms of the first order always mechanically indivisible, but thenceforth chemically divisible, and, in their turn, constituting all the other orders of composition by a series of analogous combinations. The principle is in such harmony with scientific conceptions in all departments, that it appeared like a happy generalization of the most familiar ideas of scientific men in every province of natural philosophy; and its universal and immediate admission took place as a matter of course.

It was observed by Berzelius that the deduction of the existence of definite proportions from this principle would be illusory if the combinations were not restricted to a very small number of atoms: for otherwise—if the number was, though limited, very great—the binary assemblages would be so multiplied that we might as well have combinations in any proportions whatever; and then the atomic theory might almost equally well represent the opposite doctrines of definite and indefinite proportions. Dalton was well aware of this; and the restrictions that he enunciated were presently declared too narrow by his successors, who found that they would not comprehend all existing combinations. His assertion was, that, in every combination, one of the immediate principles always enters for a single atom, and the other generally for a single atom also, and always for a very small number, rarely exceeding six. Taken with the expansion proposed by his successors, the atomic conception evidently represents the entire doctrine of definite proportions. But it is the theory of successive multiples, derived from the primary doctrine, which especially distinguishes Dr. Dalton's influence upon numerical chemistry. From the ground of his doctrine he easily saw that if two substances can combine in various distinct proportions, the ponderable quantities of the one which correspond, in the different compounds, to the same weight in the other, must naturally follow the series of whole numbers, since these compounds will have resulted from the union of one atom of the second substance with one, two, three, etc., of the first: and this constitutes a principal element, then perceived for the first time, of the theory of chemical proportions.

Extension by Berzelius. Berzelius followed, with his vast experimental study of the whole of the important points concerned in numerical chemistry, the different parts of which he has done more than any other chemist to develop and systemize. He first perfected Richter's law, so as to connect it closely with the atomic theory; by which it became susceptible of the extension given to it by Berzelius himself, to all compounds of the second order. But the most important new knowledge has arisen from his numerical study of compounds of the first order. By comparing the composition of the metallic sulphurets and that of the corresponding oxydes, he discovered a law, analogous to Richter's in regard to the salts. This law—that the quantity of sulphur of the first is always proportionate to the quantity of oxygen combined with a like weight of the base in the second—is now regarded, by induction, as applicable to all the compounds of the first order to which the same degree of chemical neutrality is assignable. And again the luminous series of the analyses of Berzelius have precisely verified in another direction the law of successive multiples discovered by Dalton in pursuance of his atomic theory.

Gay-Lussac followed, with the valuable numerical analyses he effected by having recourse to gaseous combinations, considered, not as to weight, but to volume. *Extension by Gay-Lussac.* He thus not only verified, in a special manner, the general principle of definite proportions, but presented it under a new aspect, which, by a wise induction, comprehends all possible cases—showing that all bodies in a gaseous state combine in invariable and simple numerical relations of volume. An accessory advantage of this achievement was that the specific gravity of the gases might be obtained with a precision often comparable to that of experimental estimate. It is necessary however to warn inquirers not to be led away, in their application of the theory of volumes to substances which have never been vaporized, from the point of view which in Gay-Lussac's application is equivalent to Dalton's, as adopted by Berzelius.

The labors of Wollaston bore a great part in establishing the doctrine of definite proportions. I do not *Wollaston's verification.* refer chiefly to his transformation of the atomic theory into that of chemical equivalents, though it has a more positive character, and tends to restrain the student from wandering after inaccessible objects, to which the first might tempt him, if not judiciously directed. The substitution would be of high value, no doubt, if it were not less a change of conception than an artifice of language. Nor have I in view the ingenious expedients by which Wollaston popularized numerical chemistry by rendering its use more clear and convenient. A greater service, in our present view, was his furnishing us with the indispensable complement of Richter's discovery, by establishing the theory in regard to the acid salts, since extended by analogy to the alkaline salts. The case of the acid salts was perhaps the most unfavorable possible for the ascertainment of the principle of invariable proportions. Wollaston effected the proof in the most satisfactory manner; and this special confirmation of the principle is considered, from its nature, the most decisive of all.

Such has been the logical and historical progress of the researches which have constituted numerical chemistry as it is now. We can represent by an invariable number, appropriated to each of the different elementary bodies, their fundamental relations of chemical equivalence, whence, by very simple formulas, immediately expressing the laws just indicated, we easily pass to the numerical composition proper to each combination. No further evidence of the truth of the doctrine is needed, than the fact that so many illustrious inquirers having attained the same view by ways which each one opened for himself, and all agreeing as to its positive application to all cases of importance, differing only as to the mode of expression of the results, in as far as the atomic theory left it indeterminate, and therefore optional. But we must glance at the difficulties thrown in the way of its application by a consideration of the aggregate of chemical phenomena, in order to form a clear idea of the final improvement of which this doctrine yet stands in need.

Scope of application of Numerical Chemistry.
Among the points which are beyond dispute, it is, first, evident, and no chemist has ever doubted it, that substances differ as much in the proportion as in the nature of their constituent principles. It is an axiom of chemical philosophy that any change whatever in the numerical composition causes a change in the whole of the specific properties, in a more marked degree as the alteration is greater. Varied and gradual above all others as are the proportions produced by the chemical phenomena proper to living bodies, they afford a striking confirmation to this universal maxim. Therefore, in the lowest stages of chemical analysis, chemists have always endeavored to assign, as a characteristic property, the proportion of the elements of each substance, as far as possible: and when this was omitted, it was on the understanding that the proposed combination admitted of only a certain proportion; as in the case of the neutral salts.

Again, it has long been acknowledged that there always exists, between any two substances, a certain minimum and maximum of reciprocal saturation, beyond or short of which all combination becomes impossible. At the utmost, certain variations, themselves restricted, have been supposed procurable. Berthollet established, more directly than any one else, the general and necessary existence of these limits of combination—one of the principal characters which distinguish it from simple mixture. It is clear that the two extreme degrees of all combination must be subject to special and invariable proportions: and, as all agree in this, all argument, about the opposite doctrines of indefinite and definite proportions is reduced to the question whether the passage from the minimum to the maximum of saturation can be effected gradually and almost imperceptibly, or whether it takes place always abruptly, through a small number of well-marked degrees.

Thirdly, the possibility and actual existence of intermediary definite proportions are admitted by all chemists, who can have no other dispute than about the greater or small generality of such a property. We have seen that the idea of neutrality must, sooner or later, bring after it that of a determinate and unchangeable proportion; and the gradual development of chemical knowledge has extended this character to more and more varied cases. Berthollet disclosed several other causes of definite proportions, which were entirely misconceived before his time, and which may meet in almost all combinations, modifying certain circumstances of the phenomenon. The precise question now is, therefore, whether, besides these determinate compounds, subject to fixed proportions, within the two limits of possible combination, there does or does not exist, in general, a continuous series of other intermediate compounds of a less marked character; in a word, whether definite proportion constitutes the rule, as is now generally supposed, or, as

Berthollet endeavored to establish, the exception. This is now the only dispute. It is no derogation from the interest of the doctrine of definite proportions to say, as some preceding considerations compel us to do, that the decision of this disputed point is not of the importance commonly supposed. The doctrine has tended to simplify the general problem of chemistry; but it must not be supposed that the solution would have been impossible without this aid:—it would have been simply more difficult and less precise. The eminent chemists who concurred in establishing the doctrine were naturally engrossed by that labor; but their successors, who find numerical chemistry constituted to their hand, must beware of losing sight in it of the true scientific aim of chemistry. They must not linger in this vestibule of the science, to the neglect of the direct construction of Chemistry itself—an enterprise scarcely begun, and to which it is high time that attention should be once more fully directed.

If we inquire, as we must do, how far the doctrine of definite proportions is irrevocably established, we shall bear in mind that the founders of numerical chemistry have accomplished that chief part which depends on an investigation of all known compounds, leaving only the question whether the doctrine is compatible with certain chemical phenomena, neglected during its formation, and remaining to be since referred to it.

The first general objection relates to the important phenomenon of dissolution, evidently possible in an infinity of different proportions. It must be acknowledged that the distinctions between the state of dissolution and that of combination, by which the difficulty has been met, afford little satisfaction. In my opinion the only effectual reply must consist in the extension of the principle of definite proportions to the phenomena of dissolution; and, difficult as it may be to do it, it does not seem to me impossible. The way is by the use of an hypothesis already proposed for other cases in which it might appear less admissible. All the successive degrees of concentration of the liquid must be regarded as simple mixtures of the small number of definite dissolutions which shall have been established, either between themselves or with the dissolvent, in the manner of habitual mixtures of water with alcohol, with sulphuric acid, etc. In any case, the positive verification of this hypothesis must be extremely delicate. Furthermore, to render the study of dissolutions fully rational, in this point of view, it is necessary to combine with it that of other analogous chemical phenomena, relating to the absorption of gases by liquids or by porous solids. All these different modes of molecular union are often energetic enough to resist influences able to destroy certain combinations, properly so called: why should they not be, like them, subject to the rule of definite proportions, if that rule is truly a fundamental law of nature? *Objection of dissolution.*

The next case, that of various metallic alloys, is very extensive, though more particular. The difficulty lies *Of metallic alloys.*

in the question whether these are cases of combination or of mixture. The state of combination has been taken for granted in the case of alloys; whereas the general application of the principle of numerical chemistry requires that they should be mixtures; while, again, it is difficult to conceive of such a mixture of solids as could resist perturbing influences which would appear to be necessarily destructive; as great changes of temperature, the influence of crystallization, etc. The question can be decided only by a series of special experiments, devised to find the general limits of the permanence of unquestionable mixtures; and the results might be extended to other questions of numerical chemistry, as of certain oxydes, on which explanations have been hazarded too freely. When a true chemical theory of mixtures is established on a proper basis of experiment, and we leave off referring to an hypothesis of mixture all cases in which combination seems susceptible of an indeterminate proportion, in order to bring them under the law of definite proportions, we shall get rid of a formidable objection to the principle of numerical chemistry.

Of organic substances. The remaining case constitutes the greatest obstacle of all in the way of the generalization of the law of definite proportions; and if it can not be surmounted, the law sinks to the rank of an empirical rule, fit for nothing more than facilitating a certain order of chemical analyses. I refer to the class, anomalous in this view, of substances called organic. And this is the effect, in fact, of the declaration of the chemists of our time, that organic substances do not come under the principle of definite proportions. It amounts to saying that the law rules all the elements, except oxygen, hydrogen, carbon, and azote. The division between inorganic and organic chemistry is merely scholastic; for all chemistry is, by its nature, homogeneous—that is, inorganic. And thus, if we admit of the enormous exception of the numerical composition of so-called organic substances, the doctrine of definite proportions is overthrown as a rational theory. As it evidently can not be founded on any *à-priori* considerations, it is only by a strict generality that it can become a rational theory.

Application of the principle of dualism. If we could not hold at once the grand principle of the dualism which pervades chemistry, and constitutes its homogeneous character, and the doctrine of definite proportions, I should not hesitate to sacrifice the latter; for it is more important for chemical progress to grasp the great principle of systematic dualism, than to advance our investigations by the use of the numerical rule. But there is not, in fact, any incompatibility between these two means of progress; and such a brief sketch of my conception on this subject as my limits allow may show how the doctrine of definite proportions can be duly generalized only by discarding organic chemistry as a separate body of doctrine, and extending the principle of dualism to all organic compounds.

If we are to include all organic compounds under one uniform system of chemistry, properly so called, we must refer to physiology, vegetable and animal, the study of the numerous secondary substances which owe their transient and variable existence to the development of vital phenomena, and which have no scientific interest except under the head of biology. We shall see, under that head, hereafter, what the precise classification is; and all that we have to do with it now is to show that it proceeds from the fundamental distinction between the state of death and that of life. The second, and most extended class of organic substances is chiefly composed of mixtures which, as such, admit of all imaginable proportions, within the limits of vital conditions. As for the substances which exhibit real combinations, we must conceive of them as subject to the law of definite proportions; but the complexity, and yet more the instability of such compounds will probably for ever forbid their being successfully studied under the numerical point of view, which is indeed of very inferior interest in biology. Even after this clearing of the field, we could not accomplish the desired generalization if we had not taken a new stand with regard to the ternary and quaternary substances contemplated by ordinary chemistry. The rigorous dualism which I have before, and in a higher view, shown to be necessary, seems to supply, naturally and finally, the needs of the doctrine of definite proportions.

As long as chemists persist in regarding organic combinations as ternary and quaternary—that is, in confounding their elementary with an immediate analysis;—while oxygen, hydrogen, carbon, and azote, are regarded as immediately united, the compounds from them which must be recognised as distinct, after the severest sifting, will be enough to constitute an invincible objection to the principle of a numerical chemistry. But if they become binary compounds of the second, or at most the third order, whose principles are formed by the direct and binary combination of those three or four elements, we find ourselves able to represent all the actual numerical varieties established by an elementary analysis, conceiving, for each degree of combination, a very small number of distinct and entirely definite proportions. In the ternary case—appropriate to compounds of a vegetable origin—their three elements may be united in three kinds of binary combinations. Combining these again, still employing at once oxygen, hydrogen, and carbon, we have three principal classes of compounds of the second order. But then, again, each term of the new compounds really corresponds to two distinct substances; and thus, while admitting only one proportion for the binary composition of these bodies, we have already provided for the numerical composition of twelve substances at present called ternary. But, further, we are compelled to suppose at least three different proportions for each binary combination; one producing perfect neutralization, and the other two the extreme limits of the reciprocal saturation; and chemical analogies indicate a much larger number of compounds. Putting those aside, we have thirty-six compounds, without going beyond the second order, by the known combination of three elements on the principle of dualism. We are also entitled now to conceive of a third possible combination between oxygen and carbon, or between carbon and hydrogen, etc., which already furnish two, after being long supposed to admit of only one. Hence, and in view of all these considerations, we may be assured that by dualism we might completely and naturally subject to the law of definite proportions eighty-one compounds of the second order, formed from oxygen, hydrogen, and carbon; and this would unquestionably more than suffice to represent the elementary analysis of all distinct substances in the range of vegetable chemistry.

Passing on to the quaternary case, characterizing what is called Animal Chemistry,—it seems as if the principal class of compounds of the second order must be more numerous than in the ternary case: but the indispensable condition of employing all the four elements at once restricts the classes to three. But when we examine the terms of the secondary compounds, we find that while two represent only one compound each, a third represents five. Thus, the three pairs of compounds yield fourteen of one proportion, and forty-two of the three proportions indicated in the last case. But applying, at each degree, the rational rule of a triple binary combination, without stopping at the inevitable gaps of our existing chemistry, we find ourselves in possession of ninety-nine compounds of the second order, now regarded as quaternary. This is probably a larger number than a rational analysis of animal substances will be found to require. Morever, as animal substances have undergone a greater degree of vital elaboration than vegetable matters, it would be philosophical to admit, with respect to them, the possibility of a higher order of composition, such as physiological combinations must eminently tend to realize. On such an hypothesis, without going beyond the third order, we might obtain ten thousand perfectly distinct compounds from these four elements, all formed by an invariable dualism, and strictly subject to the law of definite proportions. It is true, nature would not permit the realization of more than a small part of these speculative combinations; but I have pursued the consequences of my conception to this extreme ideal limit, to show how abundant are the rational resources supplied by this new theory for the generalization of the laws of numerical chemistry. If this view is not followed up, or some equivalent one proposed, it is evident that we must give up the doctrine of definite proportions as a law of natural philosophy, and return to Berthollet's theory, merely enlarging the cases of fixed proportions which he admitted. In the present state of the question there is no other choice. But my theory having been, not instituted for this destination, but naturally arrived at by another way, and with higher views, and proceeding from established principles, to meet the needs of chemical philosophy, seems to me to be presumptively entitled to a future, and perhaps speedy realization.

This account of the present aspects of the doctrine of definite proportions will enable any one to judge of its real progress from its institution to this day; of the conditions which must be fulfilled before its principle can be converted into a great law of nature; and of the rational course which alone can lead to such a final constitution of numerical chemistry.

CHAPTER IV.

THE ELECTRO-CHEMICAL THEORY.

FROM the beginning of modern chemistry, the chemical influence of electricity manifested itself unequivocally in many important phenomena, and, above all, in the grand experiment of the recomposition of water by the direct combination of oxygen with hydrogen, effected by the aid of the electric spark. But the special attention of chemists was not strongly drawn toward this agency till Volta's immortal discovery disclosed its principal energy, in rendering the electric action at once more complete, more profound, and more continuous. Since that time, various series of general phenomena have taught us that electricity is a chemical agent more universal and irresistible than heat itself, both for decomposition and combination. The danger now is of exaggerating the relation it bears to the general system of chemical science. Though chemistry is united to physics by this agency more than by any other, it must yet be remembered that the two sciences are distinct, and that there should be no no confounding of chemical with electrical properties. In order to ascertain with precision what are the relations of chemistry with electrology, we must briefly review the gradation of ideas which have led up to the present electro-chemical theory, as systemized by Berzelius. *Relation of Electricity to Chemistry.*

The first important chemical effect of the Voltaic influence was the decomposition of water, established by Nicholson in 1801. It was a necessary result of examination into the action of the pile, without any chemical intention. It confirmed a truth before well known: but it had a high chemical value, as revealing the chemical energy of the instrument; and it thus constituted the starting-point of electro-chemical research. We may even refer to this origin the first attempts at founding a general theory of electro-chemical phenomena; for the conception offered by Grothuss, to explain Nicholson's observation by the electric polarity of molecules, contains the germ of all the essential ideas which have expanded, according to the requisitions of the phenomena, into the present electro-chemical theory. *History of the case. Nicholson's discovery.*

The analytical power of the Voltaic pile having been once discovered, it was natural for the chemists to apply the new agent to the decomposition of substances which had hitherto resisted all known means. The first series of attempts produced, after a few years, the brilliant discovery by the illustrious Davy, of the analysis of the alkalies, properly so called, and the earths. Lavoisier's theory, which showed that every salifiable base must be a result of the combination of oxygen with some metal, had foreshown this analysis; but no chemical means had sufficed to effect it. It was believed in, in spite of Berthollet's discovery of the true composition of ammonia; and the brilliant result obtained by Davy was an easy consequence of a discovery completely prepared for. M. Gay-Lussac soon followed, with a more difficult but less striking achievement—the confirmation, by a purely chemical process, of the electrical analysis of potash.

Davy's discovery.

Nicholson's observation having originated electro-chemistry, and Davy's given it a great impulse, the next step was to investigate the chemical influence of electricity, in a scientific view. This was effectually, though indirectly, determined by Davy's great feat; for by it chemistry was proved to have achieved the most important, and hitherto inaccessible analyses: and, in fact, the science has not since made any essential acquisition. Electro-chemical action was presently and permanently subjected to direct and regular study; and it was irrevocably constituted a fundamental part of chemical science when Berzelius accomplished his series of investigations on the Voltaic decomposition of all the salts, and then of the principal oxydes and acids. It was in consequence of these researches that the habitual consideration of electric properties has assumed a growing importance in the chemical study of all substances, which are now scientifically divided into the classes of electro-negatives and electro-positives. It was thus the privilege of Berzelius to be the first to conceive of the electro-chemical theory under an entirely systematic form.

Berzelius's extension.

One condition remained to be fulfilled, to give its due scientific character to this new branch of chemistry. The Voltaic action had thus far been regarded only analytically:— it must be regarded synthetically also. This was done by the labors of Becquerel, who fully established the synthetical influence of electricity, fitly applied; and who, moreover, employed it in effecting new and valuable combinations, hitherto impracticable. From this procedure arose the necessity of modifying the method of experimentation. The apparatus which was powerful enough to decompose was much too powerful to combine, because it would probably decompose the immediate principles which were intended to be combined; and thence arose the method of employing the protracted action of very feeble electric powers,—every advantage being given to the disposition of the substances to be acted upon.

Synthetical process.

M. Becquerel did this very successfully by operating with a single Voltaic element, and seizing each substance in the state called *nascent*, which is agreed upon as most favor-

Becquerel's.

able to combination. This change of procedure is the distinctive honor of M. Becquerel. He not only determined direct combinations, not before obtainable, but exhibited in others, which were obtainable, the remarkable property of clearly manifesting their geometrical structure, through the slowness and regularity of their gradual formation;—a character especially marked in the case of certain metallic sulphurets, some oxydes, and several salts. It does not lie within our province to point out the results of this method in regard to the natural history of the globe, in explaining a great number of mineral origins, when the time for such concrete questions shall have arrived. It is more in our way to observe the importance of these labors in bringing up chemical synthesis to something like harmony with the progress of analysis,—favored as the latter pursuit had been by the ease with which we destroy, in comparison with the difficulty with which we recreate. Once more, these researches of M. Becquerel have completed the general constitution of electro-chemistry, which, being henceforth at once synthetical and analytical, can only expand in one of these two directions, however great the improvements remaining to be attained.

Such has been the filiation of electro-chemical discoveries, since the beginning of our century. A little attention to the great phenomenon which was the original subject of the electro-chemical theory will show how this study has gradually led to a new fundamental conception for the whole of Chemistry.

It has been said, through all periods of chemical research, that the study of Combustion must be the central point of the science. So it was thought in the ancient theological period of the science; and also in the more recent metaphysical stage, when combustibility was called phlogiston, and regarded as an intangible materialized entity: and the advent of the positive period of chemistry was marked by the establishment of Lavoisier's new theory of combustion. In our day it is the recognised necessity of modifying this theory that has especially led to the electric conception of chemical phenomena. *Study of combustion.*

The pneumatic theory of combustion of Lavoisier had two entirely different objects in view; objects which are too commonly confounded, but which we must be careful to keep apart. First, the analysis of the general phenomenon of combustion: and, secondly, the explanation of the effects of heat and light, which is, in the eyes of the vulgar, the more important of the two. Both were treated in the most admirable manner possible in the existing state of knowledge; and in the characteristics of positivity and rationality Lavoisier's theory has not since been surpassed. All combustion, abrupt or gradual, was regarded as consisting in the combination of the combustible body with oxygen, whence, when the body was simple, must result an oxyde, generally susceptible of becoming the basis of a salt, and, if the oxygen was preponderant, a true acid, the principle of a certain kind of salts. *Lavoisier's theory.*

As for the disengagement of heat and light, it was attributed, in a general way, to the condensation of the oxygen, and in an accessory way, to that of the combustible body, in this combination.

First division. The first part of this anti-phlogistic theory has a much more philosophical character than the second. It was eminently rational to analyze the phenomenon of combustion, so as to seize whatever was common to all cases. The conclusions drawn might be too general; and were afterward proved to be so; but whatever would stand the test of time must form a body of indestructible truth, constituting an essential part of chemical science through all future revolutions. The case is different with the explanation of heat and light. The question is not, like the first, of a chemical, but a physical nature; and, whatever may be its final solution, it can not affect chemical conceptions. It would have been wiser to abstain from offering any general explanation of the effects of heat and light through such a supposition as that of a condensation, which does not necessarily take place, and which is, in fact, found to be often absent. Lavoisier hoped to attach the thermological effect to the great law discovered by Black, of the disengagement of heat proper to the passage of any body from one state to another more dense; but such a connection could not be established on the ground of phenomena not invariably present, or indisputably manifested. However, it would be too much to expect a perfect scientific reserve in discoverers who bring out the scientific truths from a region of metaphysical phantasies. It is from their followers that we have a right to demand it; and we are compelled to charge upon the chemists who have been eager to substitute the electro-chemical theory for the anti-phlogistic theory, properly so called, a want of care in constructing explanations analogous to those which are dismissed as insufficient. To justify this charge, we must review the proofs of the imperfection of Lavoisier's theory,—still regarding it under the two divisions just exhibited.

Berthollet's limitation. Berthollet saw presently that Lavoisier's method of analysis of combustion must be modified. One of the chief consequences of this analysis was that every acid and every salifiable base must be a result of combustion; that is, of the combination of any element with oxygen; whereas, Berthollet discovered that one of the most marked of the alkalies, ammonia, is formed of hydrogen and azote alone, without any oxygen; and soon after, he proved that sulphuretted hydrogen gas, in which also there is no oxygen, nevertheless presents all the essential properties of real acid. These facts have since been confirmed in every possible way, and especially by the elecric method; and the exceptions, both as to alkalies and acids, have become so multiplied, that the investigation and comparison of them have given that high character of generality to the study of alkalies and acids which belongs to it in our time.

Moreover, the primitive theory of combustion has been gradually modified by the discovery that a rapid disengagement of heat and light is not always an indication of a combination with oxygen. Chlorine, sulphur, and several other bodies, even non-elementary, have been found to occasion true combustion. And again, the phenomenon of fire is no longer attributed exclusively to any special combination, but, in general, to all chemical action at once very intense and vivid.

It does not follow that because Lavoisier's discoveries have parted with some of their character of generality, they have lost any of their direct value: and such alteration of views as there is relates chiefly to artificial phenomena, while the natural facts remain securely established. Thus, though there are acids and alkalies without oxygen, it is unquestionable that the greater number of them, and especially the most powerful, are oxygenated: and again, if oxygen be not indispensable to combustion, it remains the chief agent, and especially in natural combustions. In natural history, the theory is applicable, almost without reserve, though it is insufficient for the severe conditions of abstract science. If the universal sovereignty of oxygen has been overthrown, it will yet be for ever the chief element of the whole chemical system.

As for the second aspect of the discovery,—the explanation of fire,—it was destroyed by the first direct examination of it. *Second division.* No new facts were required for its overthrow, but merely a more scientific appreciation of common phenomena. It will not even serve naturalists for their concrete purposes in any degree, having never really explained the most ordinary effects. The required condensation is found to be only occasionally present, and often absent in the most important cases; so that if it were not for the connection of this aspect of the theory with a sounder one, it would be inconceivable how it could have held its ground up to a recent period,—busy as the chemists were with other theoretical speculations. After finding that in cases where condensation was supposed, expansion exists instead; and that where we find vivid combustion, there should, by the theory, be a great cooling, we are brought to the reflection that if the fire on our hearths was not a matter of daily fact to us, its existence must become doubtful, or be disbelieved, through those very explanations by which the phenomenon has been proposed to be established. To my mind, this is a clear indication that the chemical production of fire does not admit, in a general way, of any rational explanation. Otherwise it appears incomprehensible that men of such genius and such science, at a time so near our own, should have been so deluded. The electric fire, now proposed for an explanation, must have been sufficiently known to Lavoisier, Cavendish, Berthollet, and others, to have served as a basis to their theory, if its preponderance over the merits of their hypothesis had been so great as is now commonly supposed. This consideration however, striking as it may be, is no dispensation from the duty of examining the electro-chemical conception, for which we have been prepared by this short account of its antecedents.

According to this theory, the fire produced in the greater number of strong chemical reactions must be attributed to a real electric discharge which takes place at the moment of combination (by the mutual neutralization, more or less complete, of the opposite electric conditions) of the two substances under consideration,—one of which must be electro-positive, and the other electro-negative. There is every reason to fear, however, that when this theory has been effectually examined, it will be found as defective in rationality as its predecessor. If electric effects are concerned in all chemical phenomena, as seems to be now agreed upon by most chemists and physicists, they must oftener be supposed than found: and the electric symptoms are most impossible to detect in precisely those chemical phenomena which have been most relied on for overthrowing the old theory. And in the cases in which electrization is evident, its chemical influence is so equivocal that some regard it as the cause, and others as the effect, of the combination. The explanation is not yet positively established for any phenomenon that has been duly analyzed: while its vague nature leaves room for fear that it will not be so radically or so speedily destroyed as its predecessor. It could be plainly shown whether the requisite condensation did or did not exist; but there is always a resource, in the more recent case, in the faint or fugitive character of the electric condition, which defies our means of positive exploration,— qualities which are far from being a ground of recommendation of a theory which is to account for very striking and intense effects. I do not desire to say that the disengagement of light and heat can never have an electric origin, any more than that it can never proceed from the condensation proposed: but I think an impartial observation would decide, that in most cases of combustion there is neither condensation nor electrization. My own view is that these vain attempts to explain the chemical production of fire proceed from the lingering metaphysical tendency to penetrate into the nature and mode of being of phenomena: and that chemical action is one of the various primitive sources of heat and light, which can not, from their nature, usually admit of any positive explanation; that is, of being referred, in this relation, to any other fundamental influence.

If our chemical science were more advanced than it is, we should not have to point out that the consideration of fire, which, however important, is only a physical accessory of chemical phenomena, can not afford a rational ground for a radical change in our conceptions of chemical action. When our predecessors regarded heat as the chief physical agent in composition and decomposition, they did not pervert such a consideration to the point of assimilating chemical to thermological effects. We are less cautious at the present day: we confound the auxiliary, or the general physical agent of the phenomenon, with the phenomenon

Difficulties of the theory.

itself, and pervert chemistry by confounding it with electrology, by irrationally assimilating chemical to electrical properties, as is seen especially in the theory of M. Berzelius. How can there be any scientific comparison between the tendency of two bodies to a mechanical adhesion after a certain mode of electrization, and the disposition to unite all their molecules, external and internal, by a true chemical action? M. Berzelius has frankly declared that cohesion, properly so called, admits of no electric explanation. Nothing is gained toward explaining the molecular connection, indissoluble by any mechanical force, in contrast with the magnetic union so easily overcome, by talking of Voltaic elements with their positive or negative pole, and their connection by the electric antagonism of the opposite poles. Such inventions give no idea whatever of molecular cohesion. Nor is affinity, or the tendency to combination, any better explained by the electro-chemical theory. Electrical phenomena, in physics, are eminently general, offering only differences of intensity in different bodies; whereas, chemical phenomena are essentially special or elective: and therefore every attempt to make chemistry, as a whole, enter into any branch of physics, is thoroughly anti-scientific. This would be enough: but we see, besides, that the smallest changes in the mode of electrization reverse the electric antagonism, and destroy the proposed electrical order of elementary bodies: we find ourselves unable to deduce the new electric properties which the theory bids us look for in the compounds of different orders; we do not know by what laws they derive their positive or negative character from the electric condition of each of the two elements; nor are we able to approach the great end of chemical science,—the prevision of the qualities of compounds by those of their constituent elements. Morever, in any case, the great body of chemical phenomena opposes insurmountable obstacles; as when oxygen, the most negative element, when entering largely into certain oxydes, finds them positive toward certain acids, into which it enters much more sparingly,—the radicals of the first being often as negative as those of the last. According to M. Berzelius's own frank declaration, organic compounds cast insuperable difficulties in the way of his arrangement of electric relations; and he alleges the transience of the combinations in that class of cases as an explanation of the anomaly: but chemical science would be impossible if compounds were not throughout considered stable till causes of decomposition arise: and if organic compounds are guarded from these, they remain chemically stable, like inorganic substances. The fundamental obstacle of the whole case—the identity of the elements, in opposition to the electric variety—can not, by any means whatever, be got over.

Leaving all these difficulties on one side, we learn nothing about chemical phenomena by likening them to electric action, for we establish thus no harmony between the pretended causes and the real effects. Every attempt seems to prove simply the auxiliary influence of electricity on chemical effects,—acting as heat does, only with a differet intensity. In fact, there are scarcely any electro-chemical combinations which can not be effected by ordinary chemical processes, without any electric indications; and the few exceptional cases still existing may, by analogy, be expected to be brought under the rule. If, in the face of all this, we were to persist in investing the electrical influence with the specific and molecular attributes of chemistry, we should merely be restoring the old entity of *affinity*, decked out with some hypothetical material attributes, which would be far from rendering it more positive: and such a procedure would be as hurtful to physics as to chemistry, by infusing new vagueness into our notions of electricity, which are at present far from being sufficiently distinct. And then might follow, as likely as not, the founding on electrology, not only the whole of chemistry, but the theories of heat, of weight, and probably, as a consequence, that of celestial mechanics. And then, if we added to this heterogeneous assemblage a confounding of the supposed nervous fluid with the pretended electric fluid, we should have attained to the show of a universal system, devoid of all scientific use, which would fall to pieces as soon as tested by real study, parting off into categories of independent doctrine, and encumbering natural philosophy with insoluble questions, which must be discarded, to enable us to begin afresh.

Thus, to sum up, the great chemical influence of electricity, like that of weight, and yet more of heat, is unquestionable; and I have endeavored to exhibit the high importance of electro-chemistry to the improvement of chemical science, of which it is one of the essential elements. But I must, once for all, reject the conception through which it has been attempted to transform all chemical into electrical phenomena. In a philosophical point of view, Lavoisier's theory appears to me, notwithstanding its serious imperfections, very superior as a scientific composition to that to which it has given place. It related directly to the aim of chemical science,—the establishment of general laws of composition and decomposition; whereas the newer theory draws attention away from it, in a vain inquiry into the intimate nature of chemical phenomena. Thus the anti-phlogistic conception has suggested numerous and important chemical discoveries, while it is very doubtful whether as much will ever be said for the electric theory. In an indirect way, however, it may operate favorably for chemical science, by its binary antagonism suggesting the extension of dualism among compounds which are as yet supposed to be more than binary. Berzelius appears to have felt this connection; and he would probably have erected dualism into a fundamental principle, but for his subjection to the old division of chemistry into organic and inorganic. Others, who are free from this entanglement, may be prepared by the electro-chemical theory for general dualism; though it is not, in principle, desirable to recur to faulty means to attain good results in an indirect way. In another view, this the-

Its position and powers.

ory may be useful,—in fixing the attention of chemists on the influence of time in the production of chemical effects; an influence remarkably manifested by many phenomena, but not, as yet, directly analyzed. Not only does time naturally increase the mass of the products of chemical reaction; it also causes formations which would not otherwise exist. The chemical theory of time is at present a blank in science; and the phenomena of electro-chemistry seem likely to enlighten us on this head,—so all-important in its connection with chemical geology, while constituting an indispensable element in the conceptions of abstract chemistry.

To complete our survey of the philosophy of Chemistry, we must turn to some considerations already suggested, on the subject of what is commonly called Organic Chemistry.

CHAPTER V.

ORGANIC CHEMISTRY.

ORGANIC Chemistry comprehends, it can not be denied, two kinds of researches, chemical and physiological, which are perfectly distinct. For instance, the study of organic acids, and especially the vegetable acids, and alcohol, ethers, etc., has a character as purely chemical as that of any inorganic substances whatever; and, on the other hand, there is no doubt of the biological character of inquiries into the composition of sap or blood, and of the analysis of the products of respiration, and many other matters included in organic chemistry. The confusion of the two orders is prejudicial to both sciences, and especially to physiology. The division of Chemistry conceals or violates essential analogies, and hinders the extension of dualism into the organic region, where it is seldom found, though, as I have shown, it is optional, in fact, throughout the whole range of chemistry; and thus the arbitrary arrangement is the chief obstacle in the way of the entire generalization of the doctrine of definite proportions. And whenever a true chemical theory shall fitly replace the antiphlogistic conception, it must necessarily comprehend all organic, as well as inorganic compounds. The most philosophical chemists are tending more and more to a recognition of the identity of the two departments; and there can be no doubt that the establishment of that identity will be an immediate consequence of a rational classification of chemical knowledge.

The confusion is more mischievous, but less felt by chemists, under the other view—the comprehension of biological phenomena among those of organic chemistry. The confusion arose from the need of chemical researches in very many physiological questions;

and these chemical researches, being usually extensive and difficult, were out of the range of the physiologists, and were taken possession of by the chemists, who annexed them to their own domain. Both classes were to blame for the vicious arrangement, and both must amend their scientific habits before a division can be effected in entire conformity to natural analogies. The physiologists have the most difficult task before them, having to qualify themselves for inquiries from which the chemists have only to abstain.

It is scarcely possible to characterize or to circumscribe the physiological part of organic chemistry, formed as it is by successive encroachments. It comprehends the chemical analysis of all the anatomical elements, solid or fluid, and that of the *products* of the organism; and if its usurpations remained unchecked, it would soon include the phenomena of what Bichat called the *organic life:* that is, the functions of nutrition and secretion, the only ones common to all living bodies, and in which the chemical point of view might well appear the natural one. In such a state of things, physiology would be reduced to the study of the functions of animal life, and to that of the development of the living being. It is easy to see what would become of biological science, if it were reduced to this fragmentary state. Chemists can not but be unfit for the rational examination of the important question of anatomy and physiology, vegetable and animal, because the research requires that comprehension of view which their studies, as chemists, preclude them from obtaining. In the anatomical relation, they are perpetually overlooking the fundamental division, established by M. de Blainville, between the true *elements* of the organism, and its simple *products;* and they take for one another, almost indifferently, the tissues, the parenchyma, and the organs. The spirit of biological investigation being unknown to them, they can neither choose their subject well, nor direct their analysis wisely. If these are grave inconveniences in anatomical questions, they are much more serious in physiological problems, properly so called, the essential conditions of which are not understood by chemists. The rational direction of physiological analysis can take place only by the subordination of the chemical to the physiological view; and therefore by the employment of chemistry by the physiologists themselves, as a simple means of investigation: It is an analogous case to that before exhibited, of the application of mathematical analysis to physical questions. If it is important that physicists should employ the instrument of analysis, instead of delivering over physical subjects to the mathematicians, to be a mere theme for analytical exercises, much more important is it that the greater diversity of view in chemistry and physiology should not be lost sight of. The irrational and incoherent studies comprised in the organic chemistry of our day give us no idea of the true nature of the aids that biology may derive from chemistry. A few instances will show the high importance of an improved organization of scientific labor.

In the anatomical order, almost all the researches of chemists have to undergo an entire revision by the physiologists, before they can be applied to the studies of the elements or the products of the organism. *Relation of Chemistry to Anatomy.* The fine series of researches of M. Chevreul on fatty bodies are perhaps the only important chemical study immediately applicable to biology, animal or even vegetable. In the chemical analysis of blood or sap, or almost any other element, a single case, taken at hazard, is usually presented as a satisfactory type, without any comparative investigation either of each species of organism in its normal state, or of the degree of development of the living being—its sex, its temperament, its mode of alimentation, the system of its exterior conditions of existence, etc., and other modifications which physiologists alone can duly estimate. Such analyses correspond to nothing in anatomy but the single case observed; and even that is seldom sufficiently characterized. Hence inevitable divergences among chemists, who choose different types, and discussions of no scientific use, as the discordance is attributed to the different analytical methods employed, instead of to the variations which physiology would have led them to anticipate. The case is the same with regard to products first secreted and then excreted, as bile, saliva, etc., which offer a still greater complication. The chemists make no inquiry about the parts from which these products are taken, or the modifications which may have been occasioned by their remaining some time after their production, etc.; and therefore the analyses of these products, however often renewed, are still incoherent and thoroughly defective. We owe to M. Raspail a full exposure of the practice of the chemists of multiplying organic principles almost without limit, from differences of character which imply no distinction of nature, but are merely marks of various degrees of elaboration of the same principle in different developments of vegetation; and even from confounding the observed substances with their anatomical envelope. It is to be regretted that M. Raspail did not complete his great service to science by founding rationally the physiological portion of organic chemistry, instead of vainly attempting to systemize organic chemistry, under the bias of our crude chemical education.

If we turn from the anatomical order of questions to the physiological, we shall find yet stronger evidences *To Physiology.* of the inaptitude of chemists for biological inquiries. All endeavors have yet failed to establish any point of general doctrine in biology; and we find ourselves merely with simple materials, which must be newly elaborated by physiologists, under the view of vitality, before they can be put to use. To give an example or two:—the experiments of Priestley, Sennebier, Saussure, and others, on the mutual chemical action of vegetables and atmospheric air, were of the highest value, as instituting positive knowledge of the vegetable economy; but the inquiry is by no means so simple as its founders naturally supposed, after having analyzed one separate portion of the phenomenon of vegetation. The absorption of carbonic acid, and the exhalation of oxygen, though very important in relation to the action of leaves, are only one aspect of the double vital motion, and can not be understood but in the physiological view of both. This action can not explain the elementary composition of vegetable substances, or determine the kind of alteration sustained by the air through vegetation, because it is, in other ways, partially *compensated by the precisely inverse action produced by the germination of seeds, the ripening of fruits, etc., and even, as regards the leaves, by the mere passage from light to darkness. It is much to have indicated the true nature of the requisite research, and to have supplied some materials for it. The rest is the business of the physiologist. If we turn to animal physiology for examples, the case is yet more striking.

After all the inquiry that has been made into the chemical phenomena of respiration, no fixed point is yet established. It was long supposed that the absorption by the lungs of atmospheric oxygen, and its transformation into carbonic acid, would explain the great phenomenon of the conversion of venous into arterial blood. But the problem is much more complicated than was supposed by the chemists who established this essential part of the phenomenon, and whose labors present the most contradictory conclusions in regard to the facts under their notice. We do not know, for instance, whether the quantity of carbonic acid formed really corresponds to the quantity of oxygen absorbed; and even the simple general difference between the inhaled and exhaled air, which is the first point to be ascertained, is far from being positively established. The atmospheric azote appears to some to be increased after respiration, while others say it is diminished, and others again that it remains the same. The disagreements about the changes in the composition of the blood are yet more marked. Perhaps the inaptitude of chemists and physicists for physiological researches is more striking still in the case of animal heat. In the early days of modern chemistry, this phenomenon seemed to be sufficiently accounted for by the disengagement of heat corresponding to the decarbonizing of the blood in the lungs, which the chemists regarded as the focus of a real combustion; but this explanation was soon found to be inadequate, even in a normal condition; and much more in various pathological cases. Uncertain as we still are as to the pulmonary influence in the process, we know that all the vital functions must concur in it, in a greater or smaller degree. There is even some reason to suppose, in direct opposition to the opinion of the chemists, that respiration, far from aiding to produce animal heat, constantly tends to diminish it. No doubt, the chemical effects occasioned by vital action must always be taken into the account in the study of animal heat; but it is only the physiologists who can deal with them in the light of the whole subject. When we have learned to combine the chemical and the physiological view, we may proceed to the formation of positive doctrine, without having to deal with preliminary obstacles; as, in regard to such questions as the general agreement between the chemical composition of living bodies, and that of the whole of their aliments; a case which constitutes one of the principal aspects of the vital state.

It is evident, at the outset, that every living body, whatever its origin, must be, in the long run, composed of the different chemical elements concerned in the substances, solid, liquid, and gaseous, by which it is habitually nourished; since, on the one hand, the vital motion subjects its parts to a continual renovation, and, on the other, we can not without absurdity suppose it capable of spontaneously producing any real element. This consideration is so far from involving any difficulty, that it might lead us to divine the general nature of the principal elements of living bodies; for animals feed in the first place on vegetable, or on other animals which have eaten vegetables; and in the second place, on air and water, which are the basis of the nutrition of plants: and thus, the organic world evidently admits of those chemical elements only which are furnished by the decomposition of air and water. When these two fluids have been duly analyzed, physiologists can, in a manner, foresee that animal and vegetable substances must be composed of oxygen, hydrogen, azote, and carbon, as chemistry taught them. It is true, such a prevision must be very imperfect, as it indicates nothing of the difference between animal and vegetable substances, nor why the latter usually contain so much carbon and so little azote. But this first glimpse, though it suggests some of the difficulty of the problem, yet indicates the possibility of establishing such a general harmony. But, when we proceed with the comparison, we encounter important objections, which are at present insoluble. The chief difficulty is that azote appears to be as abundant in the tissues of herbivorous as of carnivorous animals, though the solid aliments of the former contain scarcely any azote. The opinions of chemists, as of Berzelius and Raspail, as to the nature of azote, do not solve the difficulty, as they cast no light upon its origin. This is one of a multitude of cases in which we can not at all explain the chemical composition of anatomical elements by that of the exterior substances from which they are unquestionably derived. Another striking case is that of the constant presence of carbonate, and, yet more, phosphate of lime in the bony tissue, though the nature of the aggregate of aliments appears to afford no room for the formation of those salts. This system of investigations, considered in its whole range, constitutes one of the most important general questions that can arise from the chemical study of life; and if it is at present hardly initiated, the backwardness is owing, not only to its eminent difficulty, but to the biologists having abandoned to the chemists a task which, under a wise organization of labor, would have belonged to themselves alone.

In order to effect a rational division of organic chemistry, and to assign its portions to chemistry on the one hand and physiology on the other, we must take our stand on the *Partition of Organic Chemistry.*

separation between the state of life and that of death; or, what comes to nearly the same thing, between the stability and instability of the proposed combinations, submitted to the influence of common agents. Among the compounds indiscriminately called organic, some owe their existence only to vital motion; they exhibit perpetual variations, and usually constitute mere mixtures: and these can not belong to chemistry, but must enter into the domain of biology—statical or dynamical, according as we study their fixed condition, or the vital succession of their regular changes. Such, for instance, are blood, lymph, fat, etc. The others, which form the more immediate principles of these, are substances essentially dead, admitting of a remarkable permanence, and presenting all the characters of true combinations, independent of life: their natural place is, evidently, in the general system of chemical science, among the substances of inorganic origin, from which they differ in no important respect. Of these, the organic acids, alcohol, albumen, etc., are examples. These are the substances which truly belong to organic chemistry; and no reason exists for their separation from analogous inorganic substances, even if no injury was done by such an arbitrary division; and there is more reason for giving the title of *organic* to them than to the theory of oxygen, hydrogen, carbon, and azote, or to the study of many other substances, acid, alkaline, saline, etc., without which chemical anatomy and physiology would be unintelligible. As for chemical phenomena truly common to all compounds of this class, in consequence of the necessary identity of their chief elements, it is certainly important to assign to them their precise relations. The most important of these phenomena relate, at present, to the interesting and very imperfect theory of the different kinds of fermentation. But the consideration of these properties does not constitute a different order from that which results from the same ground in the case of many other compounds, purely inorganic. The property of fermentation, however important, has not a higher scientific value than that of burning, and has no more right to an exclusive classification. It is admitted that too much was attributed to combustion formerly, in regard to inorganic substances; and we may be attributing too much now to fermentation, or any other common property, among so-called organic bodies. We can not yet assign their proper place to these compounds in the rational system of chemical science; but we are able to affirm that, in that system, they must be more or less separated from each other, and interposed among substances called inorganic. Nothing more than this is needed to settle the question of the maintenance or the suppression of organic chemistry as a distinct body of doctrine. In applying the principle which I have proposed, to ascertain to which science any question belongs, it is enough to inquire whether chemical knowledge will serve the purposes of the research, or whether any biological considerations enter into it. The proposing such an alternative is, in fact, making the classification.

It is not our business to treat of any special application of this principle; but it is desirable to point out that in this partition of organic chemistry, its two portions are very unequally divided— the study of vegetable substances contributing most to chemistry, and that of animal substances to biology. At the first glance, we might suppose the difference to be the other way; for the importance of chemical considerations is really much greater with regard to living vegetables than animals, whose chemical functions are, except among the lowest orders of the zoological hierarchy, subordinated to a superior order of new vital actions. Yet, in virtue of the higher degree of vital elaboration that matter undergoes in the animal than in the vegetable organism, the chemical part of animal physiology presents a much greater extent and complexity than the vegetable, in which, for instance, the whole series of the phenomena of digestion is absent, and in which assimilation and secretion are much simplified. But, on account of the superior elaboration, and of the greater number of elements, animal substances are much less stable than vegetable: they rarely remain separate from the organism; and, at the same time, the new immediate principles proper to them are so few that their very existence has been questioned. Vegetation is evidently the chief source of true organic compounds, which are derived thence by the animal organism, and modified by it, either through their mutual combinations or new external influences. Thus, the true domain of chemical science must necessarily be more extended by the study of vegetable than by that of animal substances.

Enough has been said about the necessity of subjecting organic compounds to the law of dualism; but there is a particular aspect, under which the importance of this conception in improving chemical theories is worth a brief notice.

In considering substances as ternary or quaternary, their multiplicity is accounted for only by the difference in the proportions of their constituent elements—their component principles being identical. Very great differences are sometimes explained by inequalities of proportion so small as to shock the spirit of chemical analysis; and in other cases, the proportions being the same, the differences remain unaccounted for;— as, for instance, in the cases of sugar and gum, in which we find the same elements, combined in the same proportions. If we extend dualism to organic compounds, this class of anomalies disappears; for the distinction between immediate and elementary analysis enables us to resolve by dualism, in the most natural manner, the general paradox of the real diversity of two substances composed of the same elements, in the same proportions. In fact, these substances, identical in their elementary, would differ in the immediate analysis, as we may understand from what was offered in my chapter on the law of definite proportions. In another connection, chemists have remarked the possibility of exactly representing the numerical composition of alcohol or ether, etc., according to the several binary formulas, radically distinct from each other, and yet finally equivalent with regard to the elementary analysis. Now, if such fictitious combinations should ever be realized, they would produce highly distinct substances, which might differ much in the aggregate of their chemical properties, and yet coincide by their elementary composition. It is only necessary to transfer the same spirit into the study of organic combinations, by the establishment of a universal dualism, to dissipate all these anomalies: and the resource may thus be happily prepared, before the cases of isomerism (as Berzelius calls this fact) have become very numerous.

Summary of the chapter. We have now seen how heterogeneous is the body of doctrine included under the name of organic chemistry; how it should be divided; what is the duty of physiologists with regard to their share of it; and how the extension of dualism will establish a natural agreement between the composition of substances and their collective characters.

Summary of the Book. With regard to Chemistry at large, I have pointed out the true spirit of the science, under a philosophical view of its present aspects, and of the indispensable conditions of its advancement. We do not want new materials so much as the rational disposition of the details which already abound; and I have offered two prominent ideas, in my survey of chemical philosophy; the fusion of all genuinely chemical studies into one body of homogeneous doctrine; and the reduction of all combinations to the indispensable conception of a dualism always optional. These two conditions, distinct but connected, have been presented as necessary to the definitive constitution of chemical science. The application of such a conception to the only part of chemical research which yet exhibits anything of a positive rationality has removed all doubt about its general fitness, by showing its spontaneous aptitude to resolve the anomalies of numerical chemistry.

With this division closes our survey of the whole of natural philosophy that relates to universal or inorganic phenomena. In the order of phenomena to which we next proceed, there is at once much more complexity, and much less established order. The study of them is scarcely yet organized; and yet, out of the speciality of the phenomena arises the most indispensable part of natural philosophy—that of which Man is first the chief object, and then Society.

BOOK V.
BIOLOGY.

CHAPTER I.

GENERAL VIEW OF BIOLOGY.

THE study of the external world and of Man is the eternal business of philosophy; and there are two methods of proceeding; by passing from the study of man to that of external nature, or from the study of external nature to that of man. Whenever philosophy shall be perfect, the two methods will be reconciled: meantime, the contrast of the two distinguishes the opposite philosophies—the theological and the positive. We shall see hereafter that all theological and metaphysical philosophy proceeds to explain the phenomena of the external world from the starting-point of our consciousness of human phenomena; whereas, the positive philosophy subordinates the conception of man to that of the external world. All the multitude of incompatibilities between the two philosophies proceed from this radical opposition. If the consideration of man is to prevail over that of the universe, all phenomena are inevitably attributed to *will*—first natural, and then outside of nature; and this constitutes the theological system. On the contrary, the direct study of the universe suggests and develops the great idea of the *laws* of nature, which is the basis of all positive philosophy, and capable of extension to the whole of phenomena, including at last those of Man and Society. The one point of agreement among all schools of theology and metaphysics, which otherwise differ without limit, is that they regard the study of man as primary, and that of the universe as secondary—usually neglecting the latter entirely. Whereas, the most marked characteristic of the positive school is that it founds the study of man on the prior knowledge of the external world.

Opposite starting-points in philosophy.

This consideration affects physiology further than by its bearing on its encyclopedical rank. In this one case the character of the science is affected by it. The basis of its positivity is its subordination to the knowledge of the external world.

Starting-point of Physiology.

Any multitude of facts, however well analyzed, is useless as long as the old method of philosophizing is persisted in, and physiology is conceived of as a direct study, isolated from that of inert nature. The study has assumed a scientific character only since the recent period when vital phenomena began to be regarded as subject to general laws, of which they exhibit only simple modifications. This revolution is now irreversible, however incomplete and however imperfect have been the attempts to establish the positive character of our knowledge of the most complex and individual of physiological phenomena; especially that of the nerves and brain. Yet, unquestionable as is the basis of the science, its culture is at present too like that to which men have been always accustomed, pursued independently of mathematical and inorganic philosophy, which are the only solid foundations of the positivity of vital studies.

There is no science with regard to which it is so necessary to ascertain its true nature and scope; because we have not only to assign its place in the scale, but to assert its originality. On the one hand, metaphysics strives to retain it; and on the other, the inorganic philosophy lays hold of it, to make it a mere outlying portion of its scientific domain. For more than a century, during which biology has endeavored to take its place in the hierarchy of fundamental sciences, it has been bandied between metaphysics and physics; and the strife can be ended only by the decision of positive philosophy, as to what position shall be occupied by the study of living bodies.

Its present imperfection. The present backwardness of the science is explained by the extreme complexity of its phenomena, and its recent date. That complexity forbids the hope that biological science can ever attain a perfection comparable to that of the more simple and general parts of natural philosophy; and, from its recent date, minds which see in every other province the folly of looking for first causes and modes of production of phenomena, still carry these notions into the study of living bodies. For more than a century, intelligent students have in physics put aside the search after the mystery of weight, and have looked only for its laws; yet they reproach physiology with teaching us nothing of the nature of life, consciousness, and thought. It is easy to see how physiology may thus be supposed to be far more imperfect than it is; and if it be, from unavoidable circumstances, more backward than the other fundamental sciences, it yet includes some infinitely valuable conceptions, and its scientific character is far less inferior than is commonly supposed.

We must first describe its domain.

Its field. There is no doubt that the gradual development of human intelligence would, in course of time, lead us over from the theological and metaphysical state to the positive by a series of logical conceptions. But such an advance would be extremely slow; and we see, in fact, that the process is much quickened by a special stimulus of one sort or another. Our historical experience, which testifies to every great advance having been made in this way, shows that the most common auxiliary influence is the need of application of the science in question. Most philosophers have said that every science springs from a corresponding art:—a maxim which, amid much exaggeration, contains solid truth, if we restrict it, as we ought, to the separation of each science from the theological and metaphysical philosophy which was the natural product of early human intelligence. <small>Relation to Medicine.</small> In this view, it is true that a double action has led to the institution of science, the arts furnishing positive data, and then leading speculative researches in the direction of real and accessible questions. But there is another side to this view. When the science has once reached a certain degree of extension, the progress of speculative knowledge is checked by a too close connection of theory with practice. Our power of speculation, limited as it is, still far surpasses our capacity for action: so that it would be radically absurd to restrict the progress of the one to that of the other. The rational domains of science and art are, in general, perfectly distinct, though philosophically connected: in the one we learn to know, and therefore to foreknow; in the other to become capable, and therefore to act. If science springs from art, it can be matured only when it has left art behind. This is palpably true with regard to the sciences whose character is clearly recognised. Archimedes was, no doubt, deeply aware of it when he apologized to posterity for having for the moment applied his genius to practical inventions. In the case of mathematics and astronomy we have almost lost sight of this truth, from the remoteness of their formation; but in the case of physics and chemistry, at whose scientific birth we may almost be said to have been present, we can ourselves testify to their dependence on the arts at the outset, and to the rapidity of their progress after their separation from them. The first series of chemical facts were furnished by the labors of art; but the prodigious recent development of the science is certainly due, for the most part, to the speculative character that it has assumed.

These considerations are eminently applicable to physiology. No other science has been closely connected with a corresponding art, as biology has with medical art,—a fact accounted for by the high importance of the art and the complexity of the science. But for the growing needs of practical medicine, and the indications it affords about the chief vital phenomena, physiology would have probably stopped short at those academical dissertations, half literary, half metaphysical, studded with episodical adornments, which constituted what was called the science little more than a century ago. The time, however, has arrived when biology must, like the other sciences, make a fresh start in a purely speculative direction, free from all entanglement with medical or any other art. And when this science and the others shall have attained an abstract completeness, then will arise the further duty, as I have indicated before, of connecting the system of the arts with that of the sciences by an intermediate order of rational conceptions. Meanwhile such an operation would be premature, because the system of the sciences is not completely formed; and, with regard to physiology especially, the first necessity is to separate it from medicine, in order to secure the originality of its scientific character, by constituting organic science as a consequence of inorganic. Since the time of Haller, this process has gone on; but with extreme uncertainty and imperfection; so that even now the science is, with a few valuable exceptions, committed to physicians, who are rendered unfit for such a charge both by the eminent importance of their proper business, and by the profound imperfection of their existing education. Physiology is the only science which is not taken possession of by minds exclusively devoted to it. It has not even a regularly-assigned place in the best-instituted scientific corporations. This state of things can not last, the importance and difficulty of the science being considered. If we would not confide the study of astronomy to navigators, we shall not leave physiology to the leisure of physicians. Such an organization as this is a sufficient evidence of the prevalent confusion of ideas about physiological science; and when its pursuit has been duly provided for, that reaction for the benefit of art will ensue which should put to flight all the fears of the timid about the separation of theory from practice. We have seen before how the loftiest truths of science concur to put us in possession of an art; and the verification of this truth, which physics and chemistry have afforded before our eyes, will be repeated in the case of physiology when the science has advanced as far.

Its object. Having provided for a speculative view of physiology, we must inquire into its object; and, as the vital laws constitute the essential subject of biology, we must begin by analyzing the fundamental idea of *life*.

Idea of Life. Before the time of Bichat, this idea was wrapped in a mist of metaphysical abstractions; and Bichat himself, after having perceived that a definition of life could be founded on nothing else than a general view of phenomena proper to living bodies, so far fell under the influence of the old philosophy as to call life a struggle between dead nature and living nature. The irrationality of this conception consists especially in its suppressing one of the two elements whose concurrence is necessary to the general idea of *life*. This idea supposes, not only a being so organized as to admit of the vital state, but such an arrangement of external influences as will also admit of it. The harmony between the living being and the corresponding *medium* (as I shall call its environment) evidently characterizes the fundamental condition of life; whereas, on Bichat's supposition, the whole environment of living beings tends to destroy them. If certain perturbations of the medium occasionally destroy life, its influence is, on the whole, preservative; and the causes of injury and death proceed at least as often from necessary and spontaneous modifications of the organism as from external influences. Moreover, one of the main distinctions between the organic and the inorganic regions is that inorganic phenomena, from their greater simplicity and generality, are produced under almost any external influences which admit of their existence at all; while organic bodies are, from their complexity, and the variety of actions always proceeding, very closely dependent on the influences around them. And the higher we ascend in the ranks of organic bodies, the closer is this dependence, in proportion to the diversity of functions; though, as we must bear in mind, the power of the organism in modifying the influences of the medium rises in proportion. The existence of the being then requires a more complex aggregate of exterior circumstances; but it is compatible with wider limits of variation in each influence taken by itself. In the lowest rank of the organic hierarchy, for instance, we find vegetables and fixed animals which have no effect on the medium in which they exist, and which would therefore perish by the slightest changes in it, but for the very small number of distinct exterior actions required by their life. At the other extremity we find Man, who can live only by the concurrence of the most complex exterior conditions, atmospherical and terrestrial, under various physical and chemical aspects; but, by an indispensable compensation, he can endure, in all these conditions, much wider differences than inferior organisms could support, because he has a superior power of reacting on the surrounding system. However great this power, it is as contradictory to Bichat's view as his dependence on the exterior world. But this notion of Man's independence of exterior nature, and antagonism to it, was natural in Bichat's case, when physiological considerations bore no relation to any hierarchy of organisms, and when Man was studied as an isolated existence. However, the radical vice of such a starting-point for such a study could not but impair the whole system of Bichat's physiological conceptions; and we shall see how seriously the effects have made themselves felt.

Next ensued the abuse, by philosophers, and especially in Germany, of the benefits disclosed by comparative anatomy. They generalized extravagantly the abstract notion of life yielded by the study of the aggregate of organized beings, making the idea of life exactly equivalent to that of spontaneous activity. As all natural bodies are active, in some manner and degree, no distinct notion could be attached to the term; and this abuse must evidently lead us back to the ancient confusion, which arose from attributing life to all bodies. The inconvenience of having two terms to indicate a single general idea should teach us that, to prevent scientific questions from degenerating into a contest of words, we must carefully restrict the term *life* to the only really living beings,—that is, those which are organized,—and not give it a meaning which would include all possible organisms, and all their modes of vitality. In this case, as in all primitive questions, the philosophers would have done wisely to respect the rough but judicious indications of popular good sense, which will ever be the true starting-point of all wise scientific speculation.

De Blainville's definition. I know of no other successful attempt to define life than that of M. de Blainville, proposed in the introduction to his treatise on Comparative Anatomy. He characterizes life as the double interior motion, general and continuous, of composition and decomposition, which in fact constitutes its true universal nature. I do not see that this leaves anything to be desired, unless it be a more direct and explicit indication of the two correlative conditions of a determinate organism and a suitable medium. This criticism however applies rather to the formula than to the conception; and the conditions are implied in the conception,—the conditions of an organism to sustain the renovation, and a medium to minister to the absorption and exhalation; yet it might have been better to express them. With this modification, the definition is unexceptionable,—enunciating the one phenomenon which is common to all living beings, and excluding all inert bodies. Here we have, in my view, the first elementary basis of true biological philosophy.

It is true, this definition neglects the eminent distinction between the *organic* and the *animal* life, and relates solely to the vegetative life; and it appears to violate the general principle of definitions,—that they should exhibit a phenomenon in the case in which it is most, and not in that in which it is least developed. But the proposed definition is shown, by these very objections, to rest upon a due estimate of the whole biological hierarchy: for the animal life is simply a complementary advancement upon the organic or fundamental life, adapted to procure materials for it by reaction upon the external world, and to prepare or facilitate its acts by sensations, locomotion, etc., and to preserve it from unfavorable influences. The higher animals, and Man especially, are the only ones in which this relation is totally subverted,—the vegetative life being destined to support the animal, which is erected into the chief end and preponderant character of organic existence. But in Man himself, this admirable inversion of the usual order becomes comprehensible only by the aid of remarkable development of intelligence and sociality, which tends more and more to transform the species artificially into a single individual, immense and eternal, endowed with a constantly progressive action upon external nature. This is the only just view to take of this subordination of the vegetative to the animal life, as the ideal type toward which civilized humanity incessantly tends, though it can never be fully realized. We shall hereafter show how this conception is related to the new fundamental science which I propose to constitute: but in pure biology, the view is unscientific, and can only lead us astray. It is not with the essential properties of humanity that biology is concerned; but with the individual in his relation to other organic beings; and it must therefore rigorously maintain the conception of animal life being subordinated to the vegetative, as a general law of the organic realm, and the only apparent exception to which forms the special object of a wholly different fundamental science. It should be added that, even where the animal life is the most developed, the organic life, besides being the basis and the end, remains common to all the tissues, while, at the same time, it alone proceeds in a necessarily continuous manner,—the animal life, on the contrary, being intermittent. These are the grounds on which M. de Blainville's definition of life must be confirmed, while, nevertheless, we may regard the consideration of animality, and even of humanity, as the most important object of biology.

This analysis of the phenomenon of life will help us to a clear definition of the science which relates to it. *Definition of Biology.* We have seen that the idea of life supposes the mutual relation of two indispensable elements,—an organism, and a suitable medium or environment. It is from the reciprocal action of these two elements that all the vital phenomena proceed;—not only the animal, but also the organic. It immediately follows that the great problem of positive biology consists in establishing, in the most general and simple manner, a scientific harmony between these two inseparable powers of the vital conflict, and the act which constitutes that conflict: in a word, in connecting, in both a general and special manner, the double idea of organ and medium with that of function. The idea of function is, in fact, as double as the other; and, if we were treating of the natural history of vital beings, we must expressly consider it so: for, by the law of the equivalence of action and reaction, the organism must act on the medium as much as the medium on the organism. In treating of the human being, and especially in the social state, it would be necessary to use the term *function* in this larger sense: but at present there will be little inconvenience in adopting it in its ordinary sense, signifying organic acts, independently of their exterior consequences.

Biology, then, may be regarded as having for its object the connecting, in each determinate case, the anatomical and the physiological point of view; or, in other words, the statical and dynamical. This perpetual relation constitutes its true philosophical character. Placed in a given system of exterior circumstances, a definite organism must always act in a necessarily determinate manner; and, inversely, the same action could not be precisely produced by really distinct organisms. We may then conclude interchangeably, the act from the subject, or the agent from the act. The surrounding system being always supposed to be known, according to the other fundamental sciences, the double biological problem may be laid down thus, in the most mathematical form, and in general terms: *Given, the organ or organic modification, to find the function or the act, and reciprocally.* This definition seems to me to fulfil the chief philosophical conditions of the science; and especially it provides for that rational prevision, which, as has been so often said, is the end of all true science; an end which abides through all the degrees of imperfection which, in any science, at present prevents its attainment. It is eminently important to keep this end in view in a science so intricate as this, in which the multitude of details tempts to a fatal dispersion of efforts upon desultory researches. No one disputes that the most perfect portions of the science are those in which prevision has been best realized; and this is a sufficient justification of the proposal of this aim, whether or not it shall ever be fully attained. My definition excludes the old division between anatomy and physiology, because I believe that division to have marked a very early stage of the science, and to be no longer sustainable. It was by the simple and easy considerations of anatomy that the old metaphysical view was discredited, and positivity first introduced into biology: but that service once accomplished, no reason remains for the separation; and the division, in fact, is growing fainter every day.

Not only does my definition abstain from separating anatomy from physiology; it joins to it another essential part, the nature of which is little known. If the idea of life is really inseparable from that of organization, neither can be severed, as we have seen, from that of a medium or environment, in a determinate relation with them. Hence arises a third elementary aspect; viz., the general theory of organic media, and of their action upon the organism, abstractedly regarded. That is what the German philosophers of our day confusedly asserted in their notion of an intermediate realm—of air and water—uniting the inorganic and organic worlds: and this is what M. de Blainville had in view in what he called the study of exterior modifiers, general and special. Unhappily, this portion which, after anatomy proper, is the most indispensable preliminary of biology, is still so obscure and imperfect that few physiologists even suspect its existence.

The definition that I have proposed aids us in describing not only the object or nature of the science, but its subject, or domain; for, according to this formula, it is not in a single organism, but in all known, and even possible organisms, that biology must endeavor to establish a constant and necessary harmony between the anatomical point of view and the physiological. This unity of subject is one of the chief philosophical beauties of biology; and, in order to maintain it, we must here avow that, in the midst of an almost infinite diversity, the study of man must always prevail, and rule all the rest, whether as starting-point or aim. Our hope, in studying other organisms, is to arrive at a more exact knowledge of man: and again, the idea of man is the only possible standard to which we can refer other organic systems. In this sense, and in this only, can the point of view of the antiquated philosophy be sustained by the deeper philosophy which is taking its place. Such is, then, the necessary consolidation of all the parts of biological science, notwithstanding the imposing vastness of its rational domain.

MEANS OF INVESTIGATION.

As for the means of investigation in this science—the first observation that occurs is that it affords a striking confirmation of the philosophical law before laid down, of the inevitable increase of our scientific resources in proportion to the complication of the phenomena in question. If biological phenomena are incomparably more complex than those of any preceding science, the study of them admits of the most extensive assemblage of intellectual means (many of them new) and develops human faculties hitherto inactive, or known only in a rudimentary state. The logical resources which are thus obtained will be exhibited hereafter. At present, we must notice the means of direct exploration and analysis of phenomena in this science.

Means of investigation.

First, Observation acquires a new extension. Chemistry admitted the use of all the five senses; but biology is, in this respect, an advance upon chemistry. We can here employ an artificial apparatus to perfect the natural sensations, and especially in the case of sight. Much needing precaution in the use, and very subject to abuse, as is this resource, it will always be eagerly employed. In a statical view, such an apparatus helps us to a much better estimate of a structure whose least perceptible details may acquire a primary importance, in various relations: and, even in the dynamical view, though much less favorable, we are sometimes enabled by these artificial means to observe directly the elementary play of the smallest organic parts, which are the ordinary basis of the principal vital phenomena. Till recently, these aids were limited to the sense of sight, which here, as everywhere else, is the chief agent of scientific observation. But some instruments have been devised in our day to assist the hearing; and, though invented for pathological investigations, they are equally fit for the study of the healthy organism. Though rough at present, and not to be compared to microscopic apparatus, these instruments indicate the improvements that may be made hereafter in artificial hearing. Moreover, they suggest, by analogy, that the other senses, not excepting even touch, may admit of such assistance, hinted to the restless sagacity of explorers by a better theory of the corresponding sensations.

Observation.

Artificial apparatus.

Next, the biologist has an advantage over the chemist in being able to employ the whole of chemical procedures, as a sort of new power, to perfect the preliminary exploration of the subject of his researches, according to the evident rule of philosophy that each doctrine may be converted into a method with regard to those that follow it in the scientific hierarchy; but never with regard to those which precede it. In anatomical observations, especially, as might be foreseen, a happy use is made of chemical procedure, to characterize with precision the different elementary tissues, and the chief products of the organism. In physiological observations, also, though they are less favorable to the use of such means, they are of real and notable efficacy—always supposing, in both cases, that they are used under the guidance of

Chemical exploration.

sound philosophy, and not overcharged with the minute numerical details which too often burden the chemical analyses of the organic tissues. One more resource may be mentioned, which was often employed by Bichat to make up for the absence or imperfection of chemical tests; the examination of alimentary effects—the substances which immediately compose organized bodies being, usually, by their nature, more or less fit for nutrition. In an anatomical view, this study may become a useful complement of the other means of investigation.

EXPERIMENT. Proceeding to the second class of means—Experiment can not but be less and less decisive, in proportion to the complexity of the phenomena to be explored: and therefore we saw this resource to be less effectual in chemistry than in physics; and we now find that it is eminently useful in chemistry in comparison with physiology. In fact, the nature of the phenomena seems to offer almost insurmountable impediments to any extensive and prolific application of such a procedure in biology. These phenomena require the concurrence of so large a number of distinct influences, external and internal, which, however diverse, are closely connected with each other, and yet within narrow limits, that, however easy it may be to disturb or suspend the process under notice, it is beyond measure difficult to effect a determinate perturbation. If too powerful, it would obviate the phenomenon: if too feeble, it would not sufficiently mark the artificial case. And, on the other hand, though intended and directed to modify one only of the phenomena, it must presently affect several others, in virtue of their mutual sympathy. Thus, it requires a highly philosophical spirit, acting with extreme circumspection, to conduct physiological experiments at all; and it is no wonder if such experiments have, with a few happy exceptions, raised scientific difficulties greater than those proposed to be solved—to say nothing of those innumerable experiments which, having no definite aim, have merely encumbered the science with idle and unconnected details.

In accordance with what has been said of the mutual relations of the organism and its environment, we must bear in mind that experiments in physiology must be of two kinds. We must introduce determinate perturbations into the medium as well as the organism; whereas the latter process has alone been commonly By affecting the attempted. If it is objected that the organism must organism. itself be disturbed by such affection of the medium, the answer is that the study of this reaction is itself a part of the experiment. It should be remarked that experimentation on the organism is much the less rational of the two methods, because the conditions of experiment are much less easily fulfilled. The first rule, that the change introduced shall be fully compatible with the existence of the phenomenon to be observed, is rendered often impracticable by the incompatibility of life with much alteration of the organs; and the second rule—that the two compared cases shall differ under only one point of view—is baffled by the mutual sympathy of the organs, which is very different from their harmony with their environment. In both lights, nothing can be imagined more futile in the way of experiment than the practice of vivisection, which is the commonest of all. Setting aside the consideration of the cruelty, the levity, and the bad moral stimulus involved in the case, it must be pronounced absurd; for any positive solution is rendered impossible by the induced death of a system eminently indivisible, and the universal disturbance of the organism under its approach.

The second class of physiological experiments appears to me much more promising; that in which the system of exterior circumstances is modified for a determinate purpose. *By affecting the Medium.* Scarcely anything has been done in this direction beyond some incomplete researches into the action of artificial atmospheres, and the comparative influence of different kinds of alimentation. We are here better able to circumscribe, with scientific precision, the artificial perturbation we produce; we can control the action upon the organism, so that the general disturbance of the system may affect the observation very slightly; and we can suspend the process at pleasure, so as to allow the restoration of the normal state before the organism has undergone any irreparable change. It is easy to see how favorable, in comparison, these conditions are to rational induction. And to these considerations may be added the one more, that under this method we can observe varying states in one individual; whereas, under the practice of vivisection, we have to observe the normal state in one individual, and the artificial in another. Thus we are justified in our satisfaction that the least violent method of experimentation is the most instructive.

As to the application of experiment in the various degrees of the biological scale;—it is easiest in the *Comparative experiment.* lower order of organisms, because their organs are simpler and fewer, their mutual sympathy is less, and their environment is more definite and less complex; and these advantages, in my opinion, more than compensate for the restriction of the field of experiment. It is true, we are remote from the human type, which is the fundamental unity of biology; and our judgment is thus impaired, especially with regard to the phenomena of animal life: but, on the other hand, we are all the nearer to the scientific constitution proper to inorganic physics, which I consider to be the ultimate destination of the art of experiment. The advantages at the other end of the scale are that the higher the organism, the more is it susceptible of modification, both from its own complexity and from the greater variety of external influences involved;—every advantage bringing with it, as we have seen, an increase of difficulty.

No one will suppose, I trust, that from anything I have said I have the slightest desire to undervalue the use of experiment in biology, or to slight such achievements as Harvey's experiments on circulation; Haller's on irritability; part of Spallanzani's on digestion and generation; Bichat's on the triple harmony between the heart, the brain, and the lungs in the superior animals; those of Legallois on animal heat; and many analogous efforts which, seeing the vast difficulty of the subject, may rival the most perfect investigations in physics. My object is simply to rectify the false or exaggerated notions of the capacity of the experimental method, misled by its apparent facility to suppose it the best method of physiological research; which it is not. One consideration remains in this connection; the consideration of the high scientific destination of pathological investigation, regarded as offering, in biology, the real equivalent of experimentation, properly so called.

Pathological Investigation. Precisely in the case in which artificial experimentation is the most difficult, nature fulfils the conditions for us; and it would surely be mistaking the means for the end to insist on introducing into the organism perturbations of our own devising, when we may find them taking place without that additional confusion which is caused by the use of artificial methods. Physiological phenomena lend themselves remarkably to that spontaneous experimentation which results from a comparison of the normal and abnormal states of the organism. The state of disease is not a radically different condition from that of health. The pathological condition is to the physiological simply a prolongation of the limits of variation, higher or lower, proper to each phenomenon of the normal organism; and it can never produce any entirely new phenomenon. Therefore, the accurate idea of the physiological state is the indispensable ground of any sound pathological theory; and therefore, again, must the scientific study of pathological phenomena be the best way to perfect our investigations into the normal state. The gradual invasion of a malady, and the slow passage from an almost natural condition to one of fully-marked disease, are far from being useless preliminaries, got rid of by the abrupt introduction of what may be called the violent malady of direct experiment: they offer, on the contrary, inestimable materials to the biologist able to put them to use. And so it is also in the happy converse case, of the return, spontaneous or contrived, to health, which presents a sort of verification of the primitive analysis. Moreover, the direct examination of the chief phenomenon is not obscured, but much elucidated, by this natural process. And again, it may be applied directly to Man himself, without prejudice to the pathology of animals, and even of vegetables. We may enjoy our power of turning our disasters to the profit of our race: and we can not but deplore the misfortune that our great medical establishments are so constituted as that little rational instruction is obtained from them, for want of complete observations and duly-prepared observers.

Here, as elsewhere, the distinction holds of the phenomena belonging to the organism or to the medium; and here, as before, we find the maladies produced from without the most accessible to inquiry. Pathological inquiry is also more suitable than experimental, to the whole biological series: and thus it answers well to extend our observations through the entire hierarchy, though our object may be the study of Man; for his maladies may receive much light from a sound analysis of the derangements of other organisms,—even the vegetable, as we shall see when we treat of the comparative process.

Again, pathological analysis is applicable, not only to all organisms, but to all phenomena of the same organism; whereas direct experimentation is too disturbing and too abrupt to be ever applied with success to certain phenomena which require the most delicate harmony of a varied system of conditions. For instance, the observation of the numerous maladies of the nervous system offers us a special and inestimable means of improving our knowledge of the laws of intellectual and moral phenomena, imperfect as are yet our qualifications for using them. There remains one other means of knowledge under this head; the examination of exceptional organizations, or cases of monstrosity. As might be anticipated, these organic anomalies were the last to pass over from the gaze of a barren curiosity to the investigation of science; but we are now learning to refer them to the laws of the regular organism, and to subject them to pathological procedures, regarding such exceptions as true maladies, of a deeper and more obscure origin than others, and of a more incurable nature;—considerations which, of course, reduce their scientific value. This resource shares with pathology the advantage of being applicable through the whole range of the biological system.

It is still necessary to insist that, in either method of experimentation, direct or indirect, artificial or natural, the elementary rules should be kept in view; first, to have a determinate aim, that is, to seek to illustrate an organic phenomenon, under a special aspect; and, secondly, to understand beforehand the normal state, and its limits of variation. In regard to the more advanced sciences, it would seem puerile to recommend such maxims as these; but we must still insist on them in biology. It is through neglect of them that all the observations yet collected on the derangements of the intellectual and moral phenomena have yielded scarcely any knowledge of their laws. Thus, whatever may be the value of the most suitable method of experimentation, we must ever remember that here, as elsewhere, and more than elsewhere, pure observation must always hold the first rank, as casting light, primarily, on the whole subject, which it is proposed to examine afterward, as a special study, with a determinate view, by the method of experimentation.

In the third place, we have to review the method of Comparison, which is so specially adapted to the study of living bodies, and by which, above all others, that study must be advanced. In Astronomy, this method is necessarily inapplicable: and it is not till we arrive at Chemistry that this third means of investigation can be used; and then, only in subordination to the two others. It is in the study, both statical and dynamical, of living bodies, that it first acquires its full development; and its use elsewhere can be only through its application here.

The fundamental condition of its use is the unity of the principal subject, in combination with a great diversity of actual modifications. According to the definition of life, this combination is eminently realized in the study of biological phenomena, however regarded. The whole system of biological science is derived, as we have seen, from one great philosophical conception; the necessary correspondence between the ideas of organization and those of life. There can not be a more perfect fundamental unity of subject than this; and it is unnecessary to insist upon the almost indefinite variety of its modifications,—statical and dynamical. In a purely anatomical view, all possible organisms, all the parts of each organism, and all the different states of each, necessarily present a common basis of structure and of composition, whence proceed successively the different secondary organizations which constitute tissues, organs, and systems of organs, more or less complicated. In the same way, in a physiological view, all living beings, from vegetable to man, considered in all the acts and periods of their existence, are endowed with a certain common vitality, which is a necessary basis of the innumerable phenomena which characterize them in their degrees. Both these aspects present as most important, and really fundamental, what there is in common among all the cases; and their particularities as of less consequence; which is in accordance with the great prevalent law, that the more general phenomena overrule the less. Thus broad and sound is the basis of the comparative method, in regard to biology.

At the first glance the immensity of the science is overwhelming to the understanding, embracing as it does all organic and vital cases, which it appears impossible ever to reduce within the compass of our knowledge: and no doubt, the discouragement hence arising is one cause of the backwardness of biological philosophy. Yet the truth is that this very magnitude affords not an obstacle, but a facility to the perfecting of the science, by means of the luminous comparison which results from it, when once the human mind becomes familiar enough with the conditions of the study to dispose its materials so as to illustrate each other. The science could make no real progress while Man was studied as an isolated subject. Man must necessarily be the type; because he is the most complete epitome of the whole range of cases: Man, in his adult and normal state, is the representative of the great scientific unity, whence the successive terms of the great biological series recede, till they terminate in the simplest organizations, and the most imperfect modes of existence. But the science would remain in the most defective state in regard to Man himself, if it were not pursued through a perpetual comparison, under all possible aspects, of the first term with all inferior ones, till the simplest was reached: and then, back again, through the successive complications which occur between the lowest type and the highest. This is the most general, the most certain, the most effectual method of studying physiological as well as anatomical phenomena. Not only is there thus a greater number of cases known, but each case is much better understood by their approximation. This would not be the case, and the problem would be embarrassed instead of simplified, if there were not a fundamental resemblance among the whole series, accompanied by gradual modifications, always regulated in their course: and this is the reason why the comparative method is appropriate to biology alone, of all the sciences, except, as we shall see hereafter, in social physics.

Complete and spontaneous as this harmony really is, no philosopher can contemplate without admiration the eminent art by which the human mind has been aided to convert into a potent means what appeared at first to be a formidable difficulty. I know no stronger evidence of the force of human reason than such a transformation affords. And in this case, as in every other in which primordial scientific powers are concerned, it is the work of the whole race, gradually developed in the course of ages, and not the original product of any isolated mind—however some moderns may be asserted to be the creators of comparative biology. Between the primitive use that Aristotle made of this method in the easiest cases —as in comparing the structure of Man's upper and lower limbs— to the most profound and abstract approximations of existing biology, we find a very extensive series of intermediate states, constantly progressive among which history can point out individually only labors which prove what had been the advance in the spirit of the comparative art at the corresponding period, as manifested by its larger and more effectual application. It is evident that the comparative method of biologists was no more the invention of an individual than the experimental method of the physicists.

There are five principal heads under which biological comparisons are to be classed. *Five kinds of comparison.*

1. Comparison between the different parts of the same organism.

2. Comparison between the sexes.

3. Comparison between the various phases presented by the whole of the development.

4. Comparison between the different races or varieties of each species.

5. Lastly, and pre-eminently, Comparison between all the organisms of the biological hierarchy.

It must be understood that the organism is always to be supposed in a normal state. When the laws of that state are fully established, we may pass on to pathological comparison, which will extend the scope of those laws: but we are not yet advanced enough in our knowledge of normal conditions to undertake anything beyond. Moreover, though comparative pathology would be a necessary application of biological science, it can not form a part of that science, but rather belongs to the future medical science, of which it must form the basis.—Again, biological comparison can take place only between the organisms, and not between them and their medium. When such comparison comes to be instituted, it will be, not as biological science, but as a matter of natural history.

The spirit of biological comparison is the same under all forms. It consists in regarding all cases as radically analogous in respect to the proposed investigation, and in representing their differences as simple modifications of an abstract type; so that secondary differences may be connected with the primary according to uniform laws; these laws constituting the biological philosophy by which each determinate case is to be explained. If the question is anatomical, Man, in his adult and normal state, is taken for the fundamental unity, and all other organizations as successive simplifications, descending from the primitive type, whose essential features will be found in the remotest cases, stripped of all complication. If the question is physiological, we seek the fundamental identity of the chief phenomenon which characterizes the function proposed, amid the graduated modifications of the series of comparative cases, till we find it isolated, or nearly so, in the simplest case of all; and thence we may trace it back again, clothed in successive complications of secondary qualities. Thus the theory of analogous existences, which has been offered as a recent innovation, is only the necessary principle of the comparative method, under a new name. It is evident that this method must be of surpassing value when philosophically applied: and also that, delicate as it is, and requiring extreme discrimination and care in its estimate and use, it may be easily converted into a hinderance and embarrassment, by giving occasion to vicious speculations on analogies which are only apparent.

Of the five classes specified above, three only are so marked as to require a notice here: the comparison between the different parts of the same organism; between the different phases of each development; and between the distinct terms of the great hierarchy of living bodies.

Comparison of parts of the same organism. The method of comparison began with the first of these. Looking no further than Man, no philosophical mind can help being struck by the remarkable resemblance that his different chief parts bear to each other in many respects,—both as to structure and function. First, all the tissues, all the apparatus, in as far as they are organized and living, offer those fundamental characteristics which are inherent in the very ideas of organization and life, and to which the lowest organisms are reduced. But, in a more special view, the analogy of the organs becomes more and more marked as that of the functions is so; and the converse; and this often leads to luminous comparisons, anatomical and physiological, passing from the one to the other, alternately. This original and simple method of comparison is by no means driven out by newer processes. It was thus, for instance, that Bichat, whose subject was Man only, and adult Man, discovered the fundamental analogy between the mucous and the cutaneous systems, which has yielded so much advantage to both biology and pathology. And again, with all M. de Blainville's mastery of the principle of the comparative method, we can not doubt the sufficiency of the analysis of the human organism to establish the resemblance he exhibited between the skull and the other elements of the vertebral column.

A new order of resources presents itself when we compare the different phases of the same organism. Its chief value is in its offering, on a small scale, and, as it were, under one aspect, the whole series of the most marked organisms of the biological hierarchy; for it is obvious that the primitive state of the highest organism must present the essential characters of the complete state of the lowest; and thus successively,—without, however, compelling us to find the counterpart of every inferior term in the superior organism. Such an analysis of ages unquestionably offers the property of realizing in an individual, that successive complication of organs and functions which characterizes the biological hierarchy, and which, in this homogeneous and compact form, constitutes a special and singular order of luminous comparisons. Useful through all degrees of the scale, it is evidently most so in the case of the highest type, the adult Man, as the interval from the origin to the utmost complexity is in that case the greatest. It is valuable chiefly in the visible ascendant period of life; for we know very little of the fœtal period; and the declining stage, which is in fact only a gradual death, presents little scientific interest: for, if there are many ways of living, there is only one natural way of dying. The rational analysis of death, however, has its own importance, constituting a sort of general corollary, convenient for the verification of the whole body of biological laws.

The popular notion of comparative biology is that it consists wholly of the last of the methods I have pointed out: and this shows how pre-eminent it is over the others; the popular exaggeration however being mischievous by concealing the origin of the art. The peculiarity of this largest application consists in its being founded on a very protracted comparison of a very extensive series of analogous cases, in which the modification proceeds by almost insensible graduated declension. The two more restricted methods could not offer a series of cases extensive enough to establish, without confirmation, the nature and value of the comparative method, though, that point once fixed, they may then come into unquestionable use. As for the value of the largest application, it demonstrates itself. There is clearly no structure or function whose analysis may not be perfected by an examination of what all organisms offer in common with regard to that structure and function, and by the simplification effected by the stripping away of all accessory characteristics, till the quality sought is found alone, whence the process of reconstruction can begin. It may even be fairly said that no anatomical arrangement, and no physiological phenomenon, can be really understood till the abstract notion of its principal element is thus reached, by successively attaching to it all secondary ideas, in the rational order prescribed by their greater or less persistence in the organic series. Such a method seems to me to offer, in biology, a philosophical character very like mathematical analysis genuinely applied; when it presents, as we have seen, in every indefinite series of analogous cases, the essential part which is common to all, and which was before hidden under the secondary specialities of each separate case. It can not be doubted that the comparative art of biologists will produce an equivalent result, up to a certain point; and especially, by the rational consideration of the organic hierarchy.

This great consideration was at first established only in regard to anatomy; but it is yet more necessary in physiology, and not less applicable, except from the difficulty of that kind of observation. In regard to physiological problems particularly, it should be remarked that not only all animal organisms, but the vegetable also, should be included in the comparison. Many important phenomena, and among others those of organic life, properly so called, can not be analyzed without an inclusion of the vegetable form of them. There we see them in their simplest and most marked condition, for it is by the great act of vegetable assimilation that brute matter passes really into the organized state, all ulterior transformations by means of the animal organization being much less marked. And thus, the laws of nutrition, which are of the highest importance, are best disclosed by the vegetable organism. The method is unquestionably applicable to all organs and all acts, without my exception; but its scientific value diminishes as it is applied to the higher apparatus and functions of the superior organisms, because these are restricted in proportion to their complexity and superiority. This is eminently the case with the highest intellectual and moral functions which below Man disappear almost entirely; or, at least, almost cease to be recognisable below the first classes of the mammifers. We can not but feel it to be an imperfection in the comparative method that it serves us least where we are most in need of all our resources; but it would be unphilosophical to deprive ourselves, even in this case, of the light which is cast upon the analysis of Man as moral, by the study of the intellectual and affective qualities of the superior animals, and of all others which present such attributes, however imperfect our management of the comparison may yet be. And we may observe that the comparative method finds a partial equivalent in the rational analysis of ages,—thus rendered more clear, extensive, and complete,—for the disadvantages which belong to the same stage of the biological hierarchy.

Thus I have presented the principal philosophical characters of the comparative method. It being the aim of biological study to ascertain the general laws of organic existence, it is plain that no course of inquiry could be more favorable than that which exhibits organic cases as radically analogous, and deducible from each other.

This study of our means of exploration has shown that our resources do indeed increase with the complexity of our subject. The two first methods—of Observation and Experiment—we have seen to acquire a large extension in the case of this science: while the third, before almost imperceptible, becomes, by the nature of the phenomena, well-nigh unbounded in its scope. We have next to examine the true rational position of Biology in the hierarchy of the fundamental sciences; that is, its relation to those that precede it, and to the one which follows it, in order to ascertain what kind and degree of speculative perfection it admits of, and what preliminary training is best adapted to its systematic cultivation. By this inquiry we shall see why we are justified in assigning to it a place between chemistry and social science.

Of the relation of Biology to social science, I need say little here, as I shall have to speak of it at length in another chapter. My task will then be to separate them, rather than to establish their connection, which it is the tendency of our time to exaggerate, through the spontaneous development of natural philosophy. None but purely metaphysical philosophers would at this day persist in classing the theory of the human mind and of society as anterior to the anatomical and physiological study of individual man. We may therefore regard this point as sufficiently settled for the present, and pass on to the relation of Biology to inorganic philosophy. *Relation of Biology to other sciences.*

It is to chemistry that Biology is, by its nature, most directly and completely subordinated. In analyzing the phenomenon of life, we saw that the fundamental acts which, by their perpetuity, characterize that state, consist of a series of compositions and decompositions; and they are therefore of a chemical nature. Though in the most imperfect organisms, vital reactions are widely separated from common chemical effects, it is not the less true that all the functions of the proper organic life are necessarily controlled by those fundamental laws of composition and decomposition which constitute the subject of chemical science. If we could conceive throughout the whole scale the same separation of the organic from the animal life that we see in vegetables alone, the vital motion would offer only chemical conceptions, except the essential circumstances which distinguish such an order of molecular reactions. The general source of these important differences is, in my opinion, to be looked for in the result of each chemical conflict not depending only on the simple composition of the bodies between which it takes place, but being modified by their proper organization; that is, by their anatomical structure. Chemistry must clearly furnish the starting-point of every rational theory of nutrition, secretion, and, in short, all the functions of the vegetative life, considered separately; each of which is controlled by the influence of chemical laws, except for the special modifications be- *To Chemistry.*

longing to organic conditions. If we now bring in again the consideration, discarded for the moment, of the animal life, we see that it could in no way alter this fundamental subordination, though it must greatly complicate its actual application: for we have seen that the animal life, notwithstanding its vast importance, can never be regarded in biology otherwise than as destined to extend and perfect the organic life, whose general nature it can not change. Such an intervention modifies, anew and largely, the chemical laws of the purely organic functions, so as to render the effect very difficult to foresee; but not the less do these laws continue to control the aggregate of the phenomena. If, for instance, a change in the nervous condition of a superior organism disturbs a given secretion, as to its energy or even its nature, we can not conceive that such an alteration can be of a random kind: such modifications, irregular as they may appear, are still submitted to the chemical laws of the fundamental organic phenomenon, which permit certain variations, but interdict many more. Thus, no complication produced by animal life can withdraw the organic functions from their subordination to the laws of composition and decomposition. This relation is so important, that no scientific theory could be conceived of in biology without it; since, in its absence, the most fundamental phenomena might be conceived of as susceptible of arbitrary variations, which would not admit of any true law. When we hear, at this day, on the subject of azote, such a doctrine as that the organism has the power of spontaneously creating certain elementary substances, we perceive how indispensable it still is to insist directly on those principles which alone can restrain the spirit of aberration.

Besides this direct subordination of biology to chemistry, there are relations of method between them. Observation and experimentation being much more perfect in chemistry, they serve as an admirable training for biological inquiry. Again, a special property of chemistry is its developing the art of scientific nomenclature; and it is in chemistry that biologists must study this important part of the positive method, though it can not, from the complexity of their science, be of so much scientific value as in chemistry. It is on the model of the chemical nomenclature that those systematic denominations have been laid down by which biologists have classified the most simple anatomical arrangements, certain well-defined pathological states, and the most general degrees of the animal hierarchy: and it is by a continued pursuit of the same method that further improvements will be effected.

We thus see why biology takes its place next after chemistry, and why chemical inquiries constitute a natural transition from the inorganic to the organic philosophy.

To Physics. The subordination of biology to Physics follows from its relation to Chemistry: but there are also direct reasons, relating both to doctrine and method, why it should be so.

As to doctrine,—it is clear that the general laws of one or more branches of Physics must be applied in the analysis of any physiological phenomenon. This application is necessary in the examination of the medium, in the first place; and the analysis of the medium is required to be very exact, on account of the strong effect of its variations on phenomena so easily modified as those of the organism. And next, the organism itself is no less dependent on those laws, relating as they do to weight, heat, electricity, etc. It is obvious that if biology is related to chemistry through the organic life, it is related to Physics by the animal life,—the most special and noble of the sensations, those of sight and hearing, requiring for the starting-point of their investigation an application of optics and acoustics. The same remark holds good in regard to the theory of utterance, and the study of animal heat and the electric properties of the organism. It remains to be wished that the biologists would study and apply these laws themselves, instead of committing the task to physicists: but they have hitherto followed too much the example of the physicists, who, as we have seen, have committed the application of mathematical analysis in their own science to the geometers; whereas, it can not be too carefully remembered that if the more general sciences are independent of the less general, which, on the other hand, must be dependent on them, the students of the higher must be unfit, in virtue of that very independence, to apply them to a more complex science, whose conditions they can not sufficiently understand. If the case was clear in regard to the intrusion of the geometers into physics, it is yet more so with regard to the intrusion of the physicists into biology; on account of the more essential difference in the nature of the two sciences. The biologists should qualify themselves for the application of the preceding sciences to their own, instead of looking to the physicists for guidance which can only lead them astray.

In regard to Method, biology is indebted to physics for the most perfect models of observation and experimentation. Observations in physics are of a sufficient complexity to serve as a type for the same method in biology, if divested of their numerical considerations, which is easily done. Chemistry, however, can furnish an almost equally good model in simple observation. It is in experimentation that biologists may find in physics a special training for their work. As the most perfect models are found in the study of physics, and the method is singularly difficult in physiology, we see how important the contemplation of the best type must be to biologists. Such is the nature of the dependence, as to doctrine and method, of biology on physics. We turn next to its relations with Astronomy; and first, with regard to doctrine.

The relation of physiology to astronomy is more important than is usually supposed. I mean something more than the impossibility of understanding the theory of weight, and its effects upon the organism, apart from the consideration of general gravitation. I mean, besides, and more specially, that *To Astronomy.*

it is impossible to form a scientific conception of the conditions of vital existence without taking into the account the aggregate astronomical elements that characterize the planet which is the home of that vital existence. We shall see more fully, in the next book, how humanity is affected by these astronomical conditions; but we must cursorily review these relations in the present connection.

The astronomical data proper to our planet are, of course, statical and dynamical. The biological importance of the statical conditions is immediately obvious. No one questions the importance to vital existence of the mass of our planet in comparison with that of the sun, which determines the intensity of gravity; or of its form, which regulates the direction of the force; or of the fundamental equilibrium and the regular oscillations of the fluids which cover the greater part of its surface, and with which the existence of living beings is closely implicated; or of its dimensions, which limit the indefinite multiplication of races, and especially the human; or of its distance from the centre of our system, which chiefly determines its temperature. Any sudden change in one or more of these conditions would largely modify the phenomena of life. But the influence of the dynamical conditions of astronomy on biological study is yet more important. Without the two conditions of the fixity of the poles as a centre of rotation, and the uniformity of the angular velocity of the earth, there would be a continual perturbation of the organic media which would be incompatible with life. Bichat pointed out that the intermittence of the proper animal life is subordinate in its periods to the diurnal rotation of our planet; and we may extend the observation to all the periodical phenomena of any organism, in both the normal and pathological states, allowance being made for secondary and transient influences. Moreover, there is every reason to believe that, in every organism, the total duration of life and of its chief natural phases depends on the angular velocity proper to our planet; for we are authorized to admit that, other things being equal, the duration of life must be shorter, especially in the animal organism, in proportion as the vital phenomena succeed each other more rapidly. If the earth were to rotate much faster, the course of physiological phenomena would be accelerated in proportion; and thence life would be shorter; so that the duration of life may be regarded as dependent on the duration of the day. If the duration of the year were changed, the life of the organism would again be affected: but a yet more striking consideration is that vital existence is absolutely implicated with the form of the earth's orbit, as has been observed before. If that ellipse were to become, instead of nearly circular, as eccentric as the orbit of a comet, both the medium and the organism would undergo a change fatal to vital existence. Thus the small eccentricity of the earth's orbit is one of the main conditions of biological phenomena, almost as necessary as the stability of the earth's rotation; and every other element of the annual motion exercises an influence, more or less marked, on biological conditions, though not so great as the one we have adduced. The inclination of the plane of the orbit, for instance, determines the division of the earth into climates, and, consequently, the geographical distribution of living species, animal and vegetable. And again, through the alternation of seasons, it influences the phases of individual existence in all organisms; and there is no doubt that life would be affected if the revolution of the line of the nodes were accelerated; so that its being nearly immoveable has some biological value. These considerations indicate how necessary it is for biologists to inform themselves accurately, and without any intervention, of the real elements proper to the astronomical constitution of our planet. An inexact knowledge will not suffice. The laws of the limits of variation of the different elements, or, at least, a scientific analysis of the chief grounds of their permanence, are essential to biological investigation; and these can be obtained only through an acquaintance with astronomical conceptions, both geometrical and mechanical.

It may at first appear anomalous, and a breach of the encyclopedical arrangement of the sciences, that astronomy and biology should be thus immediately and eminently connected, while two other sciences lie between. But, indispensable as are physics and chemistry, astronomy and biology are, by their nature, the two principal branches of philosophy. They, the complements of each other, include in their rational harmony the general system of our fundamental conceptions. The solar system and man are the extreme terms within which our ideas will for ever be included. The system first, and then Man, according to the positive course of our speculative reason: and the reverse in the active process: the laws of the system determining those of Man, and remaining unaffected by them. Between these two poles of natural philosophy the laws of physics interpose, as a kind of complement of the astronomical laws; and again, those of chemistry, as an immediate preliminary of the biological. Such being the rational and indissoluble constitution of these sciences, it becomes apparent why I insisted on the subordination of the study of Man to that of the system, as the primary philosophical characteristic of positive biology.

Though in the infancy of the human mind, when it was in its theological state, and in its youthful metaphysical stage, the order of these sciences was reversed, there was a preparation for the true view. Through all the fanciful notions of the ancient philosophy about the physiological influence of the stars, we discern a strong though vague perception of some connection between vital and celestial phenomena. Like all primitive intuitions of the understanding, this one needed only rectification by the positive philosophy; under the usual condition, however, of being partially overthrown in order to be reorganized. But modern students, finding no astronomical conditions in the course of their anatomical and physiological observations, have discarded the idea of them altogether,—as if it were possible for facts to bear immediate testimony to the conditions without which they could not exist, and which do not admit of a moment's suspension! Such an order of primitive conditions is however now established beyond dispute. In order to prevent any return to vicious or exaggerated notions about the physiological influence of the stars, it is enough to bear in mind two considerations: first, that the astronomical conditions of vital existence are comprised within our own planetary system; and, secondly, that they relate, not directly to the organism, but to its environment, affecting as they do the constitution of our globe.

In regard to method,—the importance of astronomical study to biologists consists, as in other cases, in its offering the most perfect model of philosophizing on any phenomena whatever; the importance of this example becoming greater in proportion to the complexity of the subordinate science, on account of the stronger temptation to discursive and idle inquiries offered by the latter. The more difficult their researches become, the more sedulous should physiologists be to refresh their positive forces at the source of positive knowledge; and, in the contemplation of the few general and indisputable conceptions which constitute this lofty science, to be on their guard against the baseless notions of a vital principle, vital forces, and entities of that character. Hitherto, all advance in positivity in biology has been obtained at the expense of its dignity, which has always been implicated with an imaginary origin of life, of sensibility, etc.: but when physiologists have learned from their study of gravitation and other primary laws how to confine themselves to true science, their subject will rise to the highest elevation that positivity admits of,—that rational prevision of events which is, as I have so often said, the end of true science:—an end to be aimed at in biology, as it is perfectly fulfilled in astronomy.

Here, too, must biologists learn the character of sound scientific hypothesis. This method is eminently wanted in so complex a study as physiology; but it has been as yet used with very little effect. The way is, undoubtedly, to determine the organ from the function, or the function from the organ. It is permissible to form the most plausible hypothesis as to the unknown function of a given organ, or the concealed organ of a manifest function. If the supposition be in harmony with existing knowledge, if it be held provisionally, and if it be capable of a positive verification, it may contribute to the progress of discovery, and is simply a use of a right of the human mind, exercised as in astronomy. The only eminent example known to me of sound hypothesis in biology is that of M. Broussais, in proposing the mucous membrane of the alimentary canal as the seat of so-called essential fevers. Whether he was mistaken or not, is not the question. His hypothesis being open to unquestionable confirmation or subversion, it gave a great impulse to the study of pathology in a positive manner: and it will stand in the history of the human mind, as the first example of the spontaneous introduction of a sound hypothetical method into the positive study of living beings: a method derived from the region of astronomy.

It remains to consider the relation of biology to mathematics.

The encroachments of the pure geometers upon the domain of biology have been attended with the same mischief, but in an aggravated form, that we have witnessed in the case of other sciences. This mischief has led physiologists to repudiate mathematics altogether, and open an impassible gulf between themselves and the geometers. This is a mistake; inasmuch as their science cannot be severed from that which is the basis of the whole of natural philosophy; and it is only through the admission of this that they can maintain the originality and independence of their scientific labors. The rational study of nature proceeds on the ground that all phenomena are subject to invariable laws, which it is the business of philosophical speculation to discover. It is needless to prove that on any other supposition, science could not exist, and our collections of facts could yield no result. In the phenomena of living bodies, as in all others, every action proceeds according to precise, that is, mathematical laws, which we should ascertain if we could study each phenomenon by itself. The phenomena of the inorganic world are, for the most part, simple enough to be calculable: those of the organic world are too complex for our management: but this has nothing to do with any difference in their nature. And this is the view which both geometers and biologists should bear in mind.

To mathematics.

If in astronomy our calculations are baffled when we pass beyond two or three essential conditions, it is evident how impracticable they must be amidst the inextricable complications of physiology. And again, this complexity prevents our ever effecting a mathematical disclosure of the elementary laws of the science. This excludes all idea of this method of philosophizing in biology; for these laws are no otherwise accessible than by the immediate analysis of their numerical effects. Now, whichever way vital phenomena are looked at, they present such endless and incessant variations in their numbers, that geometers are baffled as completely as if those degrees were entirely arbitrary. Even numerical chemistry is inapplicable to bodies whose molecular composition varies incessantly; and this is precisely the distinguishing character of living organisms. However hurtful may have been the incursions of the geometers, direct and indirect, into a domain which it is not for them to cultivate, the physiologists are not the less wrong in turning away from mathematics altogether. It is not only that without mathematics they could not receive their due preliminary training in the intervening sciences: it is further necessary for them to have geometrical and mechanical knowledge, to understand the structure and the play of the complex apparatus of the living, and especially the animal organism. Animal mechanics, statical and dynamical, must be unintelligible to those who are ignorant of the general laws of rational mechanics. The laws of equilibrium and motion are, as we saw when treating of them, absolutely universal in their action, depending wholly on the energy, and not at all on the nature of the forces considered: and the only difficulty is in their numerical application in cases of complexity. Thus, discarding all idea of a numerical application in biology, we perceive that the general theorems of statics and dynamics must be steadily verified in the mechanism of living bodies, on the rational study of which they cast an indispensable light. The highest orders of animals act, in repose and motion, like any other mechanical apparatus of similar complexity, with the one difference of the mover, which has no power to alter the laws of motion and equilibrium. The participation of rational mechanics in positive biology is thus evident. Mechanics can not dispense with geometry; and besides, we see how anatomical and physiological speculations involve considerations of form and position, and require a familiar knowledge of the principal geometrical laws which may cast light upon those complex relations.

In regard to method, the necessity of recurring to a perfect model of reasoning, the more earnestly in proportion to the complexity of the science concerned, is applicable in regard to Mathematics, as to Astronomy; only with still greater urgency. In mathematics we find the primitive source of rationality; and to mathematics must the biologists resort for means to carry on their researches. If biologists have hitherto not done this, but contented themselves with what is called logic, apart from all determinate reasoning, much of the fault is chargeable upon the indifference of geometers about duly organizing the whole of mathematical knowledge. The imperfect and inadequate character of the elementary treatises on mathematics that have hitherto been given to the world quite accounts for the neglect of the fundamental logical properties of mathematical science by even intelligent minds. It accounts also for the exaggerations of some philosophers, who maintain that, far from preparing the intellectual organ for the rational interpretation of nature, a mathematical education rather tends to develop a spirit of sophistical argumentation and illusory speculation. Such an abuse, however, can not affect the real value of mathematics as a means of positive education; but rather exhibits the necessity of a philosophical renovation of the whole system of mathematical instruction. Whatever advantage can be attributed to logic in directing and strengthening the action of the understanding is found in a higher degree in mathematical study, with the immense added advantage of a determinate subject, distinctly circumscribed, admitting of the utmost precision, and free from the danger which is inherent in all abstract logic—of leading to useless and puerile rules, or to vain ontological speculations. The positive method, being everywhere identical, is as much at home in the art of reasoning as anywhere else: and this is why no science, whether biology or any other, can offer any kind of reasoning, of which mathematics does not supply a simpler and purer counterpart. Thus, we are enabled to eliminate the only remaining portion of the old philosophy which could even appear to offer any real utility; the logical part, the value of which is irrevocably absorbed by mathematical science. Hither, then, must biologists come, to study the logical art so as to apply it to the advancement of their difficult researches. In this school must they learn familiarly the real characters and conditions of scientific evidence, in order to transfer it afterward to the province of their own theories. The study of it here, in the most simple and perfect cases, is the only sound preparation for its recognition in the most complex.

The study is equally necessary for the formation of intellectual habits; for obtaining an aptitude in forming and sustaining positive abstractions, without which the comparative method can not be used in either anatomy or physiology. The abstraction which is to be the standard of comparison must be first clearly formed, and then steadily maintained in its integrity, or the analysis becomes abortive: and this is so completely in the spirit of mathematical combinations, that practice in them is the best preparation for it. A student who can not accomplish the process in the more simple case may be assured that he is not qualified for the higher order of biological researches, and must be satisfied with the humbler office of collecting materials for the use of minds of another order. Hence arises another use of mathematical training—that of testing and classifying minds, as well as preparing and guiding them. Probably as much good would be done by excluding the students who only encumber the science by aimless and desultory inquiries, as by fitly instituting those who can better fulfil its conditions.

There seems no sufficient reason why the use of scientific fictions, so common in the hands of geometers, should not be introduced into biology, if systematically employed, and adopted with sufficient sobriety. In mathematical studies, great advantages have arisen from imagining a series of hypothetical cases, the consideration of which, though artificial, may aid the clearing up of the real subject, or its fundamental elaboration. This art is usually confounded with that of hypotheses; but it is entirely different, inasmuch as in the latter case the solution alone is imaginary; whereas, in the former, the problem itself is radically ideal. Its use can never be in biology comparable to what it is in mathematics; but it seems to me that the abstract character of the higher conceptions of comparative biology renders them susceptible of such treatment. The process would be to intercalate, among different known organisms, certain purely fictitious organisms, so imagined as to facilitate their comparison, by rendering the biological series more homogeneous and continuous; and it might be that several might hereafter meet with more or less of a realization among organisms hitherto unexplored. It may be possible, in the present state of our knowledge of living bodies, to conceive of a new organism capable of fulfilling certain given conditions of existence. However that may be, the collocation of real cases with well-imagined ones, after the manner of geometers, will doubtless be practised hereafter, to complete the general laws of comparative anatomy and physiology, and possibly to anticipate occasionally the direct exploration. Even now, the rational use of such an artifice might greatly simplify and clear up the ordinary system of pure biological instruction. But it is only the highest order of investigators who can be intrusted with it. Whenever it is adopted, it will constitute another ground of relation between biology and mathematics.

We have now gone over all those grounds—both of doctrine and of method. Of the three parts of mathematics, Mechanics is connected with biology in the scientific point of view; and geometry in the logical: while both rest upon the analytical theories which are indispensable to their systematic development.

This specification of the relations of biology determines its rank in the hierarchy of sciences. From this again we learn the kind and degree of perfection of which biology is susceptible; and, more directly, the rational plan of preliminary education which it indicates.

Condition and prospects of the science. If the perfection of a science were to be estimated by the means of its pursuit, biology would evidently excel all others; for we can concentrate upon it the whole of the resources of observation and of reasoning offered by all the others, together with some of high importance appropriate to itself. Yet, all this wealth of resources is an insufficient compensation for the accumulated obstacles which beset the science. The difficulty is not so much in its recent passage into the positive state as in the high complexity of its phenomena. After the wisest use of all our resources, this study must ever remain inferior to all the departments of inorganic philosophy, not excepting chemistry itself. Still, its speculative improvement will be greater than might be supposed by those who are unaware how incomplete and barren is the accumulation of observations and heterogeneous conceptions which now goes by the name of the science. All that has yet been done should be regarded as a preliminary operation—an ascertainment and trial of means, hitherto provisional, but henceforth to be organized. Such an organization having really taken place among a few qualified investigators, the state of the science may be regarded as very satisfactory. As for the direct establishment of biological laws, the few positive ideas that we have obtained justify the expectation that the science of living bodies may attain to a real co-ordination of phenomena, and therefore to their prevision, to a greater or smaller extent.

Requisite Education. As for the requisite education—as it comprehends the study of the preceding sciences, from mathematics downward, it is clearly of a more extensive and difficult order than any hitherto prescribed. But the time saved from the useless study of words, and from futile metaphysical speculations, would suffice for all the purposes of the regenerated science, which discards these encumbrances.

If, next, we look at the reaction of the science on the education of the general mind, the first thing that strikes us is that the positive study of Man affords to observers the best test and measure of the mental power of those who pursue the study. In other sciences, the real power of the inquirer and the value of his acquisitions are concealed from popular estimate by the scientific artifices which are requisite for the pursuit; as in the case of mathematics, whose hieroglyphic language is very imposing to the uninitiated: so that men of extremely small ability, rendering very doubtful services, have obtained a high reputation for themselves and their achievements. But this can hardly take place in biology; and the preference which popular good sense has accorded to the study of Man as a test of scientific intelligence is therefore well-grounded. Here the most important phenomena are common to all; and the race may be said to concur in the study of Man: and, the more difficult and doubtful the ascertainment of general laws in so complex a science, the higher is the value of individual and original meditation. When these laws become better known, this originality will yield some of its value to the ability which will then be requisite for their application. The moral world will, under all future, as under all past circumstances, regard the knowledge of human nature as the most indubitable sign, and the commonest measure, of true intellectual superiority.

The first intellectual influence of the science is in perfecting, or rather developing, two of the most important of our elementary powers, which are little required by the preceding sciences;—the arts of comparison and of classification, which, however necessary to each other, are perfectly distinct. Of the first, I have said enough; and of the second I shall speak hereafter; so that I have now only to indicate its function in biology.

The universal theory of philosophical classifications, necessary not only to aid the memory but to perfect scientific combinations, can not be absent from any branch of natural philosophy: but it is incontestable that the full development of the art of classification was reserved for biological science. As we have seen before, each of our elementary powers must be specially developed by that one of our positive studies which requires its most urgent application, and which, at the same time, offers it the most extended field. Under both aspects, biology tends, more than any other science, to favor the spontaneous rise of the general theory of classifications. First, no other so urgently claims a series of rational classifications, on account both of the multiplicity of distinct but analogous beings, and of the necessity of organizing a systematic comparison of them in the form of a biological hierarchy; and next, the same characteristics which demand these classifications facilitate their spontaneous establishment. The multiplicity and complexity are not, as might at first appear, obstacles to the systematic arrangement of subjects: on the contrary, they are aids, as the diversity of their relations offers a greater number of analogies, more extensive and easy to lay hold of. This is the reason why the classification of animals is superior to that of vegetables; the greater variety and complexity of animal organisms affording a better hold for the art of classifying. And thus we see that the very difficulties of the science are of a nature at once to require and permit the most marked and spontaneous development of the general art of classification; and hither must the student in every other department of science resort, to form his conceptions of this all-important method. Here alone can geometers, astronomers, physicists, and even chemists learn the formation of natural groups, and their rational co-ordination; and yet more, the general principle of the subordination of characteristics, which constitutes the chief artifice of the method. The biologists alone, at this day, can be in habitual possession of clear and positive ideas in these three relations.—Each of the fundamental sciences has, as we have already so often seen, the exclusive property of specially developing some one of the great logical procedures of which the whole positive method is composed; and it is thus that the more complex, while dependent on the simpler, react on their superiors by affording them new rational powers and instruments. In this view of the hierarchical character and unity of the system of human knowledge, it becomes clear that the isolation still practised in the organization of our positive studies is as hurtful to their special progress as to their collective action upon the intellectual government of the human race.

Influence of Biology upon the Positive spirit. Looking now to the higher functions of this science—its influence upon the positive spirit, as well as method—we have only to try it by the test proposed before;—its power of destroying theological conceptions in two ways:—by the rational prevision of phenomena, and by the voluntary modification of them which it enables Man to exercise. As the phenomena of any science become more complex, the first power decreases, and the other increases, so that the one or the other is always present to show, unquestionably, that the events of the world are not ruled by supernatural will, but by natural laws. Biological science eminently answers to this test. While its complexity allows little prevision, at present, in regard to its phenomena, it supplies us with a full equivalent, in regard to theological conceptions, in the testimony afforded by the analysis of the conditions of action of living bodies. The natural opposition of this species of investigation to every kind of theological and metaphysical conception is particularly remarkable in the case of intellectual and affective phenomena—the positivism of which is very recent, and which, with the social phenomena that are derived from them, are the last battle-ground, in the popular view, between the positive philosophy and the ancient. In virtue of their complexity, these phenomena are precisely those which require the most determinate and extensive concurrence of various conditions, exterior and interior; so that the positive study of them is eminently fitted to expose the futility of the abstract explanations derived from the theological or metaphysical philosophy. Hence, we easily understand the marked aversion which this study is privilege to arouse among different sects of theologians and metaphysicians. As the labors of anatomists and physiologists disclose the intimate dependence of moral phenomena on the organism and its environment, there is something very striking in the vain efforts of followers of the old philosophies to harmonize with these facts the illusory play of supernatural influences or psychological entities. Thus has the development of biological science put the positive philosophy in possession of the very stronghold of the ancient philosophy. The same effect becomes even more striking in the other direction, from biological phenomena being, beyond all others, susceptible of modification from human intervention. We have a large power of affecting both the organism and its environment, from the very considerable number of the conditions which concur in their existence: and our voluntary power of disturbing phenomena, of suspending, and even destroying them, is so striking as to compel us to reject all idea of a theological or metaphysical direction. As in the other case, of which indeed it is a mere extension, this effect is most particularly marked in regard to moral phenomena, properly so called, which are more susceptible of modification than any others. The most obstinate psychologist could not well persist in maintaining the sovereign independence of his intellectual entities, if he would consider that the mere standing on his head for a moment would put a complete stop to the course of his own speculations. Much as we may wish that, in addition to these evidences, we had that of an extensive power of scientific prevision in biology, such a power is not needed for the conclusions of popular good sense. This prevision is not always baffled: and its success in a few marked cases is sufficient to satisfy the general mind that the phenomena of living bodies are subject, like all others, to invariable natural laws, which we are prevented from interrupting in all cases only by their extreme complexity.

But, moreover, positive biology has a special conquest of its own over the theological and metaphysical systems, by which it has converted an ancient dogma into a new principle. In chemistry the same thing occurred when the primitive notion of absolute creation and destruction was converted into the precise conception of perpetual decomposition and recomposition. In Astronomy, the same thing occurred when the hypothesis of final causes and providential rule gave place to the view of the solar system as the necessary and spontaneous result of the mutual action of the principal masses which compose it. Biology, in its close connection with astronomy, has completed this demonstration. Attacking, in its own way, the elementary dogma of final causes, it has gradually transformed it into the fundamental principle of the conditions of existence, which it is the particular aptitude of biology to develop and systemize. It is a great error in anatomists and physiologists —an error fatal to both science and theology—to endeavor to unite the two views. Science compels us to conclude that there is no organ without a function, and no function without an organ. Under the old theological influences, students are apt to fall into a state of anti-scientific admiration when they find the conditions and the fulfilment coincide—when, having observed a function, anatomical analysis discloses a statical position in the organism which allows the fulfilment of the function. This irrational and barren admiration is hurtful to science, by habituating us to suppose that all organic acts are effected as perfectly as we can imagine, thus repressing the expansion of our biological speculations, and inducing us to admire complexities which are evidently injurious: and it is in direct opposition to religious aims, as it assigns human wisdom as the rule and even the limit of the divine, which, if such a parallel is to be established, must often appear to be the inferior of the two. Though we can not imagine radically new organisms, we can, as I showed in my suggestion about the use of scientific fictions, conceive of organizations which should differ distinctly from any that are known to us, and which should be incontestably superior to them in certain determinate respects. The philosophical principle of the conditions of existence is in fact simply the direct conception of the necessary harmony of the statical and the dynamical analyses of the subject proposed. This principle is eminently adapted to the science of biology, which is continually engaged in establishing a harmony between the means and the end; and nowhere else, therefore, is seen in such perfection, that double analysis, statical and dynamical, which is found everywhere.

These, then, are the philosophical properties of positive biology. To complete our review of the science as a whole, we have only to note briefly the division and rational co-ordination of its parts.

Distribution of the science. It does not fall within the scope of this work to notice several branches of positive biological knowledge, which are of extreme importance in their own place, but secondary in regard to the principles of positive philosophy. We have no concern here with pathology, and the corresponding medical art; nor with natural history, and the corresponding art of the education of organisms. These are naturally, and not untruly, called biological studies: but we must here confine the term strictly to the speculative and abstract researches which are the foundation of the science. The interior distribution of the science, thus regarded, is this.

The speculative and abstract study of the organism must be divided, first, into statics and dynamics; according as we are seeking the laws of organization or those of life: and again, statical biology must be divided into two parts, to which M. de Blainville has given the name, in regard to animals, of *zootomy* and *zootaxy*, according as we study the structure and composition of individual organisms, or construct the great biological hierarchy which results from the comparison of all known organisms. It would be easy to modify M. de Blainville's terms so as to make them common to animals and vegetables. Dynamical biology, to which we may give the name *bionomy*, as the end and aim of the whole set of studies, evidently admits of no analogous subdivision. The general name of Biology thus includes the three divisions, biotomy, biotaxy, and pure bionomy, or physiology properly so called.

Their definition exhibits their necessary dependence; and thereby determines also their philosophical co-ordination. While it is universally allowed that anatomical ideas are indispensable to physiological studies, because the structure must be known before its action can be judged of, the subordination of bionomy to biotaxy is not so well understood. Yet it is easy to see that the place of any organism in the scale must be known before its aggregate phenomena can be effectually studied: and again, the consideration of this hierarchy is indispensable to the use of that grandest instrument of all,—the comparative method. Thus, from every point of view, the double relation of dynamical to statical biology is unquestionable.

The two divisions of statical biology are less clearly marked; and it even appears as if, in regard to them, we were involved in a vicious circle: for if, on the one hand, the rational classification of living beings requires the antecedent knowledge of their organization, it is certain, on the other hand, that anatomy itself, like physiology, can not be studied, in regard to all organisms, without an antecedent formation of the biological hierarchy. Thus we must admit a consolidation of the respective advancements of biotomic and biotaxic studies, through their intimate connection. In such a case, as a separation and determinate co-ordination are required by our understandings, it appears to me that we can not hesitate to make a dogmatic arrangement,—placing the theory of organization before that of classification,—for the last is absolutely dependent on the first; while the first could meet some wants, though in a restricted way, without the second. In a word, none but known organisms can be classified; whereas they all can and must be studied, to a certain extent, without being mutually compared. And again, there is no reason why, in a systematic exposition of anatomical philosophy, we should not borrow directly from biotaxy its construction of the organic hierarchy; an anticipation which involves much less inconvenience than severing the complete study of structure.—However, it most be always borne in mind that any system will have to undergo a general revision, with a view to bringing out the essential relations of its parts: the relations, not only of the two sections of statical biology, but of both to the dynamical. This consideration goes far to diminish the importance assigned to these questions of priority: and the only reason why such a revision appears more necessary in biology than in the other science is, that there is a profounder accordance between its departments than we find in theirs.

The interior distribution of these three departments is determined, as usual, by the order of dependence of phenomena, on the ground of their relative generality. Thus, the theory of the organic life precedes that of the animal: and the theory of the highest functions and organs of Man terminates the biological system.

Where to begin. It has often been a question whether, in studying each organ and function in the whole scale, it is best to begin at the one end or the other;—to begin with Man or the simplest known organism. I do not consider this question so all-important as it is often supposed, as all qualified inquiries admit the necessity of using the two methods alternately, whichever is taken first: but I think that a distinction should be made between the study of the organic and that of the animal life. The functions of the first being chemical, it is less necessary to begin with Man; and I think there may be a scientific advantage in studying the vegetable organism first, in which that kind of functions is the more pure and more marked, and therefore the more easily and completely studied: but every investigation, anatomical or physiological, relating to animal life, most be obscure if it began elsewhere than with Man, who is the only being in which such an order of phenomena is immediately intelligible. It is evidently the obvious state of Man, more and more degraded, and not the indecisive state of the sponge, more and more improved, that we should pursue, though the animal series, when we are analyzing any of the constituent characters of animality. If we seem to be by this procedure deserting the ordinary course of passing from the most general and simple subject to the most particular and complex, it is only to conform the better to the philosophical principle which prescribes that very course, and which leads us from the most known subjects to the least known. In all cases but this, the usual course is the fittest, in biological studies.

Here we conclude our review of biological science as a whole. The extent to which I have carried out the survey will allow us to consider its separate portions very briefly. In doing so, I shall follow the order just laid down, passing from the simple considerations of pure anatomy to that positive study of phenomena of the intellect and the affections, as the highest part of human nature, which will carry us over from biology to Social Physics,—the final object of this work.

CHAPTER II.

ANATOMICAL PHILOSOPHY.

It was during the second half of the last century that Daubenton and Vicq-d'Azyr achieved the extension of the statical study of living bodies to the whole of known organisms; and the lectures and writings of Cuvier carried on, and spread abroad, the regenerating influence of this great view. But, indispensable as was this conception to the development of anatomical science, it could not complete the character of statical biology without the aid and addition of Bichat's grand idea of the general decomposition of the organism into its various elementary tissues; the high philosophical importance of which appears to me not yet to be worthily appreciated. *Development of Statical biology.*

The natural development of comparative anatomy would, no doubt, have disclosed this analysis to us sooner or later: but how slow the process would have been we may judge by what we see of the reluctance of comparative anatomists to abandon the exclusive study of systems of organs while unable to deny the preponderant importance of the study of the tissues. Of all changes, those which relate to method are the most difficult of accomplishment; and perhaps there is no example of their resulting spontaneously from a regular advance under the old methods, without a direct impulsion from a new original conception, energetic enough to work a revolution in the system of study. Biology must, from its great complexity, be more dependent on such a necessity than any other science.

Though zoological analysis furnishes the best means of separating the various organic tissues, and especially of giving precision to the true philosophical sense of this great notion, pathological analysis offered a more direct and rapid way to suggest the first idea of such a decomposition, even regarding the human organism alone. When pathological anatomy had been once founded by Morgagni, it was evident that in the best-marked maladies no organ is ever entirely diseased, and that the alterations are usually confined to some of its constituent parts, while the others preserve their normal condition. In no other way could the distinction of the elementary tissues have been so clearly established. By the co-existence in one organ of sound and impaired tissues, and, again, by different organs being affected by similar maladies, in virtue of the disease of a common tissue, the analysis of the chief anatomical elements was spontaneously indicated, at the same time that the study of the tissues was shown *Process of discovery of the tissues.*

to be more important than that of the organs. It is not consistent with my objects to go further into this: but it was necessary to show that we owe to pathological analysis the perception of this essential truth. It was Pinel who suggested it to Bichat, by his happy innovation of studying at once all the diseases proper to the different mucous membranes. Bichat then, while knowing nothing of the study of the organic hierarchy, carried off from the students of comparative anatomy the honor of discovering the primitive idea which is most indispensable of all to the general advancement of anatomical philosophy. His achievement consisted in rationally connecting with the normal condition a notion derived from the pathological condition, in virtue probably of the natural reflection that if the different tissues of the same organ could each be separately diseased in its own way, they must have, in their healthy condition, distinct modes of existence, of which the life of the organ is really composed. This principle was entirely overlooked before Bichat published the treatise in which he established the most satisfactory *à-posteriori* development of it: and it is now placed beyond all question. The only matter of regret is that, in creating a wholly new aspect of anatomical science, Bichat did not better mark its spirit by the title he gave it. If he had called it *abstract* or *elementary* instead of *general* anatomy, he would have indicated its philosophical function, and its relation to other anatomical points of view.

Combination with comparative anatomy. The anatomical philosophy began to assume its definitive character from the very recent time when the human mind learned to combine the two great primitive ideas of the organic hierarchy, and of Bichat's discovery, which applies the universal conception to the statical study of living bodies. These combined ideas are necessarily the subject of our present examination. Putting aside the irrational distinctions, still too common among biologists, of many different kinds of anatomy, we must here recognise only one scientific anatomy, chiefly characterized by the philosophical combination of the comparative method with the fundamental notion of the decomposition of the organs into tissues. It is apparently strange that, after Bichat's discovery, comparative anatomists, with Cuvier at their head, should have persisted in studying organic apparatus in its complex state, instead of beginning with the investigation of these tissues, pursuing the analysis of the laws of their combinations into organs, and ending with the grouping of those organs into apparatus, properly so called: but not even Cuvier's great name can now prevent the application of the comparative method to the analysis of tissues, throughout the whole biological hierarchy. The work, though at present neither energetically nor profoundly pursued, is begun, and will reform the habitual direction of anatomical speculation.

Bichat's studies related to Man alone; and his method of comparison bore only upon the simplest and most restricted cases of all; the comparison of parts and that of ages. His principle must, therefore, necessarily undergo some transformations, to fit it for a more extensive application. The most important of these improvements, especially in a logical view, appears to me to consist in the great distinction introduced by M. de Blainville between the true anatomical *elements* and the simple *products* of the organism, which Bichat had confounded. We saw before the importance of this distinction in the chemical study of organic substances: and we meet it again now, face to face, as an anatomical conception.

<small>Elements and Products.</small>

We have seen that life, reduced to the simplest and most general notion of it, is characterized by the double continuous motion of absorption and exhalation, owing to the reciprocal action of the organism and its environment, and adapted to sustain during a certain time, and within certain limits of variation, the integrity of the organization. It results from this that, at every instant of its existence, every living body must present, in its structure and decomposition, two very different orders of principles: absorbed matters in a state of assimilation, and exhaled matters in a state of separation. This is the ground of the great anatomical distinction between organic elements and organic products. The absorbed matters, once completely assimilated, constitute the whole of the real materials of the organism. The exhaled substances, whether solid or fluid, become, from the time of their separation, foreign to the organism, in which they can not generally remain long without danger. Regarded in a solid state, the true anatomical elements are always necessarily continuous in tissue with the whole of the organism: and again, the fluid elements, whether stagnant or circulating, remain in the depths of the general tissue, from which they are equally inseparable: whereas the products are only deposited, for a longer or shorter time, on the exterior or interior surface of the organism. The differences are not less characteristic, in a dynamical view. The true elements alone must be regarded as really living: they alone participate in the double vital motion: and they alone grow or decrease by absorption or exhalation. Even before they are finally excreted, the products are already essentially dead substances, exhibiting the same conditions that they would manifest anywhere else, under similar molecular influences.

The separation of the elements from the products is not always easy to effect, when, as frequently happens, they combine in the same anatomical arrangement to concur in the same function. All products are not, like sweat, urine, etc., destined to be expelled without further use in the organic economy. Several others, as saliva, the gastric fluid, bile, etc., act as exterior substances, and in virtue of their chemical composition in preparing for the assimilation of the organic materials. It is difficult to fix the precise moment when these bodies cease to be products and become elements; the moment, that is, when they pass from the inorganic to the organic state,—from death to life. But these difficulties arise from the imperfection of our analysis, and not from any uncertainty in the principle of separation. It may be observed, however, that there are circumstances in which products, and particularly among the solids, are closely united to true anatomical elements in the structure of certain apparatus, to which they supply essential means of improvement. Such are, for instance, the greater number of epidermic productions, the hair, and eminently the teeth. But even in this case, a sufficiently delicate dissection, and a careful analysis of the whole of the function will enable us to ascertain, with entire precision, how much is organic and how much inorganic in the proposed structure. Such an investigation was not prepared for when Bichat confounded the teeth with the bones, and concluded the epidermis and the hair to be tissues, of a piece with the cutaneous tissue; but the rectification which ensued was all-important; as enabling us to define the idea of tissue or *anatomical element,* which is the preferable term. It was through comparative anatomy that the rectification took place; for the study of the biological series showed that the inorganic parts, which in Man appear inseparable from the essential apparatus, are, in fact, only simple means of advancement, gradually introduced at assignable stages of the ascending biological series.

If we assert that, in the order of purely anatomical speculations, the study of products must be secondary to that of elements, it will not be supposed that we undervalue the study of products. This study is of extreme importance in physiology, whose principal phenomena would be radically unintelligible without it; and without it, pathological knowledge must come to a stand. As results, they indicate organic alterations; and as modified, they exhibit the origin of a great number of those alterations. In fact, the knowledge of them is much promoted by their separation from the anatomical elements, which withdraw the attention of biologists from the real claims of the whole class of products.

The consideration of products being once dismissed to its proper place, anatomical analysis has assumed its true character of completeness and clearness. Thus we may undertake now what was before impossible—an exact enumeration of anatomical elements. And again, these tissues can be classified according to their true general relations; and may even be reduced to a single tissue, modified by determinate laws. These two are the other chief transformations undergone by the great anatomical theory of Bichat, through the application of the comparative method: and these two we now proceed to review.

Vitality of the organic fluids. The first is connected with the great question of the vitality of the organic fluids, about which our ideas are far from being, I think, sufficiently settled. Every living body consists of a combination of solids and fluids, the respective proportions of which vary, according to the species, within very wide limits. The very definition of the vital state supposes this conjunction; for the double motion of composition and decomposition which characterizes life could not take place among solids alone;

and, on the other hand, a liquid or gaseous mass not only requires a solid envelop, but could admit of no real organization. If the two great primitive ideas of Life and Organization were not inseparable, we might imagine the first to belong to fluids, because they are so readily modified; and the other to solids, as alone capable of structural formation: and here, under another view, we should find the necessary harmony of the two elements. The comparison of types in the biological series confirms, in fact, the general rule that the vital activity increases with the preponderance of fluid elements in the organism; while a greater persistence of the vital state attends the preponderance of solids. This has long been regarded as a settled law by philosophical biologists, in studying the series of ages alone. These considerations seem to show that the controversy about the vitality of fluids rests, like many other famous controversies, on a vicious proposal of the problem, since such a mutual relation of the solids and the fluids excludes at once both humorism and solidism. Discarding, of course, the products from the question, there can be no doubt that the fluid elements of the organism manifest a life as real as that of the solids. The founders of modern pathology, in their reaction against the old humorism, have not paid sufficient attention, in the theory of diseases, to the direct and spontaneous alterations of which the organic fluids, especially the blood, are remarkably susceptible, in virtue of the complexity of their composition. It would appear, from a philosophical point of view, very strange if the most active and susceptible of the anatomical elements did not participate, primarily or consecutively, in the perturbations of the organism. But, on the other hand, it is not less certain that the fluids, animal and vegetable, cease to live as soon as they have quitted the organism; as, for instance, the blood after venesection. They then lose all organization, and are in the condition of products. The vitality of the fluids, considered separately, constitutes then an ill-defined and therefore interminable question.

A truly positive inquiry, however, arises out of the question—the inquiry as to which of the immediate principles of a fluid are vital; for it can not be admitted that all are so indiscriminately. Thus, the blood being chiefly composed of water, it would be absurd to suppose such an inert vehicle to participate in the life of the fluid; but then, which of the other constituents is the seat of life? Microscopic anatomy gives us the answer,—that it resides in the globules, properly so called, which are at once organized and living. However valuable such a solution would be, it can be regarded at present only as an attempt; for it is admitted that these globules, though determinate in form, shrink more and more as the arterial blood passes through an inferior order of vessels,—that is, as it approaches its incorporation with the tissues; and that, at the precise moment of assimilation, there is a complete liquefaction of the globules. It would thus appear that we must cease to regard the blood as living at the very moment when it accomplishes its chief act of vitality. Before any decision can be made, we must have the counter proof,—the acknowledgment that true globules are exclusively characteristic of living fluids, in opposition to those which, as simple products, are essentially inert, and hold in suspension various solids, which make them difficult to be distinguished from true globules, notwithstanding the determinate form of the latter. Microscopical observations are too delicate, and sometimes deceptive, to admit at present of the irreversible establishment of this essential point of anatomical doctrine.

The statical study of living bodies would form but a very incomplete introduction to the dynamical, if the fluids were left out of the investigation of the organic elements, however much remains to be desired in our knowledge of them. The omission of them in Bichat's treatise leaves a great gap. Still, as the anatomy of solids must always take precedence of that of fluids, Bichat chose the true point of departure, though he did not undertake the whole subject. It must be added that the examination of the fluids is so much the more difficult of the two as to be well nigh impracticable. In an anatomical sense, it is impracticable: and the only two methods,—microscopal and chemical examination,—are impaired by the rapid disorganization which ensues when the fluids quit the organism. The chemical method is in itself the more valuable of the two: but, besides that the chemists habitually confound the elements and the products, they have always examined the former in a more or less advanced state of decomposition: and being unaware of this, they have offered only the most false and incoherent notions of the molecular constitution of the organized fluids. In such a state of things, it is only by a full preparation, from the study of the solid elements, that anything can be done in the study of the fluids. It is almost needless to say that by the same rule which prescribes this order, we should study fluids in the order of their increasing liquefaction,—taking the fatty substances first, then the blood and other liquids, and lastly the vaporous and gaseous elements, which will always be the least understood.

Classification of the tissues. The order of inquiry being thus settled, the next subject is the rational classification of the tissues, according to their anatomical filiation. It was not by such a study as Bichat's,—that of Man alone,—that anything certain could become known of such obscure differences as those of the fundamental tissues. In order to obtain such knowledge, the study of the whole biological series is indispensable.

ORGANIC. Cellular tissue. The first piece of knowledge thus obtained is that the cellular tissue is the primitive and essential web of every organism; it being the only one that is present through the whole range of the scale. The tissues which appear in Man so multiplied and distinct, lose all their characteristic attributes as we descend the series, and tend to merge entirely in the general cellular tissue, which remains the sole basis of vegetable, and perhaps of the lowest animal organization. The fact harmonizes well with the philosophical account of the basis of life, in its last degree of simplicity; for the cellular tissue is eminently fitted, by its structure, for absorption and exhalation. At the lower end of the series, the living organism, placed in an unvarying medium, does nothing but absorb and exhale by its two surfaces, between which are ever oscillating the fluids destined for assimilation, and those which result from the contrary process. For so simple a function as this, the cellular tissue suffices. It remained to be ascertained under what laws the original tissue becomes gradually modified so as to engender all the others, with those attributes which at first disguise their common derivation: and this is what Comparative anatomy has begun to establish, with some distinctness.

The characteristic modifications of the tissue are of two prominent classes: the first, more common and less profound, are limited to the simple structure: the other class, more special, and more profound, affect the composition itself.

Of the first order the prominent case is that of the dermous tissue, properly so called, which is the basis of the general organic envelope, exterior and interior. The modification here is mere condensation, differently marked, in regard to animal organisms, according as the surface is, as in exterior surfaces, more exhalant than absorbent, or, as in interior surfaces, more absorbent than exhalant. Even this first transformation is not rigorously universal; and we must ascend the scale a little way to find it clearly characterized. Not only in some of the lowest of the animal organisms, are the exterior and interior essentially alike, so that the two surfaces may be interchanged, but, if we go a little lower, we find no anatomical distinction between the envelope and the whole of the organism, which is uniformly cellular. *Dermous tissue.*

By an increasing condensation of the parent tissue, three distinct but inseparable tissues proceed from the derma, all of which are destined to an important, though passive office in the animal economy, either as envelopes protecting the nervous organs, or as auxiliaries of the locomotive apparatus. These are the fibrous, cartilaginous, and bony tissues, ranged by Bichat in their rational order, and named by M. Laurent, in their combination, the *sclerous* tissue. The different degrees of consolidation here arise from the deposition in the cellular network of a heterogeneous substance, organic or inorganic, the extraction of which leaves no doubt as to the nature of the tissue. When, on the other hand, by a last direct condensation, the original tissue becomes itself more compact, without being incrusted by a foreign substance, we recognise a new modification, in which impermeability becomes compatible with suppleness, which is the characteristic of the serous, or (as M. Laurent calls it) the *kystous* tissue, the office of which is to interpose between the various mobile organs, and to contain liquids, both circulating and stagnant. *Sclerous tissue.* *Kystous tissue.*

The second order of transformations exhibits two secondary kinds of tissue which distinguish the animal organism, and which appear at about the same degree *ANIMAL. Muscular and nervous tissues.*

of the scale—the muscular and the nervous tissues. In each there is an anatomical combination of the fundamental tissue with a special organic element, semi-solid and eminently vital, which, having long gone by the name of *fibrine*, in the first case, has suggested the corresponding name of *neurine* (given us by M. de Blainville) for the other. Here the transformation of the parent tissue is so complete, that it would be difficult to establish, and yet more to detect it in the higher organisms; but the analogies of comparative anatomy leave no doubt, and only make us wish that we could understand with more precision the mode of anatomical union of the muscular and nervous substances with the cellular tissue.

Passing on to the chief subdivision of each of the secondary tissues, the first consideration is of the general position, which is always related to a modification, greater or smaller, of the structure itself. Comparative analysis shows us that in the case of both the muscular and the nervous system, the organization of the tissue becomes more special and elevated, exactly in proportion to its deeper position between the exterior and interior surfaces of the animal envelope. Thence arises the rational division of each of these systems into superficial and profound. This distinction is more especially remarkable with regard to the nervous system, arranged, first, in the form of filaments, and afterward that of ganglions, with or without external apparatus.

This is the family of tissues, the study of which forms the basis of anatomical analysis. It would be departing from my object to inquire into the laws of composition under which the ascent is made from this primary study to that of porous substances, and thence on to the theory of the organs, and then to that of systems of organs, which would lead us on to physiological analysis. I have fulfilled the aim of this section in exhibiting the methodical connection of the four degrees of anatomical speculation, about which no real uncertainty exists. Deeper than this we can not go. The last term in our abstract, intellectual decomposition of the organism is the idea of tissue. To attempt the passage from this idea to that of molecule, which is appropriate to inorganic philosophy, is to quit the positive method altogether: and those who do so, under the fancy that they may possibly establish a notion of organic molecules, and who gave that vain search the name of transcendental anatomy, are in fact imitating the chemists in a region into which Chemistry must enter in its own shape where admissible at all, and are asserting in other words that, as bodies are formed of invisible molecules, animals are formed of animalcules. This is simply an attempt, in the old spirit, to penetrate into the nature of existences, and to establish an imaginary analogy between orders of phenomena which are essentially heterogeneous. It is little creditable to the scientific spirit of our time that this aberration should call for exposure and rebuke, and that it should need to be asserted that the idea of tissue is, in organic speculation, the logical equivalent of the idea of molecule in inorganic speculation.

We here find ourselves in possession of a sufficient basis of anatomical science, while we need yet a more complete and profound combination of the ideas of comparative and textural anatomy. This want will be supplied when we become universally familiarized with the four analytical degrees, complementary to each other, which must henceforth be recognised and treated as the basis of anatomical speculation.

CHAPTER III.

BIOTAXIC PHILOSOPHY.

After the statical analysis of living bodies, there must be a hierarchical co-ordination of all known, or even possible organisms, in a single series, which must serve as a basis for the whole of biological speculations. The essential principles of this philosophical operation are what I have now to point out.

We have already seen that it is the distinction of biological science to have developed the theory of classifications, which, existing in all sciences, attains its perfection when applied to the complex attributes of the animal organisms. *Comparative anatomy of vegetables and animals.* In all ages, the vegetable organism was the direct subject of biological classification; but it was pursued on the principles furnished by the consideration of animals, whence the type was derived which guided philosophical speculation in the case. It could not be otherwise, so marked and incontestable as are the distinctions among animal organisms: and even the zoological classification of Aristotle, imperfect as it is, is infinitely superior to anything which could then have been attempted with regard to vegetables. This natural original classification has been rather rectified than changed by the labors of modern times; while that of vegetables has met with an opposite fate. As a fact, the first successful attempts in the animal region long preceded the establishment of the true principles of classification; whereas, it was only by a laborious systematic application of these principles that it has been possible, even within a century, to effect any rational co-ordination in the vegetable region, so little marked, in comparison, are the distinctions in the latter case. The natural result was that the animal realm, used as a type, became more and more attended to, till the improvements in zoological classification have gone so far as perhaps to lead us to fear that the vegetable organism, owing to its great simplicity, can never become subject to a much better classification than that in which it was left in the last century. The labors of the reformers of that time are very far indeed from having been useless; only, what they undertook for the vegetable kingdom has turned rather to the profit of the animal—an inevitable circumstance, since the property which rendered the animal kingdom the natural type of the taxonomical series must adapt it to receive all the improvements arising from the general principles of the theory. The character of the theory could not but remain incomplete, however, as long as the vegetable classification continued to be regarded as the chief end of the research; and the classification became rational only when it was seen that the vegetable region was the further end of the series, in which the most complex animal organism must hold the first place; an order of arrangement under which the vegetable organism will be more effectually studied than it ever was while made an object of exclusive investigation. All that is needed is that naturalists should extend to the whole series the anatomical and physiological considerations which have been attached too exclusively to animal organisms; and this will certainly be done now that the human mind is fairly established at the true point of view, commanding the fundamental theory of natural classification.

Animal anatomy our subject. These prefatory remarks indicate our theme. We must have the whole series in view; but the animal region must be our immediate and explicit subject—both as furnishing the rational bases of the general theory of classification, and as exhibiting its most eminent and perfect application.

Division of the natural method. The subject divides itself into two parts: the formation of natural groups, and their hierarchical succession—a division necessary for purposes of study, though the two parts ultimately and logically coalesce.

Natural group. In contemplating the groups, the process is to class together those species which present, amidst a variety of differences, such essential analogies as make them more like each other than like any others—without attending, for the present, to the gradation of the groups, or to their interior distribution. If this were all, the classification must remain either doubtful or arbitrary, as the circumscription of each group could seldom be done so certainly as inevitably to include or exclude nothing that might not belong to another group: and great discordance was therefore observed in the early division into *orders, families*, and even *genera*. But the difficulty disappears on the foundation of the fundamental hierarchy, which rigorously assigns its place to each species, and clearly defines the ideas of *genera, families*, and *classes*, which henceforth indicate different kinds of decomposition, effected through certain modifications of the principle which graduates the whole series. The animal realm, especially in its higher parts, is as yet the only one in which the successive degrees have admitted a fully scientific description. The rough classification into natural groups was an indispensable preparation for the marshalling into a series of the immeasurable mass of materials presented by nature. The groups being thus separated, and the study of their interior distribution postponed, the innumerable throng of organic existences became manageable. This great benefit has misled botanists into the supposition that the formation of these groups is the most scientific part of the natural method—otherwise than as a preliminary process. The regular establishment of natural families offers, no doubt, great facilities to scientific study, by enabling a single case to serve for a whole group: but this is a wholly different matter from the value of the natural method, regarded, as it must henceforth be, as the highest rational means of the whole study, statical and dynamical, of the system of living bodies; and the great condition of which is that the mere position assigned to each body makes manifest its whole anatomical and physiological nature, in its relation to the bodies which rank before or after it. These properties could never belong to any mere establishment of natural families, if they could be grouped with *Co-ordination of them.* a perfection which is far from being possible; for the arbitrary arrangement of the families, and the intermediate decomposition of each of them into species, would destroy all aptitude for comprehensive anatomical or physiological comparison, and open the way for that search after partial and secondary analogies which we see to be so mischievous in the study of the vegetable kingdom at this day.

The Natural Method, then, is philosophically characterized by the general establishment of the organic hierarchy, reduced, if desired, to the rational co-ordination of genera and even of families, the realization of which is found only in the animal region; and there only in an initiatory state. And the co-ordination proceeds under three great laws, which are these: *Three laws of co-ordination.* first, that the animal species present a perpetually increasing complexity, both as to the diversity, the multiplicity, and the speciality of their organic elements, and as to the composition and augmenting variety of their organs and systems of organs. Secondly: that this order corresponds precisely, in a dynamical view, with a life more complex and more active, composed of functions more numerous, more varied, and better defined. Thirdly: that the living being thus becomes, as a necessary consequence, more and more susceptible of modification, at the same time that he exercises an action on the external world, continually more profound and more extensive. It is the union of these three laws which rigorously fixes the philosophical direction of the biological hierarchy, each one dissipating any uncertainty which might hang about the other two. Hence results the possibility of conceiving of a final arrangement of all living species in such an order as that each shall be always inferior to all that precede it, and superior to all that follow it, whatever might otherwise, from its nature, be the difficulty of ever realizing the hierarchical type to such a degree of precision as this.

All adequate inquirers are now agreed upon this conception as the starting-point of biological speculation; and I need not therefore stop to take notice of any prior controversies, except one, which is noticeable from its having tended to illustrate and advance the principle of the natural method. I refer to the discussion raised by Lamarck, and maintained, though in an imperfect manner, by Cuvier, with regard to the general permanence of organic species. The first consideration in this matter is, that whatever may be the final decision of this great biological question, it can in no way affect the fundamental existence of the organic hierarchy. Instead of there being, as Lamarck conjectured, no real zoological series, all animal organisms being identical, and their characteristics due to external circumstances, we shall see, by a closer examination, that the hypothesis merely presents the series under a new aspect, which itself renders the existence of the scale more clear and unquestionable than before; for the whole zoological series would then become, in fact and in speculation, perfectly analogous to the development of the individual; at least, in its ascending period. There would be simply a long determinate succession of organic states, gradually deduced from each other in the course of ages by transformations of growing complexity, the order of which, necessarily linear, would be precisely comparable to that of the consecutive metamorphoses of hexapod insects, only much more extended. In brief, the progressive course of the animal organism, which is now only a convenient abstraction, adapted to facilitate thought by abridging discourse, would thus be converted into a real natural law. This controversy, then, in which Lamarck showed by far the clearer and profounder conception of the organic hierarchy, while Cuvier, without denying, often misconceived it, leaves, in fact, wholly untouched the theory of the biological series, which is quite independent of all opinion about the permanence or variation of living species.

Question of permanence of organic species.

The only attribute of this series which could be affected by this controversy is the continuity or discontinuity of the organic progression: for, if we admit Lamarck's hypothesis, in which the different organic states succeed each other slowly by imperceptible transitions, the ascending series must evidently be conceived of as rigorously continuous; whereas, if we admit the stability of living species, we must lay down as a fundamental principle the discontinuousness of the series, without pretending, either, to limit, *à priori*, in any way the small elementary intervals. This is the question to be considered; and thus restricted, the discussion is of extreme importance to the general advancement of the Natural Method, which will be, in fact, much more clearly described if we are able to regard the species as essentially stable, and the organic series, therefore, as composed of distinctly separate terms, even at its highest stages of development; for the idea of *species*, which is the principal biotaxic unity, would no longer allow any scientific definition, if we must admit the indefinite transformation of different species into each other, under the sufficiently-prolonged influence of circumstances sufficiently intense. However certain might be the existence of the biological hierarchy, we should have almost insurmountable difficulty in realizing it; and this proves to us the high philosophical interest which belongs to this great question.

Lamarck's reasoning rested on the combination of these two incontestable but ill-described principles: first, the aptitude of any organism (and especially an animal organism) to be modified to a conformity to the exterior circumstances in which it is placed, and which solicit the predominant exercise of some special organ, corresponding to some faculty become requisite; and secondly, the tendency of direct and individual modifications to become fixed in races by hereditary transmission, so that they may increase in each new generation, if the environment remains unaltered. It is evident that if this double property is admitted without restriction, all organisms may be regarded as having been produced by each other, if we only dispose the environment with that freedom and prodigality so easy to the artless imagination of Lamarck. The falseness of this hypothesis is now so fully admitted by naturalists that I need only briefly indicate where its vice resides.

We need not stop to object to the immeasurable time required for each system of circumstances to effect such an organic transformation; nor yet to expose the futility of imagining organic environments, purely ideal, which are out of all analogy with existing media. We may pass on to the consideration that the conjecture rests on a deeply erroneous notion of the nature of the living organism. The organism and the medium must doubtless be mutually related; but it does not follow that either of them produces the other. The question is simply of an equilibrium between two heterogeneous and independent powers. If all possible organisms had been placed in all possible media, for a suitable time, the greater number of them would necessarily disappear, leaving those only which were accordant with the laws of the fundamental equilibrium; and it is probably by a series of eliminations like this that a biological harmony has become gradually established on our globe, where we see such a process now for ever going on. But the whole conception would be overthrown at once if the organism could be supposed capable of modification, *ad infinitum*, by the influence of the medium, without having any proper and indestructible energy of its own.

Though the solicitation of external circumstances certainly does change the primitive organization by developing it in some particular direction, the limits of the alteration are very narrow: so that, instead of wants creating faculties, as Lamarck would have us believe, those wants merely develop the powers to a very inconsiderable degree, and could have no influence at all without a primitive tendency to act upon. The disappearance of the superior races of animals before the encroachments of Man, shows how limited is the power of the organism to adapt itself to an altered environment; even the human barbarian gives way to civilized Man; and yet the power of adaptation is known to be greatest in the highest organisms, whereas the hypothesis of Lamarck would require the fact to be the other way. In a statical view, too, this conception would compel us to regard the introductory animal as containing, at least in a rudimentary state, not only all the tissues (which might be admissible, reducible as they are to the cellular tissue), but all the organs and systems of organs; which is incompatible with anatomical comparison. Thus, in every view is Lamarck's conception condemned: and it even tends to destroy the philosophical balance between the two fundamental ideas of organization and life, by leading us to suppose most life where there is least organization. The lesson that we may learn from it is to study more effectually the limits within which, in each case, the medium may modify the organism, about which a very great deal remains to be learned: and meantime, there can scarcely be a doubt, especially after the luminous exposition of Cuvier, that species remain essentially fixed through all exterior variations compatible with their existence.

Cuvier's argument rests upon two chief considerations, complementary to each other;—the permanence of the most ancient known species; and the resistance of existing species to the most powerful modifying forces: so that, first, the number of species does not diminish; and next, it does not increase. We go back for evidence to the descriptions of Aristotle, twenty centuries ago: we find fossil species, identical with those before our eyes: and we observe in the oldest mummies even the simple secondary differences which now distinguish the races of men. And, as to the second view, we derive evidence from an exact analysis of the effects of domestication on races of animals and vegetables. Human intervention, affording as it does the most favorable case for alteration of the organism, has done nothing more, even when combined with change of locality, than alter some of the qualities, without touching any of the essential characters of any species; no one of which has ever been transformed into any other. No modification of race, nor any influences of the social state, have ever varied the fundamental and strongly-marked nature of the human species. Thus, without straying into any useless speculations about the origin of the different organisms, we rest upon the great natural law that living species tend to perpetuate themselves indefinitely, with the same chief characteristics, through any exterior changes compatible with their existence. In non-essentials the species is modified within certain limits, beyond which it is not modified, but destroyed. To know thus much is good: but we must remember that it teaches us nothing, with any completeness, of the kind of influence exercised by the medium on the organism. The rational theory of this action remains to be formed: and the laying down the question was the great result of the Lamarck controversy, which has thus rendered an eminent service to the progress of sound biological philosophy.

We may now proceed on the authorized conception that the great biological series is necessarily discontinuous. The transitions may ultimately become more gradual, by the discovery of intermediate organisms, and by a better directed study of those already known: but the stability of species makes it certain that the series will always be composed of clearly distinct terms separated by impracticable intervals. It now appears that the preceding examination was no needless digression, but an inquiry necessary to establish, in the hierarchy of living bodies, this characteristic property, so directly involved in the rational establishment of the hierarchy itself.

Having surveyed the two great conception of Natural Groups, and the biological series, which together constitute what is called the Natural Method, we must now notice two great logical conditions of the study. *Two logical conditions of the study.* The first, or primordial, is the principle of the subordination of characters: the other, the final, prescribes the translation of the interior characters into exterior, which, in fact, results from a radical investigation of the same principle.

From the earliest use of the natural method, even before the investigation had passed on from the natural groups to the series of them, it was seen that the taxonomic characters must be not only numbered but weighed, according to the rules of a certain fundamental subordination which must exist among them. *Subordination of characteristics.* The only subordination which is strictly scientific, and free from all arbitrary intermixture, is that which results from a comparative analysis of the different organisms: this analysis is of recent date, and even yet is adequately applied only in the animal region; and thus the subordination was no more than barely conceived of before the institution of comparative anatomy, and the weighing of attributes is closely connected with the conception of the organic hierarchy. The subordination of taxonomic characters is effected by measuring their respective importance according to the relation of the corresponding organs to the phenomena which distinguish the species under study,—the phenomena becoming more special as we descend to smaller subdivisions. In short, here as elsewhere, the philosophical task is to establish a true harmony between statical conditions and dynamical properties; between ideas of life and ideas of organization, which should never be separated in our scientific studies but in order to their ulterior combination. Thus our aim, sometimes baffled but always hopeful, is to subordinate the taxonomic characters to each other, without the admission of anything arbitrary into any arrangement of importance. We thus meet with gaps in our schedules which we should avoid, or be insensible to, under an arbitrary system; but we may subdue our natural impatience under this imperfection by accustoming our minds to regard rational classification as a true science, continually progressive, always perfectible, and therefore always more or less imperfect, like all positive science.

In conducting the process of comparison, the characters must be admitted without restriction, in virtue of their positive rationality, however inconvenient to man- *Procedure, in use of the Natural Method.*

age and difficult to verify. This is the foundation of the proposed classification. The next step is to discard from the collection those whose verification would be too difficult, substituting for them some customary equivalents. Without this second process, which is as yet inadequately appreciated, the passage from the abstract to the concrete would be inextricably embarrassed. The anatomist and physiologist may be satisfied with a definition of groups which will not suit the zoologist, and still less the naturalist. The kind of transformations required is easily specified. First, it is clear how important it is to discard the characters which are not permanent, and those which do not belong to the various natural modifications of the species under study. They can be admitted only as provisional attributes till true equivalents, permanent and common, have been discovered. But the very nature of the problem indicates that the aim of the chief substitution should be to replace all the interior characters by exterior: and it is this which constitutes the main difficulty, and, at the same time, the highest perfection of this final operation. When such a condition is fulfilled, on the basis of a rational primitive classification, the natural method is irrevocably constituted, in the plenitude of its various essential properties, as we now find it in the case of the animal kingdom.

This transformation appears to be necessarily possible; for, as a chief characteristic of animality is action upon the external world and corresponding reaction, the most important primitive phenomena of animal life must take place at the surface of separation between the organism and its environment; and considerations with regard to this envelope, its form, consistence, etc., naturally furnish the principal distinctions of the different animal organizations. The interior organs, which have no direct and continuous relation to the medium, will always be of the highest importance among vegetative phenomena, the primitive and uniform basis of all life: but they are of secondary consequence in considering the degrees of animality; so that the interior part of the animal envelope, by which various materials for use are elaborated, is less important, in a taxonomic point of view, than the exterior part, which is the seat of the most characteristic phenomena. Accordingly, the transformation of interior into exterior zoological characters is not merely an ingenious and indispensable artifice, but a simple return from the distraction of an overwhelming mass of facts to a direct philosophical course of investigation. When therefore we see a recourse to interior characteristics in the study of the animal series, we must recognise the truth that not only is the classification as yet unfinished, but that the operation is imperfectly conceived of; that the inquirer has not ascended, through sound biological analysis, to the original source of analogies empirically discovered.

After thus ascertaining the nature of the Natural Method, we must, before quitting the subject, glance at the mode of its application in the co-ordination of the biological series, condensed into its principal masses.

The most general division of the organic world is into the animal and vegetable kingdoms; a division which remains as an instance of thorough discontinuity, in spite of all efforts to represent it as an artificial arrangement. *Division of animal and vegetable kingdoms.* The deeper we go in the study of the inferior animals, the more plainly we perceive that locomotion, partial at least, and a corresponding degree of general sensibility, are the predominant and uniform characters of the entire animal series. These two attributes are even more universal in the animal kingdom than the existence of a digestive canal, which is commonly regarded as its chief exclusive characteristic: a predominance which would not have been assigned to this attribute of the organic life but for its being an inevitable consequence, and therefore an unquestionable test, of the double property of locomotion and sensibility; to which we must, in consequence, assign the first place. Such a transformation, however, relates only to moveable animals; so that for the rest, we should have still to seek some other yet more general indication of universal animality, if we must despair of finally discovering in it every direct anatomical condition of these two animal properties. As for the case of certain gyrating plants which appear to manifest some signs of these properties, our imperfect analysis of their motions discloses no true character of animality, since we can discover no constant and immediate relation with either exterior impressions or the mode of alimentation.

Next to the division of the two organic kingdoms comes the question of the rational hierarchy of the animal kingdom, by itself. *Hierarchy of the animal kingdom.* In the place of the irrational considerations, so much relied on formerly, of abode, mode of nutrition, etc., we now rest upon the supreme consideration of the greater or less complexity of the organism, of its relative perfection, speciality, elevation; in short, of the degree of animal *dignity*, as M. Jussieu has well expressed it. The next preparatory step was in the anatomical field, to determine the successive degrees of animality proper to the different organs. The combination of the two great inquiries,—into the bases of the zoological hierarchy as residing in the organization, and into the rank of the organs in their relation to life,—has furnished, since the beginning of this century, the first direct and general sketch of a definitive graduation of the animal kingdom. Henceforth, it became admitted that, as the nervous system constitutes the most animal of the anatomical elements, the classification must be directed by it; other organs, and, yet more, inorganic conditions, being recurred to only on the failure of the chief in the most special subdivisions; and the substitutes being employed according to their decreasing animality. Whatever share other zoologists may have contributed by their labors to the formation of this theory, it is to M. de Blainville that the credit of it especially belongs: and it is by his classification that we must proceed in estimating the application of the Natural Method to the direct construction of the true animal hierarchy.

Attribute of symmetry. The happiest innovation which distinguishes this zoological system is that it attributes a high taxonomic importance to the general form of the animal envelope, which had before been neglected by naturalists, and which offers the most striking feature, in regard to description, in the symmetry which is the prevailing character of the animal organism. We must here reserve the case of the non-symmetrical animals; and this shows that the idea is insufficiently analyzed as yet. The principle is perhaps saved, or the difficulty distanced, by the fact that in these animals no trace can be discovered of a nervous system; but there is a sufficient want of precision and clearness to mark this as a case reserved for further analysis. We shall not wonder at this imperfection if we remember how erroneous were the notions, no further back than two generations ago, about very superior orders of animals—the whole of the radiated, a part of the mollusks, and even of the lower articulated animals. Among the orders thus restricted, there are two kinds of symmetry, the most perfect of which relates to a plane, and the other to a point, or rather to an axis: hence the further classification of animals into the duplicate and the radiated. It is impossible to admire too much the exactness with which an attribute, apparently so unimportant, corresponds with the aggregate of the highest biological comparisons, which are all found spontaneously converging toward this simple and luminous distinction. Still it remains empirical in its preponderance; and we yet need a clear and rational explanation, both physiological and anatomical, of the extreme necessary inferiority of the radiated to the duplicate animals, which, by their nature, must be nearer to Man, the fundamental unity in zoology.

Division of duplicate animals. Taking the duplicate animals, or *artiozoaries*, their order is again divided according to the consistence of the envelope—whether it is hard or soft—and therefore more or less fit for locomotion. This is, in fact, a protraction of the last consideration, as symmetry must be more marked in the case of a hard than of a soft covering. The two great attributes of animality—locomotion and sensation—establish profound and unquestionable differences, anatomical and physiological, between these two cases; and we may easily connect them rationally with this primitive distinction, and perceive how they exhibit inarticulated animals as necessarily inferior to the articulated.

Division of articulated animals. The articulated animals must next be distinguished into two great classes, according to the mode of articulation; whether under the envelope, by a bony skeleton, or a cartilaginous one, in the lowest degree; or whether the articulation is external, by the consolidation of certain horny parts of the envelope, alternating with the soft parts. The inferiority of this latter organization, especially with regard to the high functions of the nervous system, must be seen at a glance. It is observable that the more imperfect development of this eminently animal system always coincides with a fundamental difference in the position of its central part, which is always above the digestive canal in vertebrated animals, and below it in those which have an external articulation.

The rational hierarchy of the chief organisms in the upper part of the animal series is, then, composed of the three great classes; the vertebrated animals, those which are articulated externally, and the mollusks; or, in scientific language, the *osteozoaries*, the *entomozoaries*, and the *malacozoaries*.

Glancing, finally, at the division of the first of these classes, I may remark that all former descriptions and definitions may merge in the consideration of the envelope. It will be enough here to refer merely to the secondary view of the envelope that of the inorganic productions which separate it from its environment. M. de Blainville has shown us how the descent from Man, through all the mammifera, the birds, the reptiles, amphibious animals, and fishes, is faithfully represented by the consideration of a cutaneous surface furnished with hair, feathers, scales, or left bare. The same determining importance of the envelope is perceived in the next order of animals, in which the descent is measured by the increasing number of pairs of locomotive appendages, from the hexapods to the myriapods, and even to the apodes, which are at the lower extremity.

Consideration of the envelope.

It is not consistent with the object of this Work to go further into a description of the animal hierarchy. My aim in giving the above details has been to fix the reader's attention on my preliminary recommendation to study the present co-ordination of the animal kingdom as an indispensable concrete explanation of the abstract conceptions which I had offered, in illustration of the natural method. We must pass by, therefore, all speculations and studies which belong to zoological philosophy, and merely observe that there is one portion of the fundamental system which remains to be constituted, and the general principles of which are as yet only vaguely perceived;—I mean the rational distribution of the species of each natural genus. This extreme and delicate application of the taxonomic theory would have been inopportune at an earlier stage of the development of the science: but the time has arrived for it to be undertaken now.

We can not but see that the natural method does not admit of anything like the perfection in the vegetable kingdom that it exhibits in even the lower stages of the animal. The families may be regarded as established, though in an empirical way; but their natural co-ordination remains almost entirely arbitrary, for want of a hierarchical principle by which to subordinate them rationally. The idea of animality yields a succession of degrees, deeply marked, so as to supply the basis of a true animal hierarchy: but there is nothing of the kind in our conception of vegetable existence. The intensity in this region is not always equal; but the character of vegetable life is homogeneous;—it is always assimilation and the contrary, continuous, and issuing in a necessary reproduction. The

Natural method applied to the Vegetable kingdom.

Difficulty of co-ordination.

mere differences of intensity in such phenomena can not constitute a true vegetable scale, analogous to the animal; and the less because the gradation is owing at least as much to the preponderant influence of external circumstances as to the characteristic organization of each vegetable. Thus, we have here no sufficient rational basis for a hierarchical comparison. A second obstacle ought to be noticed—serious enough, though of less importance than the first;—that each vegetable is usually an agglommeration of distinct and independent beings. The case does not resemble that of the polypus formation. In the compound structure of the lowest animal orders a scientific definition is still possible. There is a vital basis common to all the animal structures which are otherwise independent of each other: but in the vegetable case, it is a mere agglomeration, such as we can often produce by grafting, and where the only common elements are inorganic parts, aiding a mechanical consolidation. There is no saying, in the present state of our knowledge, how far such a system may extend, without being limited by any organic condition, as it seems to depend on purely physical and chemical conditions, in combination with exterior circumstances. Faintly marked as the original organic diversity is by nature, it is evident how all rational subordination of the vegetable families in a common hierarchy is impeded by the coalescent tendency just noticed.

The principal division, which is the starting-point of M. Jussieu's classification, is the only beginning of a true co-ordination in the vegetable kingdom. It consists in distinguishing the vegetables by the presence or the absence of seminal leaves; and when they are present, by their having several or only one. For the successive passage from those which have several to those that have none may be regarded as a continuous descent, like that of the biological series, though less marked. Such a view has been verified by the investigation of the organs of nutrition, according to the discovery of Desfontaines—as yet the only eminent example of a large and happy application of comparative anatomy to the vegetable organism. By this concurrence of the two modes of comparison—of the reproductive and nutritive arrangements—this proposition has taken its rank among the most eminent theorems of natural philosophy. But this beginning of a hierarchy remains obviously insufficient—the numerous families in each of the three divisions remaining under a purely arbitrary arrangement, which we can hardly hope to convert into a rational one. The interior distribution of species, and even of genera, within each family must be radically imperfect, as the requisite taxonomical principles can not be applied to their arrangement till the difficulty of the co-ordination of the families—a difficulty much less, but as yet insurmountable—has been overcome. The Natural Method has, therefore, as yet yielded no other result, as to the vegetable kingdom, than the more or less empirical establishment of families and genera. We can not be surprised that it has not yet excluded the use of artificial methods, and above all that of Linnæus—true as it is that, up to our time, the co-ordination of the vegetable kingdom was the field for the application of the Natural Method. It should ever be remembered, however, that the Natural Method is not merely a means of classification, but an important system of real knowledge as to the true relations of existing beings: so that even if it should be disused for the purposes of descriptive botany, it would not the less be of high value for the study of plants, the comparative results of which would be fixed and combined by it. In the present condition and prospects of the science, as to the establishment of a vegetable series, we must take the whole vegetable kingdom together as the last term of the great biological series—as the last of the small number of essential modes of organization which (when the subdivisions are disregarded) are markedly separated from each other in the classification that gives us the logical command of the study of living beings. Applied first to the vegetable kingdom, the natural method is now seen to be the means by which the animal realm, the type of all our knowledge of organic life, is to be perfected. To whatever orders of phenomena natural classification is to be applied, here its theory must first be studied; and hence it is that biological science bears so important a part in the advancement of the whole of the positive method. It is much that a considerable progress has been made in ranging, in a due order of dignity, the immense series of living beings, from Man to the simplest plant: but, moreover, this theory of classification is an indispensable element of the whole positive method; an element which could not have been developed in any other way, nor even otherwise appreciated.

CHAPTER IV.

ORGANIC OR VEGETATIVE LIFE.

We have to pass on to dynamical biology, which is very far indeed from having attained the clearness and certainty of the statical department of the science. Important as are the physiological researches of recent times, they are only preliminary attempts, which must be soundly systemized before they can constitute a true dynamical biology. The minds which are devoted to mathematical, astronomical, and physical studies, are not of a different make from those of physiologists; and the sobriety of the former classes, and the extravagance of the latter, must be ascribed to the definite constitution of the simpler sciences and the chaotic state of physiology. The melancholy condition of this last is doubtless owing in part to the vicious education of those who cultivate it, and who go straight to the study of the most com-

Condition of Dynamical Biology.

plex phenomena without having prepared their understandings by the practice of the most simple and positive speculation: but I consider the prevalent license as due yet more to the indeterminate condition of the spirit of physiological science. In fact, the two disadvantages are one; for if the true character of the science were established, the preparatory education would immediately be rectified.

This infantine state of physiology prescribes the method of treating it here. I can not proceed, as in statical biology, to an analytical estimate of established conceptions. I can only examine, in pure physiology, the notions of method; that is, the mode of organization of the researches necessary to the ascertainment of the laws of vital phenomena. The progress of biological philosophy depends on the distinct and rational institution of physiological questions, and not on attempts, which must be premature, to resolve them. Conceptions relating to method are always important in proportion to the complexity of the phenomena in view; therefore are they especially valuable in the case of vital phenomena; and above all, while the science is in a nascent state.

Vital phenomena: their division. Though all vital phenomena are truly interconnected, we must, as usual, decompose them for purposes of speculative study, into those of greater and those of less generality. This distinction answers to Bichat's division into the organic or vegetative life, which is the common basis of existence of all living bodies; and animal life, proper to animals, but the chief characters of which are clearly marked only in the higher part of the zoological scale. But, since Gall's time, it has become necessary to add a third division—the positive study of the intellectual and moral phenomena which are distinguished from the preceding by a yet more marked speciality, as the organisms which rank nearest to Man are the only ones which admit of their direct exploration. Though, under a rigorous definition, this last class of functions may doubtless be implicitly included in what we call the animal life, yet its restricted generality, the dawning positivity of its systematic study, and the peculiar nature of the higher difficulties that it offers, all indicate that we ought, at least for the present, to regard this new scientific theory as a last fundamental branch of physiology; in order that an unseasonable fusion should not disguise its high importance, and alter its true character. These, then, are the three divisions which remain for us to study, in our survey of biological science.

Theory of Organic Media. Before proceeding to the analysis of organic or vegetative life, I must say a few words on the theory of organic media, without considering which, there can be no true analysis of vital phenomena.

This new element may be said to have been practically introduced into the science by that controversy of Lamarck, already treated of, about the variation of animal species through the prolonged influence of external circumstances. It is our business here to exclude from the researches thus introduced, everything but what concerns physiology, properly so called, reduced to the abstract theory of the living organism. We have seen that the vital state supposes the necessary and permanent concurrence of a certain aggregate of external actions with the action of the organism itself: it is the exact analysis of these conditions of existence which is the object of the preliminary theory of organic media; and I think it should be effected by considering separately each of the fundamental influences under which the general phenomenon of life occurs. It can not be necessary to point out the importance of the study of this half of the dualism which is the condition of life; but I may just remark on the evidence it affords of the subordination of the organic to the inorganic philosophy; the influence of the medium on the organism being an impracticable study as long as the constitution of the medium is not exactly known.

The exterior conditions of the life of the organism are of two classes—physical and chemical; or, in other words, mechanical and molecular. Both are indispensable; but the first may be considered, from their more rigorous and sensible permanence, the most general—if not as to the different organisms, at least as to the continued duration of each of them. *Exterior conditions of Life.*

First in generality we must rank the action of Weight. There is no denying that Man himself must obey, whether as weight or projectile, the same mechanical laws that govern every other equivalent mass; and by reason of the universality of these laws, weight participates largely in the production of vital phenomena, to which it is sometimes favorable, sometimes opposed, and scarcely ever indifferent. There is great difficulty in the analysis of its effects, because its influence can not be suspended or much modified for the purpose; but we have ascertained something of them, both in the normal and the pathological states of the organism. In the lower, the vegetable portion of the scale, the physiological action of weight is less varied but more preponderant, the vital state being there extremely simple and least removed from the inorganic condition. The laws and limits of the growth of vegetables appear to depend essentially on this influence, as is proved by Mr. Knight's experiments on germination, as modified by a quicker or slower motion of rotation. Much higher organisms are subject to analogous conditions, without which we could not explain, for instance, why the largest animal masses live constantly in a fluid sufficiently dense to support almost their whole weight, and often to raise it spontaneously. However, the superior part of the animal series is least fit for the ascertainment of the physiological influence of weight, from its concurrence with a great number of heterogeneous actions; but this again enables us to study it in a variety of vital operations: for there is scarcely a function, organic, animal, or even intellectual, in which we may not point out the indispensable intervention of weight, which specially manifests itself in all that relates to the stagnation or movement of *Mechanical conditions. Weight.*

fluids. It is, therefore, much to be regretted that a subject so extended and important has not been studied in a rational spirit and method.

Pressure. The next mechanical condition,—pressure, liquid or gaseous,—is an indirect consequence of weight. Some few scientific results have here been obtained, from the facility with which pressure may be modified by artificial or natural circumstances. There are limits in the barometrical scale, outside of which no atmospherical animal,—Man or any other,—can exist. We can not so directly verify such a law in the case of aquatic animals: but it would seem that in proportion to the density of the medium must be the narrowness of the vertical limits assignable to the abode of each species. Of the relation between these intervals and the degree of organization, it must be owned, however, that we have no scientific knowledge, our ideas being, in fact, wholly confused as to the inferior organisms, and especially in the vegetable kingdom. Though, through many difficulties and complexities, the science is in a merely nascent state, some inquiries, such as those relating to the influence of atmospheric pressure on the venous circulation, and recent observations on its co-operation in the mechanism of standing and moving, etc., show that biologists are disposed to study this order of questions in a rational manner.

Motion and rest. Among the physical conditions, and perhaps first among them, the physiological influence of motion and rest should be investigated. Amidst the confusion and obscurity which exist on this subject, I think we may conclude that no organism, even the very simplest, could live in a state of complete immobility. The double movement of the earth, and especially its rotation may probably be as necessary to the development of life as to the periodical distribution of heat and light. Too much care, however, can not be taken to avoid confounding the motion produced by the organism itself with that by which it is affected from without; and the analysis had therefore better be applied to communicated than spontaneous motion. And as rotary motion tends, by the laws of mechanics, to disorganize any system, and therefore, eminently, to trouble its interior phenomena, it is this kind of motion which may be studied with the best result; for which object we should do well to investigate, in a comparative way, the modifications undergone by the principal functions from the organism being made to rotate in such a gradual variety as is compatible with a normal state. The attempt has as yet been made only with plants, and for another purpose; while, in the case of the superior animals, including Man, we have only incomplete and disjointed observations, scarcely transcending mere popular notions.

Thermological action. After the mechanical influences, we reach one which affects structure—the thermological action of the medium. It is the best known of all; for nothing is plainer than that life can exist only within certain limits of the thermometrical scale, and that there are limits affecting every family, and even every living race; and again, that the distribution of organisms over our globe takes place in zones sufficiently marked, as to differences of heat, to furnish thermometrical materials to the physicists, in a general way. But amidst the multitude of facts in our possession, all the essential points of rational doctrine are still obscure and uncertain. We have not even any satisfactory series of observations about the thermometrical intervals corresponding to the different organic conditions;—much less any law relating to such a harmony, which has never, in fact, been connected with any other essential biological character. This great gap exists as much with regard to the successive states of the same organism as to the scale of organisms. The necessary revision might be best applied first to the lowest states; as the egg and the lowest organisms appear able to sustain wider differences of temperature than those of a higher order; and several philosophical biologists have even believed that life may have been always possible on our planet, notwithstanding the different systems of temperature through which its surface has successively passed. On the whole, the sum of our analysis may seem to disclose, amidst many anomalies, a general law: that the vital state is so subordinated to a determinate thermometrical interval, as that this interval perpetually diminishes as life becomes more marked,—in the case both of the individual and of the series. If even this general law is not yet scientifically established, it may be supposed how ignorant we are of the modifications produced in the organism by variations of external temperature, within the limits compatible with life. There has even been much confusion between the results of abrupt and gradual changes of temperature, though experiment has shown that graduation vastly expands the limits within which the human organism can exist; and again, between the influence of external, and the organic production of vital heat. This last great error shows that even the laying down the question remains to be done. The same thing may be said of the other exterior conditions, such as light and electricity, of which all that we know in this connection is that they exert a permanent influence needful for the production and support of life. *Light, Electricity, etc.* Besides the confusion and uncertainty of our observations, we have to contend with the inferiority of our knowledge of those branches of physics, and with the mischief of the baseless hypothesis which we before saw to infest the study of them. While physicists talk of fluids and ethers, avowing that they do so in an artificial sense, for purposes of convenience, physiologists speak of them as the real principles of two orders of exterior actions indispensable to the vital state. Till reform becomes substantial and complete in the study of light and electricity, these will remain the exterior conditions of vitality of which our knowledge is the most imperfect.

Passing on from the physical to the chemical conditions of the medium, we find our amount of knowledge scarcely more satisfactory. In strict generality, this study relates to the physiological influence of air and water, the *Molecular conditions. Air and water.*

mingling of which in various degrees, constitutes the common medium necessary to vitality. As M. de Blainville remarked, they must not be considered separately, as in a physical or chemical inquiry, but in that mixture which varies only in the proportions of its elements. This might be anticipated from our knowledge of the chemical constitution of living bodies, the essential elements of which are found only in the combination of air and water: but we have physiological evidence also, which shows that air deprived of moisture, and water not aerated, are fatal to vital existence. In this view there is no difference between atmospheric and aquatic beings, animal and vegetable, but the unequal proportion of the two fluids; the air, in the one case, serving as a vehicle for vaporized water; and the water, in the other case, conveying liquefied air. In both cases, water furnishes the indispensable basis of all the organic liquids: and the air the essential elements of nutrition. We know that the higher mammifera, including Man, perish when the air reaches a certain degree of dryness, as fishes do in water which has been sufficiently deprived of air by distillation. Between these extreme terms there exists a multitude of intermediaries in which moister conditions of air and more aerated states of water correspond with determinate organisms; and the observation of Man in the different hygrometrical states of atmospheres shows how, in the individual case, physiological phenomena are modified within the limits compatible with the vital state. If we may say that the question has been laid down in this inquiry, it is only in a vague and obscure way. Besides our ignorance of the varying proportions, we have none but the most confused notions about the way in which each fluid participates in the support of life. Oxygen is the only element of the air about which we have made any intelligent inquiry, while physiologists entertain the most contradictory notions about azote: and the uncertainty and obscurity are still greater with regard to water. In this state of infancy, it can be no wonder that the science offers as yet no law as to the influence of the medium on the organism—even in regard to the question whether a certain condition of existence becomes more or less inevitable as the organism rises in the scale.

Study of specifics. The study of the influence of specifics does not enter here, on account, of course, of the absence of generality; but it should be just pointed out that, reduced as is the number of substances called specifics, there are still enough,—as aliments, medicines, and poisons,—to afford a hint of what might be learned by an exploration of them, in regard to the harmony between the organic world and the inorganic. The very quality of their operation, that it is special and discontinuous, and therefore not indispensable, indicates the experimental method in this case, as being certain, well circumscribed, and very various. This study may then be regarded as a needful appendix, completing the preliminary biological doctrine which I have called that of organic media, and offering resources which are proper to it, and can not be otherwise obtained. Unhappily, this complement is in even a more backward state than the more essential portions, notwithstanding the multitude of observations, unconnected and unfinished, already assembled in this path of research.

If such is the state of preliminary knowledge, it is clear how little has yet been learned of the laws of life themselves. The inquiry has gone through revolutions, as other questions have, before reaching the threshold of positivity, in our day. From the impulse given by Descartes, the illustrious school of Boerhaave arose in physiology, which exaggerated the subordination of biology to the simpler parts of natural philosophy so far as to assign to the study of life the place of appendix to the general system of inorganic physics. From the consequent reaction against this absurdity arose the theory of Stahl, which may be considered the most scientific formula of the metaphysical state of physiology. The struggle has since lain between these two schools, —the strength of the metaphysical one residing in its recognition of physiology as a distinct science, and that of the physico-chemical, in its principle of the dependence of the organic on the inorganic laws, as daily disclosed more fully by the progress of science. The effect of this improved knowledge has been to modify the conceptions of metaphysical physiology: the formula of Barthez, for instance, representing a further departure from the theological state than that of Stahl; as Stahl's already did than that of Van Helmont, though the same metaphysical entity might be in view when Van Helmont called it the *archeus*, and Stahl the *soul*, and Barthez the *vital principle*. Stahl instituted a reaction against the physico-chemical exaggerations of Boerhaave; but Barthez established, in his preliminary discourse, the characteristics of sound philosophizing, and exposed the necessary futility of all inquisition into causes and modes of production of phenomena, reducing all real science to the discovery of their laws. For want of the requisite practice in the positive method, the scheme of Barthez proved abortive; and, after having proposed his conception of a vital principle as a mere term to denote the unknown cause of vital phenomena, he was drawn away by the prevalent spirit of his time to regard the assumed principle as a real and complex existence, though profoundly unintelligible. Ineffectual as his enterprise proved, its design with regard to the advancement of positive science can not be mistaken. The progressive spirit is still more marked in the physiological theory of Bichat, though we find entities there too. These entities, however, show a great advance, as a determinate and visible seat is assigned to them. The vital forces of Bichat, however, still intervene in phenomena, like the old specific entities introduced into physics and chemistry, in their metaphysical period, under the name of faculties or occult virtues, which Descartes so vigorously hunted down, and Molière so happily ridiculed. Such is the character of the supposed *organic sensibility*, by which, though a mere term, Bichat endeavored to *explain* physiological phenomena, which he thus merely reproduced under another name: as when, for instance, he thought he had accounted for the successive flow of different liquids in one canal by saying that the organic sensibility of the canal was successively in harmony with each fluid, and in antipathy to the rest. But for his untimely death, however, there can be no doubt that he would have issued into an entire positivity. His treatise on General Anatomy, though appearing a very few years after his treatise on Life and Death, is a great advance upon it; and even in the construction of his metaphysical theory of vital forces he certainly first introduced, under the title *properties of tissue*, a conception of the highest value, destined to absorb all ontological conceptions, and to prepare for the entire positivity of the elementary notions of physiology. The thing required is to substitute *properties* for *forces*; and Bichat's treatment of tissue fulfilled this condition with regard to a very extensive class of effects: and thus his theory, while it amended the metaphysical doctrine of Stahl and Barthez, opened the way to its entire reformation by presenting at once the germ and the example of purely positive conceptions. This is now the state of physiological philosophy in the minds of the majority of students; and the conflict between the schools of Stahl and Boerhaave,—between the metaphysical and the physico-chemical tendency,—remains at the point to which it was brought up by the impulse communicated by Bichat. It would be hopeless to look to the oscillations of this antagonism for an advance in science. If the one doctrine prevailed, science would be in a state of retrogression; if the other, in a state of dissolution; as in our social condition, in the conflict of the two political tendencies, the retrograde and the revolutionary. The progress of physiology depends on the growth of positive elementary conceptions, such as will remand to the domain of history the controversy from which nothing more is to be expected. Abundant promise of such an issue now appears: the two schools have annulled each other; and the natural development of the science has furnished means for its complete institution to be begun. This I look upon as the proper task of the existing generation of scientific men, who need only a better raining to make them adequate to it. If, from its complexity, physiology has been later than other sciences in its rational formation, it may reach its maturity more rapidly from the ground having been cleared by the pursuit of the anterior sciences. Many delays were occasioned in their case by transitory phases which were not understood in the earlier days of positivity, and which need never again arrest experienced investigation. It may be hoped that physiologists will spare the science the useless and humbling delay in the region of metaphysical hypothesis which long embarrassed the progress of physics.

Philosophical character of Physiology. The true philosophical character of physiology consists, as we have seen, in establishing an exact and constant harmony between the statical and the dynamical points of view,—between the ideas of organization and of life,

—between the notion of the agent and that of the act; and hence arises the obligation to reduce all abstract conceptions of physiological *properties* to the consideration of elementary and general phenomena, each of which conveys the idea of a determinate seat. In the other words, the reduction of *functions*, to corresponding *properties* must be regarded as the simple consequence of decompounding the general life into the different functions,—discarding all notions about *causes*, and inquiring only into *laws*. Bichat's conception of the properties of tissue contains the first germ of this renovated view; but it only indicates the nature of the philosophical operation, and contains no solution of the problem. Not only is there a secondary confusion between the properties of tissue and simple physical properties, but the principle of the conception is vitiated by the irrational distinction between the properties of tissue and vital properties; for no property can be admitted in physiology without its being at once vital and belonging to tissue. In endeavoring to harmonize the different degrees of physiological and of anatomical analysis, we may lay down the philosophical principle that the idea of *property* which indicates the last term of the one must correspond with *tissue*, which is the extreme term of the other; while the idea of *function*, on the other hand, corresponds to that of *organ:* so that the successive ideas of function and of property present a gradation of thoughts similar to that which exists between the ideas of organ and of tissue, except that the one relates to the act and the other to the agent. This relation appears to me to constitute an incontestable and important rule in biological philosophy; and on it we may establish the first great division among physiological properties. We have seen how in anatomy there is a division between the fundamental, generating tissue, the cellular, and the secondary tissues which result from the combination of certain substances with this original web; and in the same way must physiological properties be divided into two groups,—the one comprising the general properties which belong to all the tissues, and which constitute the proper life of the cellular tissue; and the other, the special properties which characterize its most marked modifications,—that is, the muscular and nervous tissues. This division, indicated by anatomy, strikingly agrees with the great philosophical distinction between the organic or vegetative and the animal life; as the first order of properties must afford the basis of that general life, common to all organized beings, to which vegetable existence is reduced; while the second relates exclusively to the special life of animated beings. Such a correspondence at once makes the principle more unquestionable, and facilitates the application of the rule.

Division of the study.

If we look at what has been done, toward the construction of this fundamental theory, we shall find that it is fairly accomplished with regard to the secondary, or animal tissues,—all the general phenomena of animal life being unanimously connected with irritability and sensibility,—these being considered as attributes each of a definite tissue: and thus, the most marked case is the best understood. But the other division,—the properties which are wholly general, belonging to the universal life, are far more important, as underlying the others; and an extreme confusion and divergence exist with regard to them. No clear and satisfactory conception of the second class can be formed while the first is left in obscurity; and thus, the science remains in a purely provisional state,—its development having taken place in an order inverse to that which its nature requires.

Two functions of the organic life. The functions which belong to the vegetative life are two,—the antagonism of which corresponds to the definition of life itself: first, the interior *absorption* of nutritive materials from the surrounding medium; whence results, after their assimilation, final nutrition; and secondly, the *exhalation* of molecules, which then become foreign bodies, to be parted with, or disassimilated, as nutrition proceeds. It appears to be an error to make digestion and circulation characteristics of animality; as we certainly find them here in the fundamental sense of both. Digestion is properly a preparation of aliment for assimilation; and this takes place in a simple and almost unvaried manner in vegetable organisms: and circulation, though nothing like what it is in animals, where there is a central organ to effect it, is not less essential in vegetative life,—the lowest organism showing the continual motion of a fluid holding in suspension or dissolved, matters absorbed or thrown out; and this perpetual oscillation, which does not require a system of vessels to itself, but may take place through the cellular tissue, is equally indispensable to animal and vegetable existence. These, then, are the two great vegetative processes, performed by properties which are provisionally supposed (after the analysis of M. de Blainville, which is open to some objections) to be three,—hygrometricity, capillarity, and retractility. This analysis shows clearly that the actions which constitute vegetable life are simple physico-chemical phenomena; physical as to the motion of the molecules inward and outward; and chemical in what relates to the successive modifications of these different substances. Under the first aspect, they depend on the properties, hygrometrical, capillary, and retractile, of the cellular tissue: under the second, and much more obscure at present, they relate to the molecular action which its composition admits. This is the spirit in which the analysis of organic phenomena should be instituted; whereas that of animal phenomena should be regarded from a wholly different point of view, as we shall see hereafter.

The study of this vegetative life is not even yet rationally organized. We have seen that, in the anatomical view, the vegetable kingdom is regarded as the last term of a unique series,—the various degrees of which differ, for the most part, more widely from each other than any one of them from this extreme term. The same conception should direct physiological speculations on the organic life, analyzed uniformly for all living beings: but this has not hitherto been even attempted. Till it is accomplished, no essential point of physiological doctrine can be established, however able may be the investigations carried on, and however valuable the materials supplied. It may be alleged that the phenomena relating to the general life may be studied in the broadest simplicity in vegetable organisms: but it is no more possible in physiology than in anatomy to interpret the extreme cases in the scale by each other without having passed through the intermediate degrees: and the dynamical case is the more difficult of the two: so that the isolated study of the organic life in vegetables can not illustrate that of the higher order of animals. And one natural consequence of this irrational isolation of the vegetable case is that chemists and physicists have engrossed researches which properly belong to biologists alone. The comparative method, which we have seen to be the characteristic resource of biological philosophy, has not as yet been duly introduced into the general study of organic life, though it is at once more indispensable, and more completely applicable than in the case of animal life. If it were consistent with the character of this Work, we could point out gaps at almost every step, and about the simplest phenomena, which must shock any inquiring mind:—the darkness, doubts, and differences about digestion: and again about gaseous digestion, or respiration;—in regard to which the most contradictory opinions are held:—divergences about the simplest preliminary phenomena of vegetative life, which show how much has to be done before we can undertake any direct investigation into the phenomena of assimilation and the converse process.

We shall find ourselves even further from satisfaction if we turn from the consideration of the *functions* of organic life to those more compound phenomena which are usually confounded with them, but which M. de Blainville has taught us to distinguish as *results* from the action of, not one organ or set of organs, as in the case of function, but of the simultaneous action of all the principal organs. Of these results, the most immediate and necessary is the continuous state of composition and decomposition which characterizes the vegetative life. Ignorant as we are of assimilation and secretion, the very questions can not have been as yet suitably laid down. No one has thought, for instance, of instituting an exact chemical comparison between the total composition of each organism and the corresponding system of alimentation: nor, conversely, between the exhaled products and the whole of the agents which had supplied or modified them; so that we can give no precise scientific account of the general phenomenon of the composition and decomposition of every organism as a necessary consequence of the concurrence of the different functions. We have at present only incomplete and disjointed materials, which have never been referred to any general fact.

Results of organic action.

State of composition and decomposition.

It is acknowledged now that all organisms have, more or less, the character which used to be ascribed

Vital heat.

to only the highest, of sustaining a determinate temperature, notwithstanding variations of heat in their environment; and this is a second result of the whole of the vegetative functions, which almost always co-exists with the first. But this important study is not only in a backward state, but ill-conceived. Besides the error before noticed, of confounding vital heat with the temperature of the medium, the fundamental character of the phenomenon appears to me to have been misconceived. Its modification by the animal functions can never be understood till it has been studied in its primitive universal manifestation in all living bodies, each of which represents a chemical centre, able to maintain its temperature against external influences, within certain limits, as a necessary consequence of the phenomena of composition and decomposition. This is doubtless the point of view from which the positive study of vital heat must be regarded; and to consider it under the modifications of animal life, is to place the accessory before the principal, and to propose views which are merely provisional, if not erroneous. In the most recent works upon this leading subject, the organic foundations are, it is true, more carefully considered: but the investigation can not be said to be duly instituted as long as the vegetable organism is not regularly introduced into it.

Electrical state. These remarks are even more applicable to the electrical study of living bodies. Here we find again, and with aggravation, the confusion between organic action and external influence, as well as the aberrations remarked on in physics about ethers and electric fluids. Here, too, we meet with the error observed upon in the last case, about the physiological origin of the phenomenon. And here, again, we are bound to conclude that a permanent electrization is ascribable to acts of composition and decomposition, notwithstanding the electrical variations of the medium. And again we find that the animal functions can only modify, by accelerating or augmenting, more or less, the fundamental phenomenon. But the electrical analysis of the organism is yet further than the thermological from being conceived of and pursued in a rational view.

Next follow the general phenomena which result in a less direct and necessary manner from the whole of the vegetative functions—the production and development of living bodies. Notwithstanding the original investigations of Harvey and of Haller, with regard to the superior animals, this investigation may be considered, owing to its complexity, to be more in the rear of a positive institution than any of the preceding. The tendency to search for causes and modes of production of phenomena, instead of for their laws, has acted with fatal effect here: and, amidst every kind of deficiency, the main cause of the obscurity of the case is, undoubtedly, that students have occupied themselves in looking for what can not be found. However, the labors of anatomists and zoologists have evidently prepared the way for a more rational study. It is even worthy of remark that some students who were most bent on the search into causes have been led on by the spread of the positive spirit, to spend their efforts on inquiries into ovology and embryology, which are assuming a more scientific character every day. Still, the preliminary requisite for the formation of doctrine—a fundamental analysis—remains unfulfilled; and the ascertainment of the laws of production and development is not, therefore, to be attempted at present. In the lowest departments of the scale, the multiplication of organisms takes place by a simple prolongation of any part of the parent mass, which is almost homogeneous; and in this extreme case, we understand the phenomenon to be analogous to every other kind of reproduction of the primitive cellular tissue. In the higher degrees of the scale, we are in the dark from the moment we depart from immediate observation; and when the simplest previsions are so radically uncertain and even erroneous as in this case, the science may be pronounced to be in a state of infancy, notwithstanding the imposing appearance of the mass of works accumulated for its illustration.

The comparative method has been applied in a yet more incomplete way to the phenomena of organic development. The question has never yet been laid down under a form common to all organisms, including the vegetable. The grave error is still committed of studying the development in the animal cases alone; so that the most eminently animal of the systems, the nervous, is represented as the first to appear in the embryo of the higher orders—a supposition adverse to the institution of any really general conception of the theory of development, and in direct opposition to one of the most constant laws of biological philosophy—the perpetual accordance between the chief phases of the individual evolution and the most marked successive degrees of the organic hierarchy; for in this last view the nervous tissue is seen to be the latest and most special transformation of the primitive tissues. The preliminary analysis of organic development is, then, still far from being conceived of in a rational spirit, governed by the high philosophical intention of reconciling, as much as possible, the various essential aspects of the science of living bodies.

To be complete, this analysis should evidently be followed by the inverse, and yet correlative study of the decline of the organism, from its maturity to its death. The general theory of death is certainly in a very backward state, since the ablest physiological researches on this subject have usually related to violent or accidental death; considered, too, in the highest organisms exclusively, and affecting functions and systems of organs of an essentially animal nature. As for the deterioration of the organic life, we have yet attained to only one initiatory philosophical glimpse, which exhibits it as a necessary consequence of life itself, by the growing predominance of the movement of exhalation over that of absorption, whence results gradually an exaggerated consolidation of the organism which was originally almost fluid, a process which, in the absence of more rapid influences, tends to produce a state of desiccation incompatible with all vital phenomena. Valuable, however, as is such a glimpse, it serves only to characterize the true nature of the question, by indicating the general direction of the researches which it requires. The important considerations relative to animal life could not be rationally introduced into such a subject till this preliminary doctrine shall have been established; as in regard to all the other points of view before examined.

Summary as this review has been, we have seen enough to be authorized to conclude that the backward state of physiological science is owing mainly to the vicious training of physiologists, and the irrational institution of their habitual labors. The circulation of the blood, the first general fact which gave birth to positive physiology, and the laws of the fall of bodies, the first acquisition of sound physics, are discoveries almost absolutely contemporaneous; and yet, what an immense inequality there is now in the progress of two sciences setting out from so similar a disclosure! Such a difference can not be attributed wholly to the greater complexity of physiological phenomena, and must have depended much also on the scientific spirit which directed their general study, to the level of which the greater number of those who cultivate it have been unable to rise. The phenomena of the vegetative life obviously require, both for their analysis and their explanation, an intimate combination of the leading notions of inorganic philosophy with physiological considerations, obtained through a thorough familiarity with the preliminary laws relating to the structure and classification of living bodies. Now, each of these inseparable conditions is, in our day, the separate property of a particular order of positive investigators. Hence we have, on the one hand, the supposed organic chemistry, a bastard study, which is only a rough first sketch of vegetable physiology, undertaken by inquirers who know nothing of the true subject of their labors; and, on the other hand, vague, incoherent, and partly metaphysical doctrines, of which physiology has been chiefly constituted by minds almost entirely destitute of the most indispensable preliminary ideas. The barren anarchy which has resulted from so vicious an organization of scientific labor would be enough of itself to testify to the direct utility of the general, and yet positive point of view which characterizes the foregoing survey.

CHAPTER V.

THE ANIMAL LIFE.

It was only by a late and long-prepared effort that the human mind could attain that state of abstraction and physiological generality necessary for the comprehension of all vital beings—from Man to the vegetable—as one series. It is only in our own day that a point of view so new and so difficult has been established; and as yet, among only the most advanced minds, even as regards the simplest general aspects of biology—in the statical study of the organism. It is not at all surprising that physiological comparison should have been first applied to the animal functions, because they first suggest its importance and possibility, however clearly it may afterward appear that the organic life at once requires and admits a larger and more indispensable application of the comparative method. Looking more closely, however, into this evident existing superiority of animal over organic physiology, we must bear in mind the distinction between the two elementary aspects of every positive study—the analysis of phenomena and their explanation. It is only with regard to the first that the animal life has been in reality better explored than the organic. It is not possible that the explanation of the most special and complex phenomena should be more advanced than that of the most simple and general, which serve as basis to the others. Such a state of the science would be in opposition to all the established laws of the human mind.

However imperfect the theory of organic phenomena still is, it is unquestionably conceived in a more scientific spirit than we find in any explanations of animal physiology. We have seen that the vegetative phenomena approach most nearly to the inorganic; and that the school of Boerhaave sinned only in exaggeration, proceeding from insufficient knowledge; and it must be by this time evident that this is the link be- *Transition from the inorganic to the organic.* tween the inorganic and the biological philosophy, by which we are enabled to regard the whole natural philosophy as forming a homogeneous and continuous body of doctrine. By a natural consequence, a wholly different view must be taken of the rational theories of animal life: that is, of the phenomena of irritability and sensibility, which offer no basis of analogy with inorganic phenomena. With regard to sensibility, no one will question this: and, as to irritability—though contraction may be seen as a movement occasioned by heat, and, yet more, by electricity, these phenomena must be carefully separated from the contractile effect of the irritable fibre which is a product of the nervous action; and especially when it is voluntary. Irritability is as radically foreign to the inorganic world as sensibility; with which, too, it is inseparably connected.

Primitive nervous properties. This double property is, then, strictly primitive in the secondary tissues, and therefore no more a subject of explanation than weight, heat, or any other fundamental physical property. Whenever we have a true theory of animal life, it will be by comparing all the general phenomena which are connected with this double property, according to their preparatory analysis, in order to discover their laws; that is, as in all other cases, their constant relations, both of succession and similitude. This will be done in order to the usual end of obtaining a rational prevision; the subject here being the mode of action of a given animal organism, placed in determinate circumstances; or, reciprocally, the animal arrangement that may be induced by any given act of animality. All attempts to explore the nature of sensibility and irritability are mere hinderances in the way of this final aim, by drawing off our attention from the laws of animality in a vain search after what can never be found.

Relation of the animal to the organic life. The true relation of the animal to the organic life must throughout be carefully kept in mind. This relation is double. The organic life first serves as the basis of the animal; and then as its general end and object. We have dwelt enough on the first, if even any one would think of contesting that, in order to move and feel, the animal must first live; and that the fundamental vegetative life could not cease without extinguishing the other. As for the second relation, it is evident that the phenomena of irritability and sensibility are directed by the general needs of the organic life, which they serve by procuring better materials, and by guarding against unfavorable influences. Even the intellectual and moral functions have usually no other primitive office. Without such a destination, these properties would either destroy the organism or themselves perish. It is only in the human species, and even there only under a high degree of civilization, that any kind of inversion of this order can be conceived of. In that case, the vegetative life is essentially subordinated to the animal, the development of which it is alone destined to aid; and this, it seems to me, is the noblest scientific notion that we can form of humanity, distinct from animality:—a transformation which can be safely considered as possible only by transferring to the whole species, or at least to society, the primitive end which, in the case of animals, is limited to the individual, or, at the utmost, to the family, as we shall see hereafter. It is only among a small number of men, and it is very far indeed from being a just matter of expectation from the whole species, that the intellect can acquire such a preponderance in the whole of the organism as to become the end and object of human existence. An exception so special, and so easy to explain in the case of Man, can not alter the universality of a consideration verified by the whole animal kingdom, wherein the animal life is seen to be always destined to perfect the organic. It is only by a scientific abstraction, necessary for purposes of progress, that we can provisionally conceive of the first as isolated from the second, which is, strictly speaking, inseparable from it under the double aspect just exhibited. Thus, as the positive theory of animality must continually rest on that of general vitality, it is indissolubly combined with the whole of inorganic philosophy, which furnishes the basis of organic physiology. In a secondary sense, the same dependence exists. We admitted, while reviewing mathematical philosophy, that the laws of equilibrium and motion operate among all orders of phenomena, being absolutely universal. Among physiological phenomena we find them accordingly; and, when contraction is produced by the irritability of the muscular fibre, all the phenomena of animal mechanics which result, whether for rest or locomotion, are dependent on the general laws of mechanics. In an inverse way the same thing takes place with regard to the functions of sensibility, in which the inorganic philosophy must intervene in connection with the primitive impression on the sentient extremities, carefully distinguished from its transmission by the nervous filament, and its perception by the cerebral organ. This impression acts through an intermediate physical apparatus, optical, acoustic, or other, the study of which according to appropriate physical laws, constitutes a chief element of the positive analysis of the phenomenon. Not only must we use the knowledge already established, but we want, for our analysis, further progress in it, and even the creation of new doctrines, as the theory of flavors, and yet more of odors, in regard to the mode of propagation of which there are doubtless several general laws, of a purely inorganic character, remaining to be established. In investigating these connections between biology and inorganic science, we find again what we saw before, that chemistry is spontaneously related to vegetable physiology, and physics especially to animal physiology; though neither could be altogether dispensed with in either department, where they are required, more or less, in combination.

To inorganic philosophy.

Our ideas of the double property of irritability and sensibility can not be truly scientific till each is irreversibly assigned to a corresponding tissue. Bichat conceived of all tissues as necessarily sensitive and irritable, but in different degrees;—an error which was natural or inevitable at a time when so little was known of tissue in the way of anatomical analysis, but one which, if maintained now, would hand over the whole science to the physico-chemical school, and efface all real distinction between the inorganic and organic departments of natural philosophy. Rational biology requires that the two properties should be inherent in determinate tissues—themselves modifications, profound and distinctly marked, of the primitive cellular tissue—that our anatomical data may be in harmony with the physiological; in other words, that the elementary ideas of tissue and of property should be in perfect corre-

Properties of tissue.

spondence. The scientific character of physiology in this direction is essentially defective among biologists in general. But the new explorations continually made show us how it was that Bichat was misled. He considered it proved that sensibility existed where there were no nerves: but further investigation proves that the symptoms of sensibility were erroneously attributed to an organ deprived of nerves, instead of being referred to the simultaneous injury of neighboring nerves; or that the nervous tissue existed, though it was difficult to find. If cases apparently contradictory still remain, it would be obviously absurd to reject on their account a conception required by the principles of rational physiology, and founded on unquestionable cases, by far more numerous and decisive than those which still seem to be exceptional. This consideration should be applied to different organisms, as well as to the different tissues of the human organism. The animals supposed to be without nerves, on which the metaphysical school has insisted so much, disappear as comparative anatomy enables us to generalize more and more the idea of nervous tissue, and to detect it in the inferior organisms. It is thus, for instance, that it has been recently found in several radiated animals. The time has come for its being established as a philosophical axiom that nerves are necessary for any degree of sensibility, the apparent exceptions being left as so many anomalies to be resolved by the future progress of anatomical analysis.

Sensibility.

The same process must be instituted with the common notions of irritability, which are still ruled by Bichat's theory. He supposed, for instance, that the contractions of the heart were determined, independently of all nervous action, by the immediate stimulus of the flow of the blood toward it; whereas, it is now established that a provision of nerves is as indispensable to the irritability of this muscle as of any other; and generally, that the great distinction laid down by Bichat, between organic and animal contractility, must be abandoned. All irritability is then necessarily animal; that is, it requires a corresponding nervous provision, whatever may be the immediate centre from which the nervous action proceeds. Much illustration of this subject is needed for its scientific use, though not for the logical sanction of a principle already placed beyond dispute. We need this further enlightenment, not only in regard to the use of the modern distinction made by many physiologists between the sensory and the motory nerves, though such a question has considerable philosophical importance; but much more in regard to another consideration, more direct and more eminent, in regard to which we are in a state of most inconvenient uncertainty and obscurity: I mean the scientific distinction which must be maintained, sooner or later, between the voluntary and involuntary motions. The doctrine of Bichat had the advantage of representing this difference, which we see, in fact, to have furnished him with his chief arguments; whereas, now that we insist on irritablity being of one kind only, and de-

Irritability.

pendent on a nervous provision, we find ourselves involved in a very delicate fundamental difficulty, the solution of which is however indispensable, to enable us to understand how all motions must not be indistinctly voluntary. For this solution we must obtain—what we certainly have not as yet—an exact co-ordination of anatomical differences with incontestable physiological differences. There can be no question that such a phenomenon as the voluntary movements of the locomotive muscles while that of the cardiac muscle remains absolutely involuntary, must admit of analysis, however difficult it may be. Here then we find a chasm among the very principles of the science, by which the positive theory of irritability is much perplexed, certain as is its principle. In almost all cases the ablest anatomist is unable to decide otherwise than by the fact itself, if any definite motion is necessarily voluntary or involuntary; which affords sufficient proof of the absence of any real law in the case. The solution will probably be obtained by an analysis of the intermediate motions, as we may call them,—those which, involuntary at first, end in becoming voluntary; or the reverse. These cases, which are very common, appear to me eminently fit to prove that the distinction between voluntary and involuntary motions arises from no radical difference of muscular irritability, but only from the mode, and perhaps the degree of innervation, modified by long habit. If this be as generally true as it seems to be in some cases of acquired control, it must be supposed that the most involuntary motions, which are those most indispensable to life, would have been susceptible of voluntary suspension (not even excepting the motions of the heart) if their incessant rigorous necessity had not hindered the contraction of suitable habits in their case. While we conclude it to be probable that the difference between the two kinds of motion proceeds indirectly from the action of the entire nervous system upon the muscular system, we can not help perceiving how greatly science stands in need of a thorough new examination into this obscure fact.

This brief survey shows us the general imperfection of the study of animality. We shall find that even in the department of the primitive analysis of its general phenomena, in which it appears so superior to that of the organic life, it is very far indeed from being yet fit for exploration by positive laws. *Present state of analysis of animality.*

In regard to irritability, first,—the mechanism of no animal movement has yet been satisfactorily analyzed, —all the chief cases being still the subject of radical controversy among equally-qualified physiologists. We retain a vicious distinction among movements, contrary to all mechanical judgment,— a distinction between the general motion which displaces the whole mass, and the partial motions which subserve the organic life,—as for the reception of aliment, or the expulsion of any residuum, or the circulation of fluids; yet the first order are partial, though their object is unlike that of the second; for, in a mechanical view, the *Movement.*

organism allows of no others. By the great laws of motion, the animal can never displace its centre of gravity by interior motion, without co-operation from its environment, any more than a steam-carriage which should work without friction on a horizontal plane, turning its wheels without result. The movements which produce locomotion are not mechanically different from those, for instance, which carry food along the alimentary canal;—the difference is in the apparatus, which, for locomotion, consists of exterior appendages, so disposed, as to cause a reaction in the medium, which produces the displacement of the whole body. Certain mollusks furnish an illustration of this, when they change their place by means of contractions of the cardiac muscle, or of intestinal muscles. The simplest notions of animal mechanics being thus obscured and corrupted in their origin, it is no wonder that the physiologists still dispute about the mechanism of the circulation, and most of the means of locomotion, as leaping, flying, swimming, etc. In the way in which they proceed, they are remote from any mutual understanding, and the most opposite opinions may be maintained with equal plausibility. It needs but a word to suggest to those who have attended to what has gone before, that this extreme imperfection results from the inadequate and faulty education of physiologists, who are too often ignorant of the inorganic science which is here directly involved. The complexity of the animal apparatus, and the impossibility of bringing the primitive moving powers under any mathematical theory, will for ever forbid the application of numerical methods: but the great laws of equilibrium and motion are applicable, through all varieties of apparatus, and are the same in animal mechanics, or celestial, or industrial, or any other mechanics whatever. Some physiologists, finding their difficulty, have handed over their study to the geometers and physicists: and these, with their habits of numerical precision, and their ignorance of anatomy, have brought out only absurd results. The remedy is, as we know, in the work being consigned to physiologists, duly prepared by a sufficient training in inorganic science. The study of animal sounds, or phonation, for instance, can not be carried on to any purpose without such knowledge as physicists have of the theory of sound; and the general production of the voice, and the differences of utterance among animals, require for their explanation a knowledge at once of acoustics and anatomy; and speech itself requires this preparation no less, while demanding other requisites with it. It is to be hoped that all experience, in each department of scientific inquiry, will convince students more and more of the folly and mischief of the anarchical parcelling out of natural philosophy; but the physiologists are those who, above all, must see the need of a better organization of scientific labor,—so remarkable as is the subordination of their particular science to all that have gone before.

The analysis of the phenomena of sensibility is not more satisfactory than that of irritability: and even less so, if we leave out of the account the great knowledge that we have obtained, by anatomical study, of the corresponding organs; a knowledge which, however, must here be connected with physiology. The least imperfect part of this study relates to the simple exterior sensations. The phenomenon of sensation is composed *Exterior sensation.* of three elements, as we have seen: the impression of the external agent or the nervous extremities, by the aid of some physical apparatus; the transmission by the nervous fibre; and the reception by the cerebral organ. The first of these suggests, like the mechanical facts we have been considering, the immediate dependence of the phenomenon on the laws of the inorganic world; as the relation of the theory of vision to optics; of the theory of hearing to acoustics, in all that concerns the mode of action proper to the apparatus of sight and hearing. And yet, more expressly than even in the case of mechanics, have these theories been delivered into the hands of the physicists, who, again, bring out results from their treatment of them which are manifestly absurd. The only difference between this case and the preceding is that the metaphysicians have kept a longer hold upon this part of animal physiology,—the theory of sensations having been abandoned to them till a very recent time. It was not indeed till Gall imparted his ever-memorable impulse to the investigation, that physiologists claimed this department at all. It is no wonder, therefore, that the positive theory of sensations is less well conceived, and more recent than that of motion: and naturally more imperfect, independently of its superior difficulty, and the backwardness of those branches of physics to which it relates. The simplest modifications of the phenomenon of vision and of hearing can not as yet be referred with certainty to determinate organic conditions; as, for instance, the adjustment of the eye to see distinctly at very various distances; a faculty which the physiologists have allowed the physicists to attribute to various circumstances of structure, always illusory or inadequate, the physiologists the while playing the part of critics, instead of appropriating a study which belongs exclusively to them. Even the limits of the function are usually very vaguely defined: that is, the kind of exterior notions furnished by each sense, abstracted from all intellectual reflection, is rarely circumscribed with any distinctness. Thus it is no wonder if we are still ignorant of almost all positive laws of sight and hearing, and even of smell and taste. —The only point of doctrine, or rather of method, that we may consider to have attained any scientific stability, is the fundamental order in which the different kinds of sensations should be studied: and this notion has been supplied by comparative anatomy rather than by physiology. It consists in classifying the senses by their increasing speciality,—beginning with the universal sense of contact, or touch, and proceeding by degrees to the four special senses, taste, smell, sight, and finally, hearing. This order is rationally determined by the analysis of the animal series, as the senses must be considered more special and of a higher kind in proportion as they disappear from the lower degrees of the zoological scale. It is remarkable that this gradation coincides with the degree of importance of the sensation in regard to sociality, if not to intelligence. Unhappily, it measures yet more evidently the increasing imperfection of the theory.—We ought not to pass over the luminous distinction introduced by Gall between the passive and the active state of each special sense. An analogous consideration to this, but more fundamental, would consist, it seems to me, in distinguishing the senses themselves as active and and passive, according as their action is, from their nature, voluntary or involuntary. This distinction seems very marked in the case of sight and hearing; the one requiring our free participation, to a certain extent, while the other affects us without our will, or even our consciousness. The more vague, but more profound influence that music influences over us, than we receive from painting, seems to be chiefly attributable to such a diversity. An analogous difference, but less marked, exists between taste and smell.

Interior sensation. There is a second class of sensations, forming the natural transition from the study of the sensations to that of the affective and intellectual functions, which all physiologists, since the time of Cabanis, and yet more, of Gall, have found it necessary to admit, to complete the study of the sensations. They are the interior sensations which relate to the satisfaction of the natural wants; and, in a pathological state, the pains produced by bodily alteration. These are still more indispensable than the first to the perfection of the organic life; and, though they procure no direct notions of the external world, they radically modify, by their intense and continuous action, the general course of intellectual operations, which are, in most animal species, entirely subordinated to them. This great department of the theory of sensations is even more obscure and unadvanced than the foregoing. The only positive notion which is fairly established in regard to it is that the nervous system is indispensable to both kinds of sensibility.

Once again we see how the extreme imperfection of doctrine here is owing to the imperfection of method; and that again, as before, to the inadequate preparation of the inquirers to whom the study belongs. It is a great thing, however, to have withdrawn the subject from the control of the metaphysicians; and some labors of contemporary physiologists authorize us to hope that the true spirit of the inquiry is at length entered into; and that the study of the sensations will be directed, as it ought to be, to develop the radical accordance between anatomical and physiological analysis.

Mode of action of animality. Having reviewed the two orders of animal functions, we must consider the complementary part of the theory of animality;—the ideas about the mode of action which are common to the phenomena of irritability and sensibility. It is true, these ideas belong also to intellectual and moral phenomena: but we must review them here, to complete our delineation of the chief aspects of the study of animal life.

The considerations about such mode of action naturally divide themselves into two classes; the one relating to the function of either motion or sensation, separately; and the other to the association of the two functions. The first may relate to either the mode or the degree of the animal phenomenon. Following this order, the first theory that presents itself is that of the intermittence of action, and consequently, that of habit, which results from it. Bichat was the first who pointed out the intermittent character of every animal faculty, in contrast with the continuousness of vegetative phenomena. The double movement of absorption and exhalation which constitutes life could not be suspended for a moment without determining the tendency to disorganization: whereas, every act of irritability or sensibility is necessarily intermittent, as no contraction or sensation can be conceived of as indefinitely prolonged; so that continuity would imply as great a contradiction in animal life as interruption in the organic. All the progress made, during the present century, in physiological anatomy, has contributed to the perfecting of this theory of intermittence. Rationally understood, it applies immediately to a very extensive and important class of animal phenomena: that is, to those which belong to the different degrees of sleep. The state of sleep thus consists of the simultaneous suspension, for a certain time, of the principal actions of irritability and sensibility. It is as complete as the organization of the superior animals admits when it suspends all motions and sensations but such are indispensable to the organic life,—their activity being also remarkably diminished. The phenomenon admits of great variety of degrees, from simple somnolence to the torpor of hybernating animals. But this theory of sleep, so well instituted by Bichat, is still merely initiated, and presents many fundamental difficulties, when we consider the chief modifications of such a state, even the organic conditions of which are very imperfectly known, except the stagnation of the venous blood in the brain, which appears to be generally an indispensable preliminary to all extended and durable lethargy. It is easy to conceive how the prolonged activity of the animal functions in a waking state may, by the law of intermittence, occasion a proportional suspension: but it is not so easy to see why the suspension should be total when the activity has been only partial. Yet we see how profound is the sleep, intellectual and muscular, induced by fatigue of the muscles alone in men who, while awake, have given very little exercise to their sensibility, interior, or even exterior. We know still less of incomplete sleep; especially when only a part of the intellectual or affective organs, or of the locomotive apparatus is torpid; whence arise dreams and various kinds of somnambulism. Yet such a state has certainly its own general laws, as well as the waking state. Some experiments, not duly attended to, perhaps justify the idea that, in animals, in which the cerebral life is much less varied, the nature of dreams becomes, to a certain point, susceptible of being directed at the pleasure of the observer, by the aid of external impressions produced, during sleep, upon the senses whose action is involuntary; and especially smell. And in the case of Man, there is no thoughtful physician who, in certain diseases, does not take into the account the habitual character of the patient's dreams, in order to perfect the diagnosis of maladies in which the nervous system is especially implicated: and this supposes that the state is subject to determinate laws, though they may be unknown. But, however imperfect the theory of sleep may still be, in these essential respects, it is fairly constituted upon a positive basis of its own; for, looked at as a whole, it is *explained*, according to the scientific acceptation of the term, by its radical identity with the phenomena of partial repose offered by all the elementary acts of the animal life. When the theory of intermittence is perfected, we shall, I imagine, adopt Gall's view of connecting it with the symmetry which characterizes all the organs of animal life, by regarding the two parts of the symmetrical apparatus as alternately active and passive, so that their function is never simultaneous: and this, as much in regard to the external senses as the intellectual organs. All this, however, deserves a fresh and thorough investigation.

Habit. The theory of Habit is a sort of necessary appendix to that of intermittence; and, like it, due to Bichat. A continuous phenomenon would be, in fact, capable of persistence, in virtue of the law of inertia; but intermittent phenomena alone can give rise to habits, properly so called: that is, can tend to reproduce themselves spontaneously through the influence of a preliminary repetition, sufficiently prolonged at suitable intervals. The importance of this animal property is now universally acknowledged among able inquirers, who see in it one of the chief bases of the gradual perfectibility of animals, and especially of Man. Through this it is that vital phenomena may, in some sort, participate in the admirable regularity of those of the inorganic world, by becoming, like them, periodical, notwithstanding their greater complexity. Thence also results the transformation—optional up to a certain point of inveteracy of habit, and inevitable beyond that point—of voluntary acts into involuntary tendencies. But the study of habit is no further advanced than that of intermittence, in regard to its analysis: for we have paid more attention hitherto to the influence of habits once contracted than to their origin, with regard to which scarcely any scientific doctrine exists. What is known lies in the department of natural history, and not in that of biology. Perhaps it may be found, in the course of scientific study, that we have been too hasty in calling this an animal property, though the animal structure may be more susceptible of it. In fact, there is no doubt that inorganic apparatus admits of a more easy reproduction of the same acts after a sufficient regular and prolonged reiteration, as I had occasion to observe in regard to the phenomenon of sound: and this is essentially the character of animal habit. According to this view, which I commend to the attention of biologists, and which, if true, would constitute the most general point of view on this subject, the law of habit may be scientifically attached to the law of inertia, as geometers understand it in the positive theory of motion and equilibrium.

In examining the phenomena common to irritability and sensibility under the aspect of their activity, physiologists have to examine the two extreme terms,—exaggerated action, and insufficient action, in order to determine the intermediate normal degree: for the study of intermediate cases can never be successfully undertaken till the extreme cases which comprehend them have been first examined.

The need of exercising the faculties is certainly the most general and important of all those that belong to the animal life: we may even say that it comprehends them all, if we exclude what relates merely to the organic life. The existence of an animal organ is enough to awaken the need immediately. We shall see, in another chapter, that this consideration is one of the chief bases that social physics derives from individual physiology. Unhappily this study is still very imperfect with regard to most of the animal functions, and to all the three degrees of their activity. To it we must refer the analysis of all the varied phenomena of pleasure and pain, physical and moral. The case of defect has been even less studied than that of excess; and yet its scientific examination is certainly not less important, on account of the theory of *ennui*, the consideration of which is so prominent in social physics,—not only in connection with an advanced state of civilization, but even in the roughest periods, in which, as we shall see hereafter, *ennui* is one of the chief moving springs of social evolution. As for the intermediate degree, which characterizes health, welfare, and finally happiness, it can not be well treated till the extremes are better understood. The only positive principle yet established in this part of physiology is that which prescribes that we should not contemplate this normal degree in an absolute manner, but in subordination to the intrinsic energy of the corresponding faculties; as popular good sense has already admitted, however difficult it may be practically to conform to the precept in social matters, from the unreflecting tendency of every man to erect himself into a necessary type of the whole species.

Need of activity.

We have now only to notice, further, the third order of considerations; the study of the association of the animal functions.

This great subject should be divided into two parts, relating to the *sympathies*, to which Bichat has sufficiently drawn the attention of physiologists, and the *synergies*, as Barthez has called them, which are at present too much neglected. The difference between these two sorts of vital association corresponds to that between the normal and the pathological states: for there is synergy whenever two organs concur simultaneously in the regular accomplishment of any function; whereas sympathy supposes a certain perturbation, momentary or

Association of the animal functions.

permanent, partial or general, which has to be stopped by the intervention of an organ not primarily affected. These two modes of physiological association are proper to the animal life, any appearance to the contrary being due to the influence of animal over organic action. The study is fairly established on a rational basis; the physiologists of our time seeming to be all agreed as to the nervous system being the necessary agent of all sympathy; and this is enough for the foundation of a positive theory. Beyond this, we have only disjointed though numerous facts. The study of the synergies, though more simple and better circumscribed, does not present, as yet, a more satisfactory scientific character, either as to the mutual association of the different motions, or as to the different modes of sensibility; or as to the more general and complex association between the phenomena of sensibility and those of irritability. And yet this great subject leads directly to the most important theory that physiology can finally present,—that of the fundamental unity of the animal organism, as a necessary result of a harmony between its various chief functions. Here alone it is that, taking each elementary faculty in its normal state, we can find the sound theory of the *Ego*, so absurdly perverted at present by the vain dreams of the metaphysicians: for the general sense of the *I* is certainly determined by the equilibrium of the faculties, the disturbance of which impairs that consciousness so profoundly in other diseases.

CHAPTER VI.

INTELLECTUAL AND MORAL, OR CEREBRAL FUNCTIONS.

The remaining portion of biological philosophy is that which relates to the study of the affective and intellectual faculties, which leads us over from individual physiology to Social Physics, as vegetative physiology does from the inorganic to the organic philosophy.

Shortcoming of Descartes. While Descartes was rendering to the world the glorious service of instituting a complete system of positive philosophy, the reformer, with all his bold energy, was unable to raise himself so far above his age as to give its complete logical extension to his own theory by comprehending in it the part of physiology that relates to intellectual and moral phenomena. After having instituted a vast mechanical hypothesis upon the fundamental theory of the most simple and universal phenomena, he extended in succession the same philosophical spirit to the different elementary notions relating to the inorganic world; and finally subordinated to it the study of the chief physical functions of the animal organism. But, when he arrived at the functions of the affections and the intellect, he stopped abruptly, and expressly constituted from them a special study, as an appurtenance of the metaphysico-theological philosophy, to which he thus endeavored to give a kind of new life, after having wrought far more successfully in sapping its scientific foundations. We have an unquestionable evidence of the state of his mind in his celebrated paradox about the intelligence and instincts of animals. He called brutes automata, rather than allow the application of the old philosophy to them. Being unable to pursue this method with Man, he delivered him over expressly to the domain of metaphysics and theology. It is difficult to see how he could have done otherwise, in the then existing state of knowledge: and we owe to his strange hypothesis, which the physiologists went to work to confute, the clearing away of the partition which he set up between the study of animals and that of Man, and consequently, the entire elimination among the higher order of investigators, of theological and metaphysical philosophy. What the first contradictory constitution of the modern philosophy was, we may see in the great work of Malebranche, who was the chief interpreter of Descartes, *History till Gall's time.* and who shows how his philosophy continued to apply to the most complex parts of the intellectual system the same methods which had been shown to be necessarily futile with regard to the simplest subjects. It is necessary to indicate this state of things because it has remained essentially unaltered during the last two centuries, notwithstanding the vast progress of positive science, which has all the while been gradually preparing for its inevitable transformation. The school of Boerhaave left Descartes's division of subjects as they found it: and if they, the successors of Descartes in physiology, abandoned this department of it to the metaphysical method, it can be no wonder that intellectual and moral phenomena remained, till this century, entirely excluded from the great scientific movement originated and guided by the impulse of Descartes. The growing action of the positive spirit has been, during the whole succeeding interval, merely critical—attacking the inefficacy of metaphysical studies—exhibiting the perpetual reconciliation of the naturalists on points of genuine doctrine, in contrast to the incessant disputes of various metaphysicians, arguing still, as from Plato downward, about the very elements of their pretended science: this criticism itself relating only to results, and still offering no objection to the supremacy of metaphysical philosophy, in the study of Man, in his intellectual and moral aspects. It was not till our own time that modern science, with the illustrious Gall for its organ, drove the old philosophy from this last portion of its domain, and passed on in the inevitable course from the critical to the organic state, striving in its turn to treat in its own way the general theory of the highest vital functions. However imperfect the first attempts, the thing is done. Subjected for half a century to the most decisive tests, this new doctrine has clearly manifested all the indications which can guaranty the indestructible vitality of scientific conceptions. Neither enmity nor irrational advocacy has hindered the continuous spread, in all parts of the scientific world, of the new system of investigation of intellectual and moral man. All the signs of the progressive success of a happy philosophical revolution are present in this case.

Positive theory of Cerebral functions. The positive theory of the affective and intellectual functions is therefore settled, irreversibly, to be this: it consists in the experimental and rational study of the phenomena of interior sensibility proper to the cerebral ganglions, apart from all immediate external apparatus. These phenomena are the most complex and the most special of all belonging to physiology; and therefore they have naturally been the last to attain to a positive analysis; to say nothing of their relation to social considerations, which must be an impediment in the way of their study. This study could not precede the principal scientific conceptions of the organic life, or the first notions of the animal life; so that Gall must follow Bichat: and our surprise would be that he followed him so soon, if the maturity of his task did not explain it sufficiently. The grounds of my provisional separation of this part of physiology from the province of animal life generally are—the eminent differences between this order of phenomena and those that have gone before—their more direct and striking importance—and, above all, the greater imperfection of our present study of them. This new body of doctrine, thus erected into a third section of physiology, will assume its true place within the boundaries of the second when we obtain a distincter knowledge of organic, and a more philosophical conception of animal physiology. We must bear in mind what the proper arrangement should be—this third department differing much less from the second than the second differs from the first.

Its proper place.

Views of Psychological systems. We need not stop to draw out any parallel or contrast between phrenology and psychology. Gall has fully and clearly exposed the powerlessness of metaphysical methods for the study of intellectual and moral phenomena: and in the present state of the human mind, all discussion on this subject is superfluous. The great philosophical cause is tried and judged; and the metaphysicians have passed from a state of domination to one of protestation—in the learned world at least, where their opposition would obtain no attention but for the inconvenience of their still impeding the progress of popular reason. The triumph of the positive method is so decided that it is needless to devote time and effort to any demonstration, except in the way of instruction; but, in order to characterize, by a striking contrast, the true general spirit of phrenological physiology, it may be useful here to analyze very briefly the radical vices of the pretended psychological method, considered merely in regard to what it has in common in the principal existing schools;—in those called the French, the German, and (the least consistent and also the least absurd of the three) the Scotch school:—that is, as far as we can talk of schools in a philosophy which, by its nature, must engender as many incompatible opinions as it has adepts gifted with any degree of imagination. We may, moreover, refer confidently to these sects for the mutual refutation of their most essential points of difference.

As for their fundamental principle of *interior observation*, it would certainly be superfluous to add anything to what I have already said about the absurdity of the supposition of a man seeing himself think. It was well remarked by M. Broussais, on this point, that such a method, if possible, would extremely restrict the study of the understanding, by necessarily limiting it to the case of adult and healthy Man, without any hope of illustrating this difficult doctrine by any comparison of different ages, or consideration of pathological states, which yet are unanimously recognised as indispensable auxiliaries in the simplest researches about Man. But, further, we must be also struck by the absolute interdict which is laid upon all intellectual and moral study of animals, from whom the psychologists can hardly be expecting any *interior observation*. It seems rather strange that the philosophers who have so attenuated this immense subject should be those who are for ever reproaching their adversaries with a want of comprehensiveness and elevation. The case of animals is the rock on which all psychological theories have split, since the naturalists have compelled the metaphysicians to part with the singular expedient imagined by Descartes, and to admit that animals, in the higher parts of the scale at least, manifest most of our affective, and even intellectual faculties, with mere differences of degree; a fact which no one at this day ventures to deny, and which is enough of itself to demonstrate the absurdity of these idle conceptions.

<small>Method. Interior Observation.</small>

Recurring to the first ideas of philosophical common sense, it is at once evident that no function can be studied but with relation to the organ that fulfils it, or to the phenomena of its fulfilment: and, in the second place, that the affective functions, and yet more the intellectual, exhibit in the latter respect this particular characteristic,—that they can not be observed during their operation, but only in their results,—more or less immediate, and more or less durable. There are then only two ways of studying such an order of functions; either determining, with all attainable precision, the various organic conditions on which they depend,—which is the chief object of phrenological physiology; or in directly observing the series of intellectual and moral acts,—which belongs rather to natural history, properly so called: these two inseparable aspects of one subject being always so conceived as to throw light on each other. Thus regarded, this great study is seen to be indissolubly connected on the one hand with the whole of the foregoing parts of natural philosophy, and especially with the fundamental doctrines of biology; and, on the other hand, with the whole of history,—of animals as well as of man and of humanity. But when, by the pretended psychological method, the consideration of both the agent and the act is discarded altogether, what material can remain but an unintelligible conflict of words, in which merely nominal entities are substituted for real phenomena? The most difficult study of all is thus set up in a state of isolation, without any one point of support in the most simple and perfect sciences, over which it is yet proposed to give it a majestic sovereignty; and in this all psychologists agree, however extreme may be their differences on other points.

<small>Doctrine. Relation between the Affective and Intellectual faculties.</small> About the method of psychology or ideology, enough has been said. As to the doctrine, the first glance shows a radical fault in it, common to all sects,—a false estimate of the general relations between the affective and the intellectual faculties. However various may be the theories about the preponderance of the latter, all metaphysicians assert that preponderance by making these faculties their starting-point. The intellect is almost exclusively the subject of their speculations, and the affections have been almost entirely neglected; and, moreover, always subordinated to the understanding. Now, such a conception represents precisely the reverse of the reality, not only for animals, but also for Man: for daily experience shows that the affections, the propensities, the passions, are the great springs of human life; and that, so far from resulting from intelligence, their spontaneous and independent impulse is indispensable to the first awakening and continuous development of the various intellectual faculties, by assigning to them a permanent end, without which—to say nothing of the vagueness of their general direction—they would remain dormant in the majority of men. It is even but too certain that the least noble and most animal propensities are habitually the most energetic, and therefore the most influential. The whole of human nature is thus very unfaithfully represented by these futile systems, which, if noticing the affective faculties at all, have vaguely connected them with one single principle, sympathy, and, above all, self-consciousness, always supposed to be directed by the intellect. Thus it is that, contrary to evidence, Man has been represented as essentially a reasoning being, continually carrying on, unconsciously, a multitude of imperceptible calculations, with scarcely any spontaneity of action, from infancy upward. This false conception has doubtless been supported by a consideration worthy of all respect,—that it is by the intellect that Man is modified and improved; but science requires, before all things, the reality of any views, independently of their desirableness; and it is always this reality which is the basis of genuine utility. Without denying the secondary influence of such a view, we can show that two purely philosophical causes, quite unconnected with any idea of application, and inherent in the nature of the method, have led the metaphysicians of all sects to this hypothesis of the supremacy of the intellect. The first is the radical separation which it was thought necessary to make be-
<small>Brutes and Man.</small> tween brutes and man, and which would have been af-

fected at once by the admission of the preponderance of the affective over the intellectual faculties; and the second was the necessity that the metaphysicians found themselves under, of preserving the unity of what they called the *I*, that it might correspond with the unity of the *soul*, in obedience to the requisitions of the theological philosophy, of which metaphysics is, as we must ever bear in mind, the final transformation. But the positive philosophers, who approach the questions with the simple aim of ascertaining the true state of things, and reproducing it with all possible accuracy in their theories, have perceived that, according to universal experience, human nature is so far from being single that it is eminently multiple; that is, usually induced in various directions by distinct and independent powers, among which equilibrium is established with extreme difficulty when, as usually happens in civilized life, no one of them is, in itself, sufficiently marked to acquire spontaneously any considerable preponderance over the rest. Thus, the famous theory of the *I* is essentially without a scientific object, since it is destined to represent a purely fictitious state. There is, in this direction, as I have already pointed out, no other real subject of positive investigation than the study of equilibrium of the various animal functions,—both of irritability and of sensibility,—which marks the normal state, in which each of them, duly moderated, is regularly and permanently associated with the whole of the others, according to the laws of sympathy, and yet more of synergy. The very abstract and indirect notion of the *I* proceeds from the continuous sense of such a harmony; that is, from the universal accordance of the entire organism. Psychologists have attempted in vain to make out of this idea, or rather sense, an attribute of humanity exclusively. It is evidently a necessary result of all animal life; and therefore it must belong to all animals, whether they are able to discourse upon it or not. No doubt a cat, or any other vertebrated animal, without knowing how to say "I," is not in the habit of taking itself for another. Moreover, it is probable that among the superior animals the sense of personality is still more marked than in Man, on account of their more isolated life; though if we descended too far in the zoological scale we should reach organisms in which the continuous degradation of the nervous system attenuates this compound sense, together with the various simple feelings on which it depends.

It must not be overlooked that though the psychologists have agreed in neglecting the intellectual and moral faculties of brutes, which have been happily left to the naturalists, they have occasioned great mischief by their obscure and indefinite distinction between intelligence and instinct, thus setting up a division between human and animal nature which has had too much effect even upon zoologists to this day. The only meaning that can be attributed to the word *instinct*, is any spontaneous impulse in a determinate direction, independently of any foreign influence. In this primitive sense, the term evidently applies to the proper and direct activity of any faculty whatever, intellectual as well as affective; and it therefore does not conflict with the term *intelligence* in any way, as we so often see when we speak of those who, without any education, manifest a marked talent for music, painting, mathematics, etc. In this way there is instinct, or rather, there are instincts in Man, as much or more than in brutes. If, on the other hand, we describe *intelligence* as the aptitude to modify conduct in conformity to the circumstances of each case,—which, in fact, is the main practical attribute of *reason*, in its proper sense,—it is more evident than before that there is no other essential difference between humanity and animality than that of the degree of development admitted by a faculty which is, by its nature, common to all animal life, and without which it could not even be conceived to exist. Thus the famous scholastic definition of Man as a *reasonable animal* offers a real no-meaning, since no animal, especially in the higher parts of the zoological scale, could live without being to a certain extent reasonable, in proportion to the complexity of its organism. Though the moral nature of animals has been but little and very imperfectly explored, we can yet perceive, without possibility of mistake, among those that live with us and that are familiar with us,—judging of them by the same means of observation that we should employ about men whose language and ways were previously unknown to us,—that they not only apply their intelligence to the satisfaction of their organic wants, much as men do, aiding themselves also with some sort of language; but that they are, in like manner, susceptible of a kind of wants more disinterested, inasmuch as they consist in a need to exercise their faculties for the mere pleasure of the exercise. It is the same thing that leads children or savages to invent new sports, and that renders them, at the same time, liable to *ennui*. That state, erroneously set up as a special privilege of human nature, is sometimes sufficiently marked, in the case of certain animals, to urge them to suicide, when captivity has become intolerable. An attentive examination of the facts therefore discredits the perversion of the word *instinct* when it is used to signify the fatality under which animals are impelled to the mechanical performance of *acts* uniformly determinate, without any possible modification from corresponding circumstances, and neither requiring nor allowing any education, properly so called. This gratuitous supposition is evidently a remnant of the automatic hypothesis of Descartes. Leroy has demonstrated that among mammifers and birds this ideal fixity in the construction of habitations, in the seeking of food by hunting, in the mode of migration, etc., exist only in the eyes of closet-naturalists or inattentive observers.

After thus much notice of the radical vice of all psychological systems, it would be departing from the object of this work to show how the intellectual faculties themselves have been misconceived. It is enough to refer to the refutation by which Gall and

Spurzheim have introduced their labors: and I would particularly point out the philosophical demonstration by which they have exhibited the conclusion that sensation, memory, imagination, and even judgment—all the scholastic faculties, in short—are not, in fact, fundamental and abstract faculties, but only different degrees or consecutive modes of the same phenomenon, proper to each of the true elementary phrenological functions, and necessarily variable in different cases, with a proportionate activity. One virtue of this admirable analysis is that it deprives the various metaphysical theories of their one remaining credit—their mutual criticism, which is here effected, once for all, with more efficacy than by any one of the mutually opposing schools.

Again, it would be departing from the object of this portion of our work to judge of the doctrines of the schools by their results. What these have been we shall see in the next book; the deplorable influence on the political and social condition of two generations of the doctrines of the French school, as presented by Helvetius, and of the German psychology, with the ungovernable *I* for its subject; and the impotence of the Scotch school, through the vagueness of what is called its doctrines, and their want of mutual connection. Dismissing all these for the present, we must examine the great attempt of Gall, in order to see what is wanting in phrenological philosophy to form it into the scientific constitution which is proper to it, and from which it is necessarily still more remote than organic, and even animal physiology.

Basis of Gall's doctrine. Two philosophical principles, now admitted to be indisputable, serve as the immovable basis of Gall's doctrine as a whole: viz., the innateness of the fundamental dispositions, affective and intellectual, and the plurality of the distinct and independent faculties, though real acts usually require their more or less complex concurrence. Within the limits of the human race, all cases of marked talents or character prove the first; and the second is proved by the diversity of such marked cases, and by most pathological states—especially by those in which the nervous system is directly affected. A comparative observation of the higher animals would dispel all doubt, if any existed in either case. These two principles—aspects of a single fundamental conception—are but the scientific expression of the results of experience, in all times and places, as to the intellectual and moral nature of Man—an indispensable symptom of truth, with regard to all parent ideas, which must always be connected with the spontaneous indications of popular reason, as we have seen in preceding cases in natural philosophy. Thus, besides all guidance from analogy, after the study of animal life, we derive confirmation from all the methods of investigation that physiology admits; from direct observation, experiment, pathological analysis, the comparative method, and popular good sense—all of which converge toward the establishment of this double principle. Such a collection of proofs secures the stability of this much of phrenological doctrine, whatever transformations others parts may have to undergo. In the anatomical view, this physiological conception corresponds with the division of the brain into a certain number of partial organs, symmetrical like those of the animal life, and, though more contiguous and mutually resembling than in any other system, and therefore more adapted both for sympathy and synergy, still distinct and mutually independent, as we were already aware was the case with the ganglions appropriate to the external senses. In brief, the brain is no longer an organ, but an apparatus of organs, more complex in proportion to the degree of animality. The proper object of phrenological physiology thence consists in determining the cerebral organ appropriate to each clearly-marked, simple disposition, affective or intellectual; or, reciprocally, which is more difficult, what function is fulfilled by any portion of the mass of the brain which exhibits the anatomical conditions of a distinct organ. The two processes are directed to develop the agreement between physiological and anatomical analysis which constitutes the true science of living beings. Unfortunately, our means are yet further from answering our aims than in the two preceding divisions of the science.

Divisions of the brain. The scientific principle involved in the phrenological view is that the functions, affective and intellectual, are more elevated, more human, if you will, and at the same time less energetic, in proportion to the exclusiveness with which they belong to the higher part of the zoological series, their positions being in portions of the brain more and more restricted in extent, and further removed from its immediate origin,—according to the anatomical decision that the skull is simply a prolongation of the vertebral column, which is the primitive centre of the nervous system. Thus, the least developed and anterior part of the brain is appropriated to the characteristic faculties of humanity; and the most voluminous and hindmost part to those which constitute the basis of the whole of the animal kingdom. Here we have a new and confirmatory instance of the rule which we have had to follow in every science; that it is necessary to proceed from the most general to the more special attributes, in the order of their diminishing generality. We shall meet with it again in the one science which remains for us to review; and its constant presence, through the whole range, points it out as the first law of the dogmatic procedure of the positive spirit.

A full contemplation of Gall's doctrine convinces us of its faithful representation of the intellectual and moral nature of Man and animals. All the psychological sects have misconceived or ignored the pre-eminence of the affective faculties, plainly manifest as it is in all the moral phenomena of brutes, and even of Man; but we find this fact placed on a scientific basis by the discovery that the affective organs occupy all the hinder and middle portions of the cerebral apparatus, while the intellectual occupy only the front portion, which, in extreme cases, is not more than a fourth, or even a sixth part of the whole. The difference between Gall and his predecessors was not in the separation of the two kinds of faculties, but that they assigned the brain to the intellectual faculties alone, regarding it as a single organ, and distributing the passions among the organs pertaining to the vegetative life,—the heart, the liver, etc. Bichat supported this view by the argument of the sympathies of these organs, under the excitement of the respective passions; but the variableness of the seat of sympathy, according to native susceptibility or to accident, is a sufficient answer to such a plea, and teaches us simply the importance of considering the influence exercised by the state of the brain upon the nerves which supply the apparatus of the organic life.

Next comes the subdivision established by Gall and Spurzheim in each of these two orders. The affective faculties are divided into the propensities, and the affections or sentiments: the first residing in the hindmost and lowest part of the brain; and the other class in the middle portion. The intellectual faculties are divided into the various perceptive faculties, which together constitute the range of observation: and the small number of reflective faculties, the highest of all, constituting the power of combination, by comparison and co-ordination. The upper part of the frontal region is the seat of these last, which are the chief characteristic attribute of human nature. There is a certain deficiency of precision in this description; but, besides that we may expect improving knowledge to clear it up, we shall find, on close examination, that the inconvenience lies more in the language than in the idea. The only language we have is derived from a philosophical period when all moral and even intellectual ideas were shrouded in a mysterious metaphysical unity, which allows us now no adequate choice of terms.

Subdivision.

Taking the ordinary terms in their literal sense, we should misconceive the fundamental distinction between the intellectual faculties and the others. When the former are very marked, they unquestionably produce real inclinations or propensities, which are distinguished from the inferior passions only by their smaller energy. Nor can we deny that their action occasions true emotions or sentiments, more rare, more pure, more sublime than any other, and, though less vivid than others, capable of moving to tears; as is testified by so many instances of the rapture excited by the discovery of truth, in the most eminent thinkers that have done honor to their race—as Archimedes, Descartes, Kepler, Newton, etc. Would any thoughtful student take occasion, by such approximations, to deny all real distinction between the intellectual and affective faculties? The wiser conclusion to be drawn from the case is that we must reform our philosophical language, to raise it, by rigorous precision, to the dignity of scientific language. We may say as much about the subdivision of the affective faculties into propensities and sentiments, the distinction being, though less marked, by no means less real. Apart from all useless discussion of nomenclature, we may say that the real difference has not been clearly seized. In a scientific view, it would suffice to say that the first and fundamental class relates to the individual alone, or, at most, to the family, regarded successively in its principal needs of preservation,—such as reproduction, the rearing of young, the mode of alimentation, of habitation, etc. Whereas, the second more special class supposes the existence of some social relations, either among individuals of a different species, or especially between individuals of the same species, apart from sex, and determines the character which the tendencies of the animal must impress on each of these relations, whether transient or permanent. If we keep this distinctive character of the two classes in view, it will matter little what terms we use to indicate them, when once they shall have acquired a sufficient fixedness, through rational use.

These are the great philosophical results of Gall's doctrine, regarded, as I have now presented it, apart from all vain attempts to localize in a special manner the cerebral or phrenological functions. I shall have to show how such an attempt was imposed upon Gall by the necessities of his glorious mission: but notwithstanding this unfortunate necessity, the doctrine embodies already a real knowledge of human and brute nature very far superior to all that had ever been offered before.

Objections. Necessity of human actions. Among the innumerable objections which have been aimed at this fine doctrine,—considered always as a whole, the only one which merits discussion here is the supposed necessity of human actions. This objection is not only of high importance in itself, but it casts new light back upon the spirit of the theory; and we must briefly examine it from the point of view of positive philosophy.

When objectors confound the subjection of events to invariable laws with their necessary exemption from modification, they lose sight of the fact that phenomena become susceptible of modification in proportion to their complexity. The only irresistible action that *Answered.* we know of is that of weight, which takes place under the most general and simple of all natural laws. But the phenomena of life and acts of the mind are so highly complex as to admit of modification beyond all estimate; and in the intermediate regions, phenomena are under control precisely in the order of their complexity. Gall and Spurzheim have shown how human action depends on the combined operation of several faculties; how exercise develops them; how inactivity wastes them; and how the intellectual faculties, adapted to modify the general conduct of the animal according to the variable exigencies of his situation, may overrule the practical influence of all his other faculties. It is only in mania, when disease interferes with the natural action of the faculties, that fatality, or what is popularly called irresponsibility, exists. It is therefore a great mistake to accuse cerebral physiology of disowning the influence of education or legislation, because it fixes the limits of their power. It denies the possibility, asserted by the ideology of the French school, of converting by suitable arrangements, all men into so many Socrates, Homers, or Archimedes; and it denies the ungovernable energy of the *I*, asserted by the German school; but it does not therefore affect Man's reasonable liberty, or interfere with his improvement by the aid of a wise education. It is evident indeed that improvement by education supposes the existence of requisite predispositions: and that each of them is subject to determinate laws, without which they could not be systematically influenced; so that it is, after all, cerebral physiology that is in possession of the philosophical problem of education. Furthermore, this physiology shows us that men are commonly of an average constitution; that is, that, apart from a very few exceptional organizations, every one possesses in a moderate degree all the propensities, all the sentiments, and all the elementary aptitudes, without any one faculty being remarkably preponderant. The widest field is thus open for education, in modifying in almost any direction organisms so flexible, though the degree of their development may remain of that average amount which consists very well with social harmony; as we shall have occasion to see hereafter.

A much more serious objection to Gall's doctrine arises out of the venturesome and largely erroneous localization of the faculties which he thought proper to propose. *Hypothetical distribution of faculties.* If we look at his position, we shall see that he merely used the right, common to all natural philosophers, of instituting a scientific hypothesis, in accordance with the theory on that subject which we examined in connection with Physics. He fulfilled the conditions of this theory; his subject being, not any imaginary fluids, ethers, or the like, but tangible organs, whose hypothetical attributes admit of positive verifications. Moreover, none of those who have criticised his localization could have proposed any less imperfect, or, probably, so well indicated. The advice of prudent mediocrity, to abstain from hypothesis, is very easy to offer; but if the advice was followed, nothing would ever be done in the way of scientific discovery. It is doubtless inconvenient to have to withdraw or remake, at a subsequent period, the hypotheses to which a science owes its existence, and which, by that time, have been adopted by inferior inquirers with a blinder and stronger faith than that of the original proposers: but there is no use in dwelling upon a liability which arises from the infirmity of our intelligence. The practical point for the future is that strong minds, prepared by a suitable scientific education, should plant themselves on the two great principles which have been laid down as the foundation of the science, and thence explore the principal needs of cerebral physiology, and the character of the means by which it may be carried forward. Nor need there be any fear that the science will be held back by such a method. Nothing prevents us, when reasoning, as geometers do, upon indeterminate seats, or positions supposed to be indeterminate, from arriving at real conclusions, involving actual utility, as I hope to show, from my own experience, in future chapters; though it is evident that it will be a great advantage to the exactness and efficacy of our conclusions, whenever the time arrives for the positive determination of the cerebral organs. Meantime, it is clear that we owe to Gall's hypothetical localization our view of the necessity of such a course; and that if he had confined himself to the high philosophical generalities with which he has furnished us, he would never have constituted a science, nor formed a school; and the truths which we see to be inestimable would have been strangled in their birth by a coalition of hostile influences.

Needed improvements. We see what is the philosophical character of cerebral physiology. We must next inquire what are the indispensable improvements that it demands.

First, we want a fundamental rectification of all the organs and faculties, as a necessary basis for all further progress. Taking an anatomical view of this matter, we see that the distribution of organs has been directed by physiological analyses alone,—usually *Anatomical basis.* imperfect and superficial enough,—instead of being subjected to anatomical determinations. This has entitled all anatomists to treat such a distribution as arbitrary and loose, because, being subject to no anatomical consideration about the difference between an organ and a part of an organ, it admits of indefinite subdivisions, which each phrenologist seems to be able to multiply at will. Though the analysis of functions no doubt casts much light on that of organs, the original decomposition of the whole organism into systems of organs, and those again into single organs, is not the less independent of physiological analysis, to which, on the contrary, it must furnish a basis. This is established in regard to all other biological studies; and there is no reason why cerebral inquiries should be an exception. We do not need to see the digestive or the respiratory apparatus in action, before anatomy can distinguish them from each other: and why should it be otherwise with the cerebral apparatus? The anatomical difficulties are no doubt much greater, on account of the resemblance and proximity of the organs in the cerebral case: but we must not give up this indispensable analysis for such a reason as that. If it were so, we must despair of conferring a special scientific character on phrenological doctrine at all; and we must abide by those generalities alone which I have just laid down. When we propose to develop the harmony between the anatomical and the physiological analysis of any case, it is supposed that each has been separately established, and not that the one can be copied from the other. Nothing, therefore, can absolve the phrenologists from the obligation to pursue the analysis of the cerebral system by a series of vigorous anatomical labors, discarding for the time all ideas of function, or, at most, employing them only as auxiliary to anatomical exploration. Such a consideration will be most earnestly supported by those phrenologists who perceive that, in determining the relative preponderance of each cerebral organ in different subjects, it is not only the bulk and weight of the organ that has to be taken into the account, but also its degree of activity, anatomically estimated, by, for instance, the energy of its partial circulation.

Next, following a distinct but parallel order of ideas, there must be a purely physiological analysis of the various elementary faculties; and in this analysis, which has to be harmonized with the other, every anatomical idea must be, in its turn, discarded. The position of phrenology is scarcely more satisfactory in this view than any other, for the distinction between the different faculties, intellectual and even affective, and their enumeration, are conceived of in a very superficial way, though incomparably more in the positive spirit than any metaphysical analyses. If metaphysicians have confounded all their psychological notions in an absurd unity, it is probable that the phrenologists have gone to the other extreme in multiplying elementary functions. Gall set up twenty-seven; which was, no doubt, an exaggeration to begin with. Spurzheim raised the number to thirty-five; and it is liable to daily increase for want of a rational principle of circumscription for the regulation of the easy enthusiasm of popular explorers. Unless a sound philosophy interposes, to establish some order, we may have as many faculties and organs as the psychologists of old made entities. However great may be the diversity of animal natures, or even of human types, it is yet to be conceived (as real acts usually suppose the concurrence of several fundamental faculties) that even a greater multiplicity might be represented by a very small number of elementary functions of the two orders. If, for instance, the whole number were reduced to twelve or fifteen well-marked faculties, their combinations, binary, ternary, quaternary, etc., would doubtless correspond to many more types than can exist, even if we restricted ourselves to distinguishing, in relation to the normal degree of activity of each function, two other degrees—one higher and the other lower. But the exorbitant multiplication of faculties is not in itself so shocking as the levity of most of the pretended analyses which have regulated their distribution. In the intellectual order, especially, the aptitudes have been usually ill-described, apart from the organs: as when a mathematical aptitude is assigned on grounds which would justify our assigning a chemical aptitude, or an anatomical aptitude, if the whole bony casket had not been previously parcelled off into irremoveable compartments. If a man could do sums according to rules quickly and easily, he had the mathematical aptitude, according to those who do not suspect that mathematical speculations require any superiority of intellect. Though the analysis of the affective faculties, which are so much better marked, is less imperfect, there are several instances of needless multiplication in that department.

To rectify or improve this analysis of the cerebral faculties, it would be useful to add to the observation of Man and society a physiological estimate of the most marked individual cases,—especially in past times. The intellectual order, which most needs revision, is that which best admits of this procedure. If, for instance, it had been applied to the cases of the chief geometers, the absurd mistake that I have just pointed out could not have been committed; for it would have been seen what compass and variety of faculties are required to constitute mathematical genius, and how various are the forms in which that genius manifests itself. One great geometer has shone by the sagacity of his inventions; another by the strength and extent of his combinations; a third by the happy choice of his notations, and the perfection of his algebraic style, etc. We might discover, or at least verify, all the real fundamental intellectual faculties by the scientific class alone. In an inferior degree it would be the same with an analogous study of the most eminent artists. This consideration, in its utmost extent, is connected with the utility of the philosophical study of the sciences, under the historical as well as the dogmatical point of view, for the discovery of the logical laws concerned: the difference being that, in this last case, we have first to determine the elementary faculties, and not the laws of their action: but the grounds must be essentially analogous.

Phrenological analysis has, then, to be reconstituted; first in the anatomical, and then in the physiological order; and finally, the two must be harmonized; and not till then can phrenological physiology be established upon its true scientific basis. Such a procedure is fairly begun, as we have seen, with regard to the two preceding divisions of our science; but it is not yet even conceived of in relation to cerebral physiology, from its greater complexity and more recent positivity.

Pathological and Comparative analysis. The phrenologists must make a much more extensive use than hitherto of the means furnished by biological philosophy for the advancement of all studies relating to living bodies: that is, of pathological, and yet more of comparative analysis. The luminous maxim of M. Broussais, which lies at the foundation of medical philosophy,—that the phenomena of the pathological state are a simple prolongation of the phenomena of the normal state, beyond the ordinary limits of variation,—has never been duly applied to intellectual and moral phenomena: yet it is impossible to understand anything of the different kinds of madness, if they are not examined on this principle. Here, as in a former division of the science, we see that the study of malady is the way to understand the healthy state. Nothing can aid us so well in the discovery of the fundamental faculties as a judicious study of the state of madness, when each faculty manifests itself in a degree of exaltation which separates it distinctly from others. There has been plentiful study of monomania; but it has been of little use, for want of a due connection and comparison with the normal state. The works that have appeared on the subject have been more literary than scientific; those who have had the best opportunity for observation have been more engaged in governing their patients then in analyzing their cases; and the successors of Pinel have added nothing essential to the ameliorations introduced by him, half a century ago, in regard to the theory and treatment of mental alienation. As for the study of animals, its use has been vitiated by the old notions of the difference between instinct and intelligence. Humanity and animality ought reciprocally to cast light upon each other. If the whole set of faculties constitutes the complement of animal life, it must surely be that all that are fundamental must be common to all the superior animals, in some degree or other: and differences of intensity are enough to account for the existing diversities,—the association of the faculties being taken into the account, on the one hand, and, on the other, the improvement of Man in society being set aside. If there are any faculties which belong to Man exclusively, it can only be such as correspond to the highest intellectual aptitudes: and this much may appear doubtful if we compare, in an unprejudiced way, the actions of the highest mammifers with those of the least-developed savages. It seems to me more rational to suppose that power of observation and even of combination exists in animals, though in an immeasurably inferior degree; the want of exercise, resulting chiefly from their state of isolation, tending to benumb and even starve the organs. Much might be learned from a study of domestic animals, though they are far from being the most intelligent. Much might be learned by comparing their moral nature now with what it was at periods nearer to their first domestication; for it would be strange if the changes that they have undergone in so many physical respects had been unaccompanied by variation in the functions which more easily than any others admit of modification. The extreme imperfection of phrenological science is manifest in the pride with which Man, from the height of his supremacy, judges of animals as a despot judges of his subjects; that is, in the mass, without perceiving any inequality in them worth noticing. It is not the less certain that, surveying the whole animal hierarchy, the principal orders of this hierarchy sometimes differ more from each other, in intellectual and moral respects, than the highest of them vary from the human type. The rational study of the mind and the ways of animals has still to be instituted,—nothing having yet been done but in the way of preparation. It promises an ample harvest of important discovery directly applicable to the advancement of the study of Man, if only the naturalists will disregard the declamation of theologians and metaphysicians about their pretended degradation of human nature, while they are, on the contrary, rectifying the fundamental notion of it by establishing, rigorously and finally, the profound differences which positively separate us from the animals nearest to us in the scale.

The two laws of action—intermission and association—require much more attention than they have yet received in connection with cerebral physiology. The law of inter- *Laws of action.*

mittence is eminently applicable to the functions of the brain—the symmetry of the organs being borne in mind. But this great subject requires a new examination, seeing that it is requisite for science to reconcile their evident intermittence with the perfect continuity that seems to be involved in the connection which mutually unites all our intellectual operations, from earliest infancy to extreme decrepitude, and which can not be interrupted by the deepest cerebral perturbations, provided they are transient. This question, for which metaphysical theories allowed no place, certainly offers serious difficulties; but its positive solution must throw great light upon the general course of intellectual acts. As for the association of the faculties, in sympathy or synergy, the physiologists begin to understand its high importance, though its general laws have not yet been scientifically studied. Without this consideration, the number of propensities, sentiments, or aptitudes, would seem to be susceptible of any degree of multiplication. For one instance, investigators of human nature have been wont to distinguish various kinds of courage, under the names of civil, military, etc., though the original disposition to brave any kind of danger must always be uniform, but more or less directed by the understanding. No doubt, the martyr who endures the most horrible tortures with unshaken fortitude rather than deny his convictions, and the man of science who undertakes a perilous experiment after having calculated the chances, might fly in the field of battle if compelled to fight for a cause in which they felt no interest; but not the less is their kind of courage the same as that of the brave soldier. Apart from inequalities of degree, there is no other difference than the superior influence of the intellectual faculties. Without the diverse cerebral synergies, either between the two great orders of faculties, or between the different functions of each order, it would be impossible to analyze the greater proportion of mental actions; and it is in the positive interpretation of each of them by such association that the application of phrenological doctrine will chiefly consist, when such doctrine shall have been scientifically erected. When the elementary analysis shall have been instituted, allowing us to pass on to the study of these compound phenomena, we may think of proceeding to the more delicate inquiry whether, in each cerebral organ, a distinct part is not especially appropriate to the establishment of these synergies and sympathies. Some pathological observations have given rise to this suspicion—the gray substance of the brain appearing more inflamed in those perturbations which affect the phenomena of the will, and the white in those which relate to intellectual operations.

If our existing phrenology isolates the cerebral functions too much, it is yet more open to reproach for separating the brain from the whole of the nervous system. Bichat taught us that the intellectual and affective phenomena, all-important as they are, constitute, in the whole system of the animal economy, only an intermediate agency between the action of the external world upon the animal through sensorial impressions, and the final reaction of the animal by muscular contractions. Now, in the present state of phrenological physiology, no positive conception exists with regard to the relation of the series of cerebral acts to this last necessary reaction. We merely suspect that the spinal marrow is its immediate organ. Even if cerebral physiology carefully comprehended the whole of the nervous system, it would still, at present, separate it too much from the rest of the economy. While rightly discarding the ancient error about the seat of the passions being in the organs of the vegetative life, it has too much neglected the great influence to which the chief intellectual and moral functions are subject from other physiological phenomena; as Cabanis pointed out so emphatically, while preparing the way for the philosophical revolution which we owe to Gall.

We have now seen how irrational and narrow is the way in which intellectual and moral physiology is conceived of and studied; and that till this is rectified, the science, which really appears not to have advanced a single step since its institution, can not make any true progress. We see how it requires, above even the other branches of physiology, the preparation of scientific habits, and familiarity with the foregoing departments of natural philosophy; and how, from its vicious isolation, it tends to sink to the level of the most superficial and ill-prepared minds, which will make it the groundwork of a gross and mischievous quackery, if the true scientific inquirers do not take it out of their hands. No inconveniences of this kind, however, should blind us to the eminent merits of a conception which will ever be one of the principal grounds of distinction of the philosophy of the nineteenth century, in comparison with the one which preceded it. *Imperfect state of Phrenology.*

Looking back, on the completion of this survey of the positive study of living bodies, we see that, imperfect as it is, and unsatisfactory as are the parts which relate to life, compared with those which relate to organization, still the most imperfect have begun to assume a scientific character, more or less clearly indicated, in proportion to the complexity of the phenomena. *Present state of Biology.*

We have now surveyed the whole system of natural philosophy, from its basis in mathematical, to its termination in biological philosophy. Notwithstanding the vast interval embraced by these two extremities, we have passed through the whole by an almost insensible gradation, finding nothing hypothetical in the transition; through chemistry, from inorganic to organic philosophy, and verifying as we proceeded the rigorous continuity of the system of the natural sciences. That system, though comprehending all existing knowledge, is, however, still incomplete, leaving a wide area to the retrograde influence of the theologico-metaphysical philosophy, to which it abandons a whole order of ideas, the most immediately applicable of all. There is yet wanting, to complete the body of positive philosophy, and to *Retrospect of Natural Philosophy* organize its universal preponderance, the subjection to it of the most complex and special phenomena of all—those of humanity in a state of association. I shall therefore venture to propose the new science of *Social Physics*, which I have found myself compelled to create, as the necessary complement of the system. This new science is rooted in biology, as every science is in the one which precedes it; and it will render the body of doctrine complete and indivisible, enabling the human mind to proceed on positive principles in all directions whatever, to which its activity may be incited. Imperfect as the preceding sciences are, they have enough of the positive character to render this last transformation possible: and when it is effected, the way will be open for their future advancement, through such an organization of scientific labor as must put an end to the intellectual anarchy of our present condition.